T0254269

Emergence

# Emergence: Contemporary Readings in Philosophy and Science

edited by Mark A. Bedau and Paul Humphreys

A Bradford Book
The MIT Press
Cambridge, Massachusetts
London, England

This book was set in Stone Serif and Stone Sans on 3B2 by Asco Typesetters, Hong Kong.

Library of Congress Cataloging-in-Publication Data

Emergence : contemporary readings in philosophy and science / edited by Mark A. Bedau and Paul Humphreys.
p. cm.
"A Bradford book."
Includes bibliographical references and (p. ) index.
ISBN 978-0-262-02621-5 (hardcover : alk. paper) — ISBN 978-0-262-52475-9 (pbk. : alk. paper)
1. Emergence (Philosophy) 2. Science—Philosophy. I. Bedau, Mark. II. Humphreys, Paul.
Q175.32.E44E44 2007
501—dc22 2007000949

# Contents

# Preface

Thirty years ago emergence was largely ignored in philosophy and science. Its ethos ran counter to the reductionist views of the time, and it seemed to invoke mystical and unexplainable levels of reality. Things have changed. Emergence is now one of the liveliest areas of research in both science and philosophy. This activity holds out great promise for understanding a wide variety of phenomena in ways that are intriguingly different from more traditional approaches. The reason for this change is complicated, but it results in part from developments in a number of vigorous and successful research programs within complexity theory, artificial life, physics, psychology, sociology, and biology. In parallel, although often driven by independent developments in the philosophy of science and the philosophy of mind, philosophers have been developing new conceptual tools for understanding emergent phenomena.

This book covers the principal approaches to emergence found in contemporary philosophy and science. All of the chapters are contemporary classics that either have played a significant role in the development of thinking about emergence or capture and refine widely held pretheoretical positions on emergence. They originally were published in widely scattered and intellectually diverse sources. This volume for the first time collects them all in one easily accessible place. We have included selections that represent most, if not all, of the major contemporary approaches to emergence. However, in emphasizing the interactions between the philosophical and scientific approaches to emergence, we are striking out deliberately in a particular direction. For entirely understandable reasons, much of the recent philosophical literature on emergence, not to mention the broader public's attention, has been motivated by an interest in whether specifically mental features, such as consciousness, emerge from brain states and properties. We have included selections from that tradition, but we believe that progress in understanding emergence will be helped by a familiarity with work in areas outside psychology and the philosophy of mind. By understanding how emergent phenomena occur and are represented in physics and artificial life, for example, those with a philosophical interest in the subject can acquire a broader perspective on what is peculiar to emergence. Conversely, the abstractness and conceptual clarity characteristic of philosophy can provide a much broader perspective from which scientists can see connections with kinds of emergence that lie outside their own disciplines.

And so this collection has a variety of intended audiences. It aims to be informative to both philosophers and scientists, but we also hope that many others, including students, will find the selections helpful and thought provoking. Most of the chapters can be understood by an intelligent reader who is not an expert in the specific discipline represented by a given author, and the third section can be used as a reference source on somewhat more specialized topics. Although we believe that our ordering provides a natural progression of ideas within each section, readers with different backgrounds no doubt will find it natural to begin with different sections. Our part introductions put the chapters into context, explain how they are connected, and pose some key questions for further exploration. The chapters in this book form only the tip of the iceberg of the emergence literature in contemporary philosophy and science, and further reading material is listed in the bibliography. Those who wish to use the collection as the basis for a course or seminar on emergence in some specific area easily can supplement our readings with more specialized and technical material.

Above all, we have endeavored to include selections that provide constructive and useful methods for understanding emergence. Throughout the introductions, we have posed questions, many of them currently lacking definitive answers. We hope that readers who work through this book will be well positioned to advance and eventually solve those problems.

Our book has an associated Web site containing supplementary material. Among other things, the site contains links to software including flocking and schooling simulations, the Game of Life, and self-organizing systems, as well as links to other reputable sites on emergence. We encourage readers to download and experiment with the simulations because many aspects of emergence have an essentially dynamic component that can only be understood through firsthand experience. The site also contains links to some classic publications that are now in the public domain, and updates about important new publications on emergence will be added periodically. As new resources arise over time, the site will grow and evolve. The Web site can be found at:

http://mitpress.mit.edu/emergence

Mark A. Bedau
Paul Humphreys

## Acknowledgments

The editors would like to thank a number of friends and colleagues from whom we have received valuable assistance while developing this anthology. For organizing the Paris conference on emergence where we met and learned of our shared perspective, we thank: Anouk Barberousse, Jacques Dubucs, and Philippe Huneman. For feedback on contents: Andreas Hutteman, Jenann Ismael, Sandy Mitchell, Norman Packard, and Bill Wimsatt. For secretarial help: Kathy Kennedy and Amanda Waldroupe. For financial support at a critical final stage: Stillman Drake funds from the Dean of Faculty at Reed College. Mark Bedau would like to thank colleagues at ProtoLife Srl, Reed College, the Santa Fe Institute, and the European Center for Living Technology for valuable and enjoyable discussions on emergence and related topics. Paul Humphreys would also like to thank members of the Institute d'Histoire et de Philosophie des Sciences et des Techniques (Paris) and of the Keswick Society for extensive discussions on emergence, complexity, and related topics.

## Sources

Grateful acknowledgment is made to the following publishers and authors:

1. "The Rise and Fall of British Emergentism" by Brian P. McLaughlin, from *Emergence or Reduction?: Essays on the Prospects of Nonreductive Physicalism*, edited by Beckerman, Flohr, and Kim. Copyright © 1992 by Walter de Gruyter GmbH & Co. Reprinted by permission of Walter de Gruyter GmbH & Co. and Brian P. McLaughlin.

2. "On the Idea of Emergence" by Carl Hempel and Paul Oppenheim, from *Aspects of Scientific Explanation and other Essays in the Philosophy of Science* by Carl Hempel. Copyright © 1965 by the Free Press. Reprinted by permission of the University of Chicago Press.

3. "Reductionism and the Irreducibility of Consciousness" by John Searle from *The Rediscovery of the Mind*. Copyright © 1992 by the MIT Press. Reprinted by permission of the MIT Press and John Searl.

4. Excerpt from "Emergence and Supervenience" by Brian P. McLaughlin, from *Intellectica 25 (1997)*, edited by Rosenthal. Copyright © 1997 by *Intellectica*. Reprinted by permission of *Intellectica* and Brian P. McLaughlin.

5. "Aggregativity: Reductive Heuristics for Finding Emergence" by William C. Wimsatt, from *Philosophy of Science 64(4): Suppl.2: S372–S384 (1997)*, edited by Dickson. Copyright © 1997 by the University of Chicago Press. Reprinted by permission of the University of Chicago Press and William C. Wimsatt.

6. "How Properties Emerge" by Paul Humphreys, from *Philosophy of Science 64 (1997)*. Copyright © 1997 by the Philosophy of Science Association. All rights reserved. Reprinted by permission of the author.

7. "Making Sense of Emergence" by Jaegwon Kim from *Philosophical Studies 95 (1999)*, edited by Cohen. Copyright © 1999 by Springer Science and Business Media. Reprinted by permission of Springer Science and Business Media and Jaegwon Kim.

8. "Downward Causation and Autonomy in Weak Emergence" by Mark A. Bedau, from *Principia Revista Internacional de Epistemologica 6 (2003)*, edited by Dutra.

Copyright © by *Principia Revista Internacional de Epistemologica.* Reprinted by permission of *Principia Revista Internacional de Epistemologica* and Mark A. Bedau.

9. Excerpt from "Real Patterns" by Daniel C. Dennett from *The Journal of Philosophy 87 (1991),* edited by Smylie. Copyright © 1991 by the Journal of Philosophy, Inc. Reprinted by permission of the Journal of Philosophy, Inc. and by Daniel C. Dennett.

10. "More is Different" by P. W. Anderson from *Science 177* (1972), edited by Kennedy. Copyright © 1972 by the American Association for the Advancement of Science. Reprinted by permission of the American Association for the Advancement of Science and P. W. Anderson.

11. Excerpt from "Emergent Colonization in an Artificial Ecology" by Andrew Assad and Norman H. Packard from *Towards a Practice of Autonomous Systems 9 Systems: Proceedings of the First European Conference on Artifical Life,* edited by Varela and Bourgine. Copyright © 1992 by the MIT Press. Reprinted by permission of the MIT Press, Andrew Assad, and Norman H. Packard.

12. Excerpt from *Micromotives and Macrobehavior* by Thomas C. Schelling. Copyright © 1978 by W.W. Norton & Company, Inc. Reprinted by permission of W.W. Norton & Company, Inc. and Thomas C. Schelling.

13. "Alternative Views of Complexity" by Herbert Simon from *The Sciences of the Artificial,* third edition. Copyright © 1996 by the MIT Press. Reprinted by permission of the MIT Press.

14. "The Theory of Everything" by Robert B. Laughlin and David Pines, from *Proceedings of the National Academy of Sciences 97 (2000).* Copyright © 2000 by the National Academy of Sciences. Reprinted by permission of the National Academy of Sciences and David Pines.

15. "Is Anything Ever New? Considering Emergence" by James Crutchfield, from *Complexity: Metaphors, Models, and Reality,* edited by Cowan, Pines, and Meltzer. Copyright © 1999 by Westview Press. Reprinted by permission of Basic Books, a member of Perseus Books, LLC and James Crutchfield.

16. "Design, Observation, Surprise! A Test of Emergence" by Edmund M. A. Ronald, Moshe Sipper, and Mathieu S. Capcarrère from *Artificial Life 5 (1999).* Copyright © 1999 by the MIT Press. Reprinted by permission of the MIT Press, Edmund M. A. Ronald, Moshe Sipper, and Mathieu S. Capcarrère.

17. "*Ansatz* for Dynamical Hierarchies" by Steen Rasmussen, Nils A. Baas, Bernd Mayer, and Martin Nillson from *Artificial Life 7 (2001).* Copyright © 2001 by the MIT Press. Reprinted by permission of the MIT Press, Steen Rasmussen, Nils A. Baas, Bernd Mayer, and Martin Nillson.

18. "Newtonianism, Reductionism, and the Art of Congressional Testimony" by Stephen Weinberg from *Nature 330 (1987),* edited by Campbell. Copyright © 1987

by Nature Publishing Group. Reprinted by permission of Nature Publishing Group and Stephen Weinberg.

19. Excerpt from *Teleology Revisited and Other Essays in the Philosophy and History of Science* by Ernest Nagel. Copyright © Sidney Nagel and Hackett Publishing Company, Inc. Reprinted by permission of Hackett Publishing Company, Inc. All rights reserved.

20. "Chaos" by James P. Crutchfield, J. Doyne Farmer, Norman H. Packard, and Robert S. Shaw, from *Scientific American 255 (1986)*, edited by Rennie. Copyright © 1986 by *Scientific American*. Reprinted by permission of *Scientific American*, James P. Crutchfield, J. Doyne Farmer, Norman H. Packard, and Robert S. Shaw.

21. "Undecidability and Intractability in Theoretical Physics" by Stephen Wolfram from *Physical Review Letters 54* (1985), edited by Basbas, Brown, Sandweiss, and Schuhmann. Copyright © 1985 by the American Physical Society. Reprinted by permission of the American Physical Society and Stephen Wolfram.

22. "Special Sciences, or the Disunity of Science as a Working Hypotheses" by Jerry Fodor from *Synthese 28 (1974)*, edited by Hendricks, Symons, and van Benthem. Copyright © 1974 by Springer Science and Business Media. Reprinted by permission of Springer Science and Business Media and Jerry Fodor.

23. Excerpt from "Supervenience" by David J. Chalmers, from *The Conscious Mind: In Search of a Fundamental Theory*. Copyright © 1996 by David J. Chalmers. Reprinted by permission of Oxford University Press, Inc. and David J. Chalmers.

24. "The Nonreductivist's Troubles with Mental Causation" by Jaegwon Kim, from *Mental Causation*, edited by Heil and Mele. Copyright © 1993 by Oxford University Press. Reprinted by permission of Oxford University Press, Inc. and Jaegwon Kim.

# Introduction

Emergence relates to phenomena that arise from and depend on some more basic phenomena yet are simultaneously autonomous from that base. The topic of emergence is fascinating and controversial in part because emergence seems to be widespread and yet the very idea of emergence seems opaque, and perhaps even incoherent. The topic has special urgency today because of the burgeoning attention to emergence in contemporary philosophy and science.

This book examines how emergence is treated in contemporary philosophy and science, and one of our goals is to facilitate informed discussions between these communities. Less insular discussions should clarify what the main categories of emergence are thought to be today, and how well they apply to the paradigm cases considered in contemporary philosophy and science. We hope that the eventual outcome will be an understanding of emergence that is both philosophically rigorous and useful in empirical science.

This general introduction to the book gives some examples of apparent emergent phenomena, calls attention to a few methodological subtleties, and then highlights some central open questions about emergence that the chapters in this book collectively address. The first section covers contemporary philosophical perspectives on emergence. Part II covers today's scientific perspectives on emergence. The last group of chapters collects contextual and background material from both philosophy and science. Each section's introductory essay discusses the chapters' unifying themes and issues.

One of the best ways to get a feel for emergence is to consider widely cited core examples of apparent emergent phenomena. The examples involve a surprising variety of cases. One group concerns certain properties of physical systems. For example, the liquidity and transparency of water sometimes are said to emerge from the properties of oxygen and hydrogen in structured collections of water molecules. As another example, if a magnet (specifically a ferromagnet) is heated gradually, it abruptly loses its magnetism at a specific temperature—the Curie point. This is an example of physical phase transitions, which often are viewed as key examples of emergence. A third example involves the shape of a sand pile. As grains of sand are added successively to the top of the pile, the pile forms a conical shape with a characteristic slope, and successive small and large avalanches of sand play an important role in preserving that shape. The characteristic sand pile slope is said to emerge from the interactions among the grains of sand and gravity.

Life itself is one of the most common sources of examples of apparent emergence. One simple case is the relationship between a living organism and the molecules that constitute it at a given moment. In some sense the organism is just those molecules, but those same molecules would not constitute an organism if they were rearranged in any of a wide variety of ways, so the living organism seems to emerge from the molecules. Furthermore, developmental processes of individual organisms are said to involve the emergence of a more mature morphology. A multicellular frog embryo emerges from a single-celled zygote, a tadpole emerges from this embryo, and eventually a frog emerges from the tadpole. In addition, evolutionary processes shaping biological lineages also are said to involve emergence. A complex, highly differentiated biosphere has emerged over billions of years from what was originally a vastly simpler and much more uniform array of early life forms. The mind is a rich source of potential examples of emergence. Our mental lives consist of an autonomous, coherent flow of mental states (beliefs, desires, memories, fears, hopes, etc.). These, we presume, somehow emerge out of the swarm of biochemical and electrical activity involving our neurons and central nervous system.

A final group of examples concerns the collective behavior of human agents. The origin and spread of a teenage fad, such as the sudden popularity of a particular hairstyle, can be represented formally in ways similar to a physical phase transition, and so seem to involve emergence. Such phenomena often informally are said to exhibit "tipping points." Another kind of case is demonstrated in a massive traffic jam spontaneously emerging from the motions of individual cars controlled by individual human agents as the density of cars on the highway passes a critical threshold. It is interesting to speculate about whether the mechanisms behind such phenomena are essentially the same as those behind certain purely physical phenomena, such as the jamming of granular media in constricted channels.

The chapters in this book are full of many other examples of apparent emergent phenomena. These examples can serve as useful guides against which to test an account of emergence. However, testing accounts with these examples is not always simple. Everything else being equal, it would count in favor of a theory of emergence if it could explain how all these examples do involve emergence. But there is no guarantee that the best theory will classify all these examples as genuine cases of emergence. When we finally understand what emergence truly is, we might see that many of the examples are only apparent cases of emergence. Indeed, one of the hotly contested issues is whether there are *any* genuine examples of emergence.

Identifying the genuine examples of emergence is possible only given an appropriate definition of emergence, but as the chapters in this book amply illustrate, the proper characterization of emergence still is contested. Finding appropriate definitions or theories of emergence with indisputable instances has obvious consequences for the scientific legitimacy of emergence. One of the most important differences between contemporary accounts of emergence and their precedents is that the earlier accounts quickly became metascientific because the examples used to illustrate emergence tended to be phenomena such as life that at the time were well beyond the realm of serious scientific understanding. Nowadays, we know much more about complex phenomena like life, so many of the plausible candidates for emergence now are well understood by science. Any adequate definition of emergence would take these into account in the sense that at least some of these examples should be included under the definition in a clear naturalistic fashion. Fashioning such a definition, however, involves an inescapable back-and-forth process, hinted at above. Definitions and theories may be sharpened to account for more examples, but also candidate examples may be abandoned because they fail to fit an otherwise convincing theory. In a similar way, we must be prepared to abandon some of our preconceptions and background beliefs about emergence if a persuasive and detailed theory of emergence calls them into question.

One small caveat is needed here. Hunting for emergence is an exciting sport, but the claim that something is emergent should be made with care and supported with persuasive evidence. Indeed, some of the articles reprinted in this collection ultimately are quite skeptical about emergence and argue that emergent phenomena, if they exist at all, are likely to be uncommon. One should not lightly abandon nonemergent, reductionist approaches that have been successful in many areas of science and philosophy. At the same time, one also should note that many of the conceptions of emergence developed and defended in this book are consistent with many common forms of reductionism.

The study of emergence is still in its infancy and currently is in a state of considerable flux, so a large number of important questions still lack clear answers. Surveying those questions is one of the best ways to comprehend the nature and scope of the contemporary philosophical and scientific debate about emergence. Grouped together here are some of the interconnected questions about emergence that are particularly pressing, with no pretense that the list is complete.

1. **How should emergence be defined?** A number of leading ideas appear in different definitions of emergence, including irreducibility, unpredictability, conceptual novelty, ontological novelty, and supervenience. Some definitions combine a number of these ideas. We should not presume that only one type of emergence exists and needs definition. Instead, different kinds of emergence may exist, so different that they fall under no unified account. Emergent phenomena might well come in fundamentally

different types that should be distinguished along various dimensions. A further issue is whether emergence should be defined only relative to a theory, or a level of analysis, or a system decomposition. The controversy about how to define emergence is exacerbated by the casual way that terms such as "emergence" and "emergent" often are used. At least two separate issues are important here: controversies about the proper definition of emergence, and controversies about the proper way to test and evaluate definitions of emergence. Perhaps the proper definition of emergence can be attained only in the context of a comprehensive theory of emergence, resulting in a definition that is implicit rather than explicit. Another possibility is that the concept of emergence is best characterized by a cluster of features such as novelty, holism, irreducibility, and so on, but that the features drawn from the cluster differ from case to case, and that what counts as novel, for example, differs with different subject matters. Given the high level of uncertainty about how to properly characterize what emergence is, it should be no surprise that many other fundamental questions remain unanswered.

2. What ontological categories of entities can be emergent: properties, substances, processes, phenomena, patterns, laws, or something else? Within the literature on emergence, different authors say that different categories of entities are emergent. There should be no presumption that these different categories are mutually exclusive; it could be that emergence applies to many or even all of them. But it is important to be clear about which of these candidates is under discussion in any given context. Emergence in one of these categories sometimes entails emergence in another, but that is not always the case. For example, it seems clear that emergent laws can link nonemergent properties, whereas a genuinely new emergent property would seem to require new, and probably emergent, laws.

3. What is the scope of actual emergent phenomena? This question partly concerns which aspects of the world can be characterized as emergent. The examples of apparent emergence above show the prevalence of the claim that emergence captures something distinctive about consciousness and about other aspects of the mind. Another common idea is that emergence is one of the hallmarks of life. But examples of apparent emergent phenomena also include the behavior of human social organizations and of nonhuman social organizations. In addition, certain kinds of physical aggregations are commonly cited as examples of emergent phenomena. The question of the scope of emergence also concerns the question of how widespread emergence is. For example, many contemporary philosophers think that emergence is a rare and special quality found only in extremely distinctive settings, such as human consciousness. Others think that emergence is quite common and ordinary, applying to a myriad of complex systems found in nature. For those who think that nothing is truly emergent, the question still arises whether this state of affairs is simply an accident or whether the very idea of emergence is incoherent.

4. Is emergence an objective feature of the world, or is it merely in the eye of the beholder? Does emergence characterize only models or descriptions or theories of nature, or does it apply also to nature itself? Is emergence only a function of how something is described or viewed or explained? Question 4 is connected to the issue of whether emergence is defined only relative to a theory or model or representation. Some maintain that emergent phenomena are real features of the world, while others maintain that emergence is merely a result of our imposing certain kinds of representation on the world, or a result of our limited abilities to comprehend correctly what the world is like. Candidates for emergent phenomena in the real world include the physical process called *spontaneous symmetry breaking*. A simple case of this can occur when a uniform body of liquid has a flat surface. If the bottom of the liquid is heated uniformly and sufficiently, the fluid breaks up into a field of different convection cells in which the liquid continually cycles between the bottom and top of the fluid. An example of emergence that might reflect merely our limited ability to understand the world is the stable patterns that emerge in John Conway's Game of Life. If the Game of Life is initialized with the now-famous R-pentomino pattern of 5 active cells, it takes 1103 iterations of the rules to arrive at a final stable pattern. The discovery of this final pattern occurred only after the game was implemented on a computer; exploring the rules of the game "by hand" was insufficient.

5. Should emergence be viewed as static and synchronic, or as dynamic and diachronic, or are both possible? This is a major division between accounts of emergence. In synchronic emergence, the emergent feature is simultaneously present with the basal features from which it emerges. By contrast, in diachronic emergence, the base precedes the emergent phenomenon which develops over time from it. If mental phenomena emerge from neural phenomena, this is generally thought to be synchronic, there being no time gap between a recollection of one's fifteenth birthday and the brain state that gives rise to the memory. The development of the traffic jam over time is a good candidate for a diachronically emergent pattern. Discussions in the philosophical literature usually focus on synchronic emergence, while those in the scientific literature often concern diachronic emergence. A further question about diachronic emergence is whether and how it applies to both discrete and continuous systems.

6. Does emergence imply or require the existence of new levels of phenomena? A great many discussions of emergence use the terminology *levels*, with the levels having three characteristic features. First, the hierarchy of levels has no precisely defined order, but instead is determined implicitly by the organizational complexity of objects. These levels tend to coincide with the domains of individual sciences. Second, each level is assumed to contain at least one kind of object and one kind of property that is not found below that level. Third, at each level kinds exist that have novel causal powers that emerge from the organizational structure of material components. Pressing

questions thus include whether this framework of levels corresponds to an objective hierarchy in the world, whether appeal to these levels is useful or misleading, and whether there are clear criteria to identify the levels.

7. In what ways are emergent phenomena autonomous from their emergent bases? Emergent phenomena are Janus faced; they depend on more basic phenomena and yet are autonomous from that base. Therefore, if emergence is to be coherent, it must involve different senses of dependence and independence. A number of different kinds of autonomy have been discussed in the literature, including the ideas that emergent phenomena are irreducible to their bases, inexplicable from them, unpredictable from them, supervenient on them, and multiply realizable in them. In addition, emergent phenomena sometimes are thought to involve the introduction of novel concepts or properties, and functionally characterized properties sometimes are thought to be especially associated with emergent phenomena. Another important question about the autonomy of emergent phenomena is whether that autonomy is merely epistemological or whether it has ontological consequences. An extreme version of the merely epistemological interpretation of emergence holds that emergence is simply a sign of our ignorance. One final issue about the autonomy of emergent phenomena concerns whether emergence necessarily involves novel causal powers, especially powers that produce "downward causation," in which emergent phenomena have novel effects on their own emergence base. One of the questions in this context is what kind of downward causation is involved, for the coherence of downward causation is debatable.

The chapters in this book provide a variety of perspectives on possible ways to construct answers to these questions. Many of the questions are discussed at greater length in the introductions to the book's three sections, where the central themes treated in the individual chapters are highlighted.

Emergence seems to arise in many of the most interesting complexities in the world we inhabit, but it is simultaneously palpable and confusing, as the questions above reflect. New advances in contemporary philosophy and science, many of which this book collects, now are converging to enable new progress on these questions, so emergence is a topic ripe for new clarifications, unifications, and other creative conclusions. This book's chapters illuminate these questions from many perspectives to help readers with framing their own answers.

# I Philosophical Perspectives on Emergence

# Introduction to Philosophical Perspectives on Emergence

This introduction describes some of the leading themes about emergence in contemporary philosophy. The philosophical community has always placed a high premium on conceptual rigor and analytical clarity, and philosophy often has a view of the larger ramifications of issues. Because the topic of emergence is conceptually subtle and multifaceted, philosophical precision is required for sustained progress in understanding what emergence is and when it occurs. Students of emergence should therefore find it instructive to study these philosophical contributions to the analysis of emergence and surrounding topics.

The significant resurgence of interest in emergence within contemporary philosophy has occurred for at least two reasons. One is that emergence once again has become an attractive position in philosophical attempts to understand the mind; chapters 3, 4, and 7 reflect this trend in different ways. As chapters 5, 6, 8 and 9 suggest, the second reason is that many philosophers also are inspired by the recent attention to emergence in contemporary science, examples of which can be seen in the chapters in part II. In addition to calling attention to some of the methodological issues that make philosophical reflection about emergence interesting and subtle, this introduction also highlights the key issues that readers might want to ponder when reading the chapters in part I.

## Leading Ideas about Emergence

A central question addressed by most chapters in this section is how to characterize emergent phenomena more precisely. Although this question has no generally accepted answer, a number of characteristic features recur in various forms within otherwise different accounts of emergence. Emergent phenomena frequently are taken to be irreducible, to be unpredictable or unexplainable, to require novel concepts, and to be holistic. This list is almost certainly incomplete, and not all of these features are present in every account of emergence, but the cluster indicates how philosophers usually think of emergence. Each of these ideas has various forms that can be distinguished, as different chapters show. Some accounts of emergence favor only one idea to the exclusion of all the others, while other accounts simultaneously embrace many. Although these ideas provide a useful guide to the emergence literature, they do not form a mutually exclusive taxonomy or exhaustive partition of possibilities; instead, they simply indicate the leading ideas about emergence.

Irreducibility is often viewed as the sense in which emergent phenomena are autonomous from the more basic phenomena that give rise to them. This irreducibility can take a variety of forms, some much weaker than others, as the chapters in this section illustrate. The failure of reduction is sometimes supplemented with a supervenience relation that holds between the emergent and the more basic phenomena. Supervenience is proposed to be the sense in which emergent phenomena, while having a distinct existence, nevertheless depend on more basic phenomena (for background material on supervenience, see chapter 23).

The independence of emergent phenomena is in some other approaches viewed as unpredictability; a state or other feature of a system is emergent if it is impossible to predict the existence of that feature on the basis of a complete theory of basic phenomena in the system. Unpredictability sometimes is taken to be a consequence of irreducibility and, as with irreducibility, unpredictability has been interpreted in a variety of ways, as the collection of views throughout this book illustrate. A closely related idea is that emergent phenomena cannot be explained given a complete understanding of more basic phenomena.

The conceptual approach maintains that a system that has reached a critical level of complexity can be described effectively only by introducing a conceptual or descriptive apparatus that is new compared to what is used for more basic phenomena. This conceptual novelty can range from the invention of a new term to the introduction of an entirely new theory. The impossibility of understanding the phenomena without this new framework is taken to be the mark of an emergent level of phenomena. For example, the need to introduce the term *liquid* in order to effectively describe the behavior of a large collection of molecules that has undergone a phase transition illustrates the idea that emergence involves conceptual novelty.

It often is noted that the term *liquid* cannot be applied to an individual molecule of $H_2O$. So, closely related to conceptual novelty is the idea that the independence of emergent phenomena involves an appeal to holism. Some properties apply only to wholes formed out of assemblies of more basic parts; it is conceptually incoherent for them to be applied to the parts. For example, an individual monomer does not have the property of elasticity, but when a number of monomers are covalently bonded into a chain, the resulting polymer is elastic; it can be stretched, and it will spring back afterward. This kind of holism often is thought to be at least a component of emergence.

Irreducibility, unpredictability, unexplainability, conceptual novelty, and holism sometimes are taken not merely at face value, but as a reflection of a deeper ontological novelty in emergent phenomena. This ontological interpretation of emergence considers emergent entities to be genuinely novel features of the world itself, arising from but also separate in some sense from more basic aspects of the system. The ontological interpretation is the source of much of the animated interest in emergence, both positive and negative.

As Brian McLaughlin recounts in chapter 1, the empiricist John Stuart Mill often is credited with the first sustained philosophical treatment of emergence, in his 1843 treatise *A System of Logic*, although he did not himself use the word *emergence*.[1] For Mill, emergence was associated with the failure of the principle of the composition of causes, which states that the effects of causes are additive. The parallelogram of forces is a standard example of the principle's application, the idea being that the total effect of two causes acting in the "mechanical mode" is simply the sum of the effects of the two causes acting alone. In using this example, Mill generalized the idea of additive effects to include vector addition as well as scalar addition. For at least the next century, philosophers struggled without much success to identify a set of operations that fitted some very general idea of additivity.

Mill did not consider the principle of the composition of causes to be universally true because he also insisted that what he called *heteropathic laws* operated in certain sciences.[2] Such laws cover processes in which the composition of causes principle is violated, resulting in something for which, to use a famous phrase, "the whole is more than the sum of the parts." The appeal to heteropathic laws nicely illustrates the role that unpredictability plays in accounts of emergence. As Mill put it, "even if we could have ascertained . . . that oxygen and hydrogen were both present when water is produced, no experimentation on oxygen and hydrogen separately, no knowledge of their laws, could have enabled us deductively to infer that they would produce water. We require a specific experiment on the two combined." That is to say, deductive methods do not provide knowledge of emergent phenomena, although inductive methods may. It is worth noting that when Mill says, "Not a trace of the properties of hydrogen or of oxygen is observable in those of their compound, water," he also seems to be appealing to the idea that emergence involves ontological novelty. Furthermore, the fact that the term *liquid* is necessary to describe water but not to describe individual molecules of $H_2O$ suggests that the idea of conceptual novelty is also present in Mill's example, although those who wish to identify water and $H_2O$ might resist this claim.

As can be seen from the date of Mill's treatise, an interest in emergence developed surprisingly late. One reason for this may be that reductionist projects have proved attractive since the earliest days of natural philosophy. The pre-Socratics thought that reality was much simpler than appearances suggest and that an appeal to water, earth, or some mixture of other elements would suffice to explain all physical phenomena. The atoms hypothesized by Democritus and Leucippus provided the building blocks for reduction and, much later when the mechanical worldview of the seventeenth century had been developed, atomism held out the promise of unifying all of physics. Nineteenth-century developments in chemistry and twentieth-century developments in molecular biology then suggested that a combinatorial approach to matter would be exceptionally fruitful in those areas as well. It took the development of detailed and accurate microtheories before examples of actual reductions became plausible, and it was only then possible to appreciate how those same detailed reductions seemed to fail in certain other situations.

## The Scope of Emergence

There is a widely held view—which we will term *the sparse view*—that if emergence exists at all, it will appear only in unusual types of phenomena that tend to fall outside the scope of normal science. One motivation for the sparse view has been the historical success of reduction mentioned above and the accompanying belief that science has been progressively successful at showing how higher-level sciences can be reduced to more basic sciences. Indeed, one reason why interest in emergence waned in the middle part of the twentieth century was the belief that large parts of chemistry had been reduced to physics and that molecular biology, itself amenable to reduction to physical chemistry, held out the promise of reducing large parts of biology. The sparse view thus has encouraged a focus on consciousness and other exotic phenomena as the most appropriate candidates for emergence. For example, Lloyd Morgan, an early British emergentist, represented reality as consisting in a number of levels that were, from the bottom, physical, chemical, vital, conscious mentality, and reflective mentality. Samuel Alexander had a somewhat different hierarchy: space-time, primary qualities, secondary qualities, life, mind, and, at the very top, Deity. Chapter 1 explains how the British Emergentist tradition developed and describes some of the main lines of approach to emergentism that remain today. Jaegwon Kim's position in chapter 7 is one contemporary representation of the sparse view.

The sparse view often is used as a criterion to test accounts of emergence, following the idea that if an account makes emergence too common, then the account must be flawed. Yet the revival of interest in emergence has resulted in part from the discovery of candidates for emergent phenomena that either occur at very basic levels of the ontological hierarchy or are quite common. An example of the first is the phenomenon of quantum entanglements and examples of the second can be found throughout complexity theory and artificial life (see part II). And so it is worth considering whether the sparse view is in fact correct and whether emergence, rather than being rare, is in fact quite common. The chapters in this section present a variety of perspectives on this issue. In particular, William Wimsatt (chapter 5), Paul Humphreys (chapter 6), Mark Bedau (chapter 8), and Daniel Dennett (chapters 9) provide a variety of reasons for thinking that emergent phenomena are quite common. For Wimsatt, emergence is clearly the rule rather than the exception in ordinary physical systems; it generally arises in some form whenever a system has properties that are not governed by conservation laws. Since so few properties are governed by conservation laws, for Wimsatt emergence abounds. For Humphreys, the existence of emergent features in many well-understood domains of physics, together with complex and fine-grained interactions between scientific domains, suggests that emergence is not confined to a few mysterious phenomena. For Bedau, emergence is associated not with most physical systems but specifically with those that have a certain kind of complexity. Nevertheless, this kind of complexity is a familiar enough part of the landscape that emergence is far from rare. Dennett similarly associates the need for higher-level theories with physical systems that are too complex to otherwise predict and explain, and these are commonplace.

One of the key decisions facing advocates of philosophical accounts of emergence is whether the account should describe only how emergence *could* occur, leaving the issue of the existence of real cases as an optional further issue, or whether it would be a serious defect in the account if the world contained no examples of that type of emergence. Metaphysics happily can develop theories of circular time, even if our universe does not contain such a thing, and so one also could develop theories of possible but not actual emergence. But an account of emergence that has actual instances is more interesting, and scientifically informed metaphysics can learn from the way that our universe is actually structured. A related question is whether a theory of emergence must be necessarily true and apply to every logically possible world, or whether it would be sufficient if the theory were merely true of the actual world. When considering these issues, one must decide whether there are sufficiently many clear and uncontroversial examples of emergence that they can be used to test theories, or whether an independently appealing theory of emergence first must be constructed and later used to identify cases of emergence.

## Irreducibility and Mystery

Irreducibility is one of the leading ideas about emergence; a failure to be reduced often is viewed as a necessary condition for something to be emergent. Reduction can mean a variety of things, but one standard view is that a phenomenon has been reduced if it has been explained in terms of phenomena that are more fundamental (according to some ordering criterion) than the original phenomenon. Yet to presuppose in this way that emergent phenomena by their very nature cannot be explained—as Samuel Alexander bluntly put it, emergence "admits of no explanation"[3]—is to risk associating emergence with mysticism. Much of the antipathy toward emergence can indeed be traced to the idea that emergence is unavoidably mysterious. To avoid this undesirable consequence, a number of contemporary philosophers hold that emergence is compatible with reduction. This raises two related questions: To what extent are emergence and reduction compatible or incompatible, and to what extent, if any, are emergent properties and objects explainable? Chapters 5, 7, 8, and 9 offer starkly different answers to these questions.

In chapter 7 Kim provides an alternative to the once-dominant account of reduction of Ernest Nagel (see chapter 19). For Kim, reduction depends on three components. First is the ability to give a property E a functional definition, that is, a definition in terms of its role as a causal intermediary between inputs and outputs; second is the ability of science to find realizers of E at the reducing level, that is, specific entities or properties that carry out the function at the lower level; and third is the ability of science to find a theory at the reducing level that explains how the realizers carry out the causal task constitutive of E. Emergent phenomena then will be those irreducible features that remain when Kim-style reducibility has been carried out across the board. Inferring from the past successes of science, Kim concludes that emergent phenomena are likely to be rare, limited at most to qualia and consciousness (recall the sparse view of emergence, discussed above).

As it became clear that there were few, if any, real examples of Nagel-style reduction, the relation of supervenience (see chapter 23 in part III) frequently was used to show how higher-level properties depend upon lower-level properties without being reducible to them. By extension, accounts of emergence using supervenience were constructed, and Brian McLaughlin presents one such in chapter 4. Supervenience accounts of nonreductive physicalism are subject to the downward causation argument (see chapter 24 in part III), which produces what seems to be an irreconcilable tension between the causal closure of the physical realm and the need for the supervenient properties to do some nonredundant causal work. Because the emergent property and the base properties need to exist simultaneously within these supervenience approaches to emergence, the downward causation argument fails to apply in cases where the base property instances go out of existence when they combine to form an emergent property instance. In chapter 6 Humphreys uses this fact to suggest that a fusion operation on properties, the application of which produces an instance of a distinctively new property that does not have the old property instances as components, can avoid the downward causation problem. He also suggests that examples of fusion may be found in quantum entangled states and, although the chapter does not mention it, the covalent bonding of molecules appears to provide additional examples. If so, then emergence is a common rather than a sparse phenomenon and, ironically, one of the very cases that led to the decline of British Emergentism will turn out to be a central case in the new emergentist tradition.

Instead of contrasting emergence with reduction, Wimsatt in chapter 5 contrasts emergence with aggregativity and writes: "Aggregative properties depend on the parts' properties in a very strongly atomistic manner, under all physically possible decompositions. It is rare indeed that all of these conditions are met. This is the complete antithesis of functional organization." As a result, he argues that emergence is a common phenomenon, and because many of these cases of emergence are explicable, the account of emergence must be compatible with reduction. An important subsidiary claim made by Wimsatt is that emergence comes in degrees, a claim that is in striking contrast to the all-or-nothing attitude toward emergence that is widely adopted elsewhere. Elsewhere Wimsatt calls the appeal to qualia and consciousness support for a kind of mystical emergence, a criticism also leveled in chapter 3 by John Searle, who argues that emergence need not involve explanatory opacity. While Searle accepts cases of emergence that can be explained by the causal interactions of their constituents (what he terms $emergent_1$), he rejects those putative cases of emergence that are inexplicable (termed $emergent_2$). Searle also provides a useful list of types of reduction, using them to argue that consciousness is causally reducible to physical processes but not ontologically reducible.

Bedau in chapter 8 contrasts weak and strong forms of emergence. He dismisses strong emergence as a mysterious metaphysical possibility for which empirical evidence is lacking, but he defends weak emergence as a nonmysterious and scientifically explainable kind of phenomenon that arises in certain complex systems. This weak emergence is compatible with certain forms of ontological and causal reduction, but its explanatory irreducibility can generate, in certain cases, other forms of ontological and causal novelty. By contrast, Dennett in chapter 9 develops a mediating position that questions the appropriateness of even asking whether higher-level patterns are real or have causal powers. Higher-level patterns in a sense might be nothing more than the more basic elements that constitute them, but at the same time there is no choice other than to treat higher-level patterns autonomously. This discussion shows the intimate connections among the questions of whether emergence is compatible with reduction, whether emergent phenomena can be explained, and whether emergent phenomena are common.

## Unpredictability, Ignorance, and Reconceptualization

Sufficient experience with past occurrences of a certain kind of emergent phenomenon often can lead to reasonably accurate inductive predictions about that kind of emergent phenomenon in the future. This is no more surprising or significant than any other rough inductive prediction. A quite different issue is whether it is possible, in principle or in practice, to predict emergent phenomena exactly, and to do so solely on the basis of theories of the base phenomena. Unpredictability formed the core of

C. D. Broad's account of emergence, described in chapter 1. The inability to predict theoretically how a process will develop over time also lies at the heart of what was called *weak emergence* above. In essence, a phenomenon is weakly emergent if it is produced by lower-level phenomena but there are no theoretical shortcuts to predicting it exactly because the lower-level process that produces it is computationally irreducible (for background on computational irreducibility, see chapter 21). Thus, a solar eclipse is not weakly emergent because such things are predictable well in advance by using celestial mechanics. But attempts to predict how proteins fold have resisted all attempts to provide a similarly compressed derivation. The best current options are to let the folding process play out in reality or to run a computer simulation of the process and see how it develops.

The suggestion that unpredictability is a leading idea about emergence raises a number of associated questions. One is whether objective criteria exist for a system's being unpredictable because of computational irreducibility, or whether this is relative to the theories we accept today. The sort of unpredictability that is due to computational irreducibility (see chapter 21) or sensitive dependence on initial conditions (see chapter 20) is interesting in part because it is an objective mathematical property of certain complex systems. Promise of a positive answer to this question resides in the lively area of contemporary research generically known as computational complexity, an area further addressed in the introduction to part III.

A second question is whether the inability to predict a phenomenon today results from a lack of theoretical knowledge that will be provided at some point in the future. In addressing this question, Carl Hempel and Paul Oppenheim in chapter 2 suggest that the novelty of emergent phenomena is a result of their being unexplainable from more basic phenomena. When this chapter originally was written, Hempel and Oppenheim considered unpredictability and inexplicability to be two sides of the same coin, and scientific explanation had to be provided in terms of scientific theories. Thus, their chapter argues that emergence is a feature that is relative to a particular theory, so that something can be called emergent only relative to the state of knowledge at a given time. Because of this, they viewed claims of emergence as professions of ignorance. Their position on the importance of theory and linguistic frameworks was shared widely within the logical empiricist tradition and their chapter reinforced the high level of skepticism about ontological emergence that prevailed in philosophy through much of the second half of the twentieth century. There is no doubt that inadequate theories can lead to claims about emergence that are later recognized to be mistaken. Life and certain chemical phenomena are two of the examples of emergence mentioned in Mill's *A System of Logic*, and it is frequently pointed out that the low point in the fortunes of twentieth-century emergentism occurred when quantum-mechanical explanations for molecular properties were discovered and molecular biology was beginning to hold out the promise of explaining many of the properties of living organisms. It is notable that these same kinds of examples once again are being treated as paradigms of a new kind of emergence in complexity theory (see part II).

Complexity theory also allows connections with emerging fields in physics that represent physical phenomena in information-theoretic terms. One common representational device is a cellular automaton. Cellular automata are extremely simple, abstract mathematical systems in which space and time are considered to be discrete. They consist of a regular grid of cells (typically infinite, arrayed in one, two, or more dimensions), each of which can be in any one of a finite number of states. Usually all the cells are governed by the same rule, which describes how the state of a cell at a given time is determined by the states of itself and its neighbors at the preceding moment. (For more background, see chapter 21 in part III.) Bedau and Dennett appeal to cellular automata in chapters 8 and 9 because they are such simple and vivid illustrations of certain kinds of emergent phenomena. The real world is much more complex than such cellular automata. Furthermore, the determinism, discrete time and space, temporal synchronies, and spatial symmetries characteristic of cellular automata are all at variance with how nature seems. So, arguments based on such examples raise the question of the extent to which their conclusions apply not only to abstract models but also to nature. This is especially true if the notion of emergence makes sense only relative to a theory or model.

At this juncture, it is natural to ask whether some cases of emergence merely are a result of reconceptualizing phenomena. Dennett addresses this question, and also how to determine the appropriate level of vocabulary to choose when describing a given phenomenon. He suggests that certain kinds of phenomena need to be reconceptualized at a higher level of description in order to treat them effectively, because the finer-grained descriptions are far too complicated. Our everyday vocabulary is full of such redescriptions—we talk of tables rather than of vast arrangements of molecules, to borrow Eddington's famous example—but the core issue for Dennett is whether higher-level patterns are real, at least in the sense that they have their own distinctive causal effects. Dennett suggests that such levels come about when we adopt the vocabulary of what he calls *the design level*, which is a compact way of describing the data using a new vocabulary that avoids the explicit bit-by-bit description at the physical level. All of this provides fodder for the picture of emergence as conceptual novelty.

The chapters in this section nicely illustrate the flourishing discussion of emergence in contemporary philosophy. As this introduction makes plain, the discussion is far from reaching a consensus. Many fundamental questions remain controversial, and there is ample room for new fundamental insights. This section should provide a broad and solid foundation on the basis of which readers can build their own creative contributions to the philosophical debate.

## Notes

1. G. H. Lewes seems to have been the first philosopher in the British Emergentist tradition to use the term "emergent." See G. H. Lewes, *Problems of Life and Mind* I, 1874, p. 98. This is also the first usage of the term in the sense of "non-resultant effect" that is cited in the *Oxford English Dictionary*. Lewes's invention is clearly a metaphorical extension of earlier unscientific uses.

2. *A System of Logic*, Book III, Chapter X, Section 4.

3. *Space, Time, and Deity*, Volume II, p. 47.

# 1 The Rise and Fall of British Emergentism

**Brian P. McLaughlin**

My aim is to examine an account of the special sciences. The account can be found in the texts of a tradition which I hereby dub "British Emergentism." This tradition began in the middle of the nineteenth century and flourished in the first quarter of this century. It began with John Stuart Mill's *System of Logic* (1843), and traced through Alexander Bain's *Logic* (1870), George Henry Lewes's *Problems of Life and Mind* (1875), Samuel Alexander's *Space, Time, and Deity* (1920), Lloyd Morgan's *Emergent Evolution* (1923), and C. D. Broad's *The Mind and Its Place in Nature* (1925), the last truly major work in the tradition; but the tradition continues even today in the work of a few authors, for example, in the work of the noted neurophysiologist Roger Sperry. In what follows, I examine the coherence and plausibility of British Emergentism's account of the special sciences; in particular, I examine its doctrine of "emergent laws" which figures prominently in the account. And I attempt to explain how the British Emergentist tradition arose and why it has fallen.

## 1.1

Without pausing for scholarly qualifications, I will now briskly present, in modern dress, an *idealized* version of the main body of British Emergentist doctrines that will concern us. I should acknowledge that the Emergentists disagreed over some relevant details. I will, however, largely ignore their disagreements. My aim is to abstract a coherent and representative body of doctrines from their texts. I will defend my formulations of the doctrines on textual grounds in due course; but for now let us turn to the doctrines themselves. Scholarship enough will come later.

To begin, British Emergentism maintains that everything is made of matter: There are, for example, no Cartesian souls, or entelechies, vital elan, or the like.[1] And it holds that matter is grainy, rather than continuous; indeed, that it bottoms-out into elementary material particles, atoms or more fundamental particles. It even allows that there may be a single kind of material particle that wholly composes every kind of material object.[2] Moreover, on its view, nothing happens, no change occurs, without some motion of elementary particles. And all motion is to the beat of the laws of mechanics.

According to British Emergentism, there is a hierarchy of levels of organizational complexity of material particles that includes, in ascending order, the strictly physical, the chemical, the biological, and the psychological level. There are certain kinds of material substances specific to each level. And the kinds of each level are wholly composed of kinds of lower-levels, ultimately of kinds of elementary material particles. Moreover, there are certain properties specific to the kinds of substances of a given level. These are the "special properties" of matter. Physics is at the base of the hierarchy of sciences in that it is concerned with the properties common to all or nearly all material substances at whatever level of organizational complexity: These include properties such as inertial and gravitational mass, and electrical charge. Physics studies the organizational relationships such properties are responsible for, e.g., gravitational attractions, electro-magnetic attractions and repulsions, etc. But the material substances of a given level participate in certain organizational relationships in virtue of their special properties too. And it is part of the business of a special science to study the organizational relationships peculiar to a specific level and to formulate the laws governing those relationships.

While these views require further explication, they will, no doubt, seem familiar enough. What is especially striking about British Emergentism, however, is its view about the causal structure of reality. I turn to that view in the following two paragraphs.

British Emergentism maintains that some special science kinds from each special science can be wholly composed of types of structures of material particles that endow the kinds in question with fundamental causal powers. Subtleties aside, the powers in question "emerge" from the types of structures in question. Chemical elements, in virtue of their minute internal structures, have the power to bond with certain others. Certain biological organisms, in virtue of their minute internal structure, have the powers to breathe, to digest food, and to reproduce (Broad 1925, pp. 78–81). And certain kinds of organisms, in virtue of the minute internal structures of their nervous systems, have "the power of cognizing, the power of being affected by past experiences, the power of association, and so on" (Broad 1925, p. 436). These powers emerge from the types of structures in question. The property of having a certain type of structure will thus endow a special science kind with emergent causal powers. Such a structure will have an emergent causal power as a matter of law, but the law will be not be "reducible to" or "derivative from" laws governing lower levels of complexity and any boundary conditions involving the arrangements of particles. The laws that attribute such powers to the types of structures in question are "emergent laws." These laws "emerge" from the laws governing lower levels of complexity and boundary conditions involving the arrangements of particles, and so are in no sense derivative from them.

Now, the exercise of the causal powers in question will involve the production of movements of various kinds. Indeed, Emergentism maintains that special kinds, in virtue of possessing certain types of minute internal structures, have the power to influence motion. And here is the striking point: They endow the kinds with the power to influence motion in ways unanticipated by laws governing less complex kinds and conditions concerning the arrangements of particles. Emergentism is committed to the nomological possibility of what has been called "downward causation."[3]

These, then, are the main doctrines of British Emergentism that will concern us. As I noted, the doctrines are *idealizations* of doctrines that can be found in the texts cited earlier. I should mention that my interpretations of these texts differ from those I have found in my studies. I was tempted for a time to plagiarize a famous twentieth century philosopher and say that the Emergentist views that I present here are ones that occurred to me while reading the Emergentist texts. But I think I can claim more than just that. The views can be found in the texts. I will, in due course, defend my interpretations of the texts. I will examine the texts of Mill, Bain, Lewes, Alexander, and Morgan in great detail in section 1.3. And in section 1.5, I will examine Broad's texts. I will give Broad's texts special attention since the main doctrines of British Emergentism receive their most mature and careful formulation there. Moreover, it is Broad's texts which have received the most attention from critics of Emergentism.

Before turning to the texts, however, some preliminary remarks are in order to set the stage for later discussion.

## 1.2

Consider the doctrine that there are fundamental powers to influence motion associated with types of structures of particles that compose certain chemical, biological, and psychological kinds. Let us see what this would imply in the framework of classical mechanics, for example. It would imply that types of structures that compose certain special science kinds can affect the acceleration of a particle in ways unanticipated by laws concerning forces exerted by pairs of particles, general laws of motion, and the spatial or spatio-temporal arrangements of particles. In a framework of forces, the view implies that there are what we may call "configurational forces": *fundamental* forces that can be exerted only by certain types of configurations of particles, and not by any types of pairs of particles. Such forces contrast with what we may call "particle-pair forces," types of forces that can be exerted by (at least some types of) pairs of elementary particles. Thus, for example, in classical mechanics, the gravitational and electro-magnetic forces are particle-pair forces. Nowadays, we speak of forces that cannot be exerted by pairs of elementary particles; I have in mind such forces as Van der Waals forces, London forces, and Dipole-Dipole forces. But these are not configurational forces. For they are not fundamental forces: They are all electro-magnetic in origin. And configurational forces are, by stipulation, fundamental forces. In section 1.4, I will argue that configurational forces can be easily accommodated in classical mechanics. But, for the moment, I want to restrict the discussion to some general observations.

Mechanics is, of course, the science of motion. The laws of motion are the laws of mechanics. And, as I said, all motion is to the beat of the laws of mechanics. In the framework of classical mechanics, the above view would have to maintain that there are *fundamental* force-laws that cite the configurational forces in question. Are the Emergentists, then, committed to the view that the laws of the special sciences are derivative from the laws of physics? The Emergentists can concede that the laws of the special sciences are derivative from the laws of *mechanics*.[4] But whether they are derivative from the laws of *physics* is another matter. The special sciences concern themselves with specific kinds of substances. Suppose that physics is understood to be solely concerned with properties common to substances at all levels of organizational complexity (properties such as, e.g., mass and charge). Then, if there are configurational forces, mechanics is *not* a branch of physics. But, one may ask: Is not mechanics, by definition, a branch of physics? There is no need to squabble over the word "physics." Suppose that mechanics is, by definition, a branch of physics. Then, it is consistent with British Emergentism that the laws of the special sciences are derivative from laws of physics. But if British Emergentism's view of the laws of the special sciences were correct, then some fundamental laws of physics would concern structural properties specific to very specific kinds of things, for example, various chemical kinds, biological kinds, and so on. And there would, moreover, be "downward causation" from the psychological, to the biological, to the chemical, to the subchemical levels of organizational complexity. For at least some kinds from each special science will be composed of types of aggregates of particles that exert configurational forces.[5]

Let us pause for a moment to ask what the Emergentist notion of emergent causal powers and laws at the chemical, biological, and psychological levels would mean in the context of nonrelativistic quantum mechanics. Schrödinger's equation is the fundamental law of nonrelativistic quantum mechanics. It governs the evolution of systems through time. It tells us that the temporal evolution of a state vector $\psi$ is determined by

$$H\psi = ih\frac{\delta\psi}{\delta t},$$

where H is the Hamiltonian operator and h is Planck's constant divided by $2\pi$. Now, to employ the equation, one must independently determine the Hamiltonian. The Hamiltonian concerns energy, rather than forces. (Quantum mechanics could, however, be recast in terms of forces, it is just that the mathematics would be considerably more complex; scalars are, of course, easier to compute with than vectors.) But on the Emergentist view in question there would be kinds of energies specific to types of structures of particles that compose certain chemical, biological, and psychological kinds.

Hereafter, I will, however, focus on the notion of configurational forces, rather than energies and the Hamiltonian. Quantum mechanics was not developed until just after the publication of *The Mind and Its Place in Nature*; and this was, as I mentioned, the last major work in the Emergentist tradition. Alexander, Morgan, and Broad lived to see the advent of quantum mechanics. But when they were writing in the Emergentist tradition, they knew nothing of Schrödinger's equation or the like.

It is, I contend, no coincidence that the last major work in the British Emergentist tradition coincided with the advent of quantum mechanics. Quantum mechanics and the various scientific advances it made possible are arguably what led to British Emergentism's fall. It is not that British Emergentism is logically incompatible with nonrelativistic quantum mechanics. It is not. Schrödinger's equation could be the fundamental equation governing motion in a world with energies that are specific to types of structures of particles that compose certain chemical, biological, and psychological kinds. But, as will become apparent, quantum mechanical explanations of chemical bonding in terms of electro-magneticism, and various advances this made possible in molecular biology and genetics—for example, the discovery of the molecular structure of DNA—make the main doctrines of British emergentism, so far as the chemical and biological are concerned at least, seem enormously implausible.[6] Given the achievements of quantum mechanics and these other scientific theories, there seems not a scintilla of evidence that there are emergent causal powers or laws in the sense in question; there seems not a scintilla of evidence that there are configurational forces; and there seems not a scintilla of evidence that there is downward causation from the psychological, biological, or chemical levels.[7]

On the current evidence, the main doctrines of British Emergentism seem "kooky." As should go without saying, however, the epistemic situation was not always this way. We should keep in mind that Mill, Bain, and Lewes, for example, knew nothing of the atom's structure. J. J. Thomson did not discover the electron, "cathode rays," until around 1897, thirteen years after Mill's death. (The existence of atoms was even resisted by some in the community of chemists, until after Einstein's 1905 paper on Brownian Motion.) And there were, in Mill's, Bain's, and Lewes's lifetimes, no even remotely plausible micro-explanations of chemical bonding. The laws concerning which chemical elements have the power to bond with which others seemed to many like "brute facts" that admitted of no explanation. To borrow a phrase from Wordsworth which was a favorite of the Emergentists, it was thought that the laws of chemical bonding had to be accepted with "natural piety."

And there was talk of chemical forces. Consider the following remarks by Walter Nernst in his 1909 review article entitled "Development of General and Physical Chemistry During the Last Forty Years":

> we are obliged to admit that during the period under consideration [c. 1865–1905] there has been no answer to the question [of the nature of chemical bonding] which really tells us anything more than we can see with our own eyes. It seems reasonably certain that we should admit the existence, not only of electrical and therefore polar forces, but of nonpolar natural forces somewhat of the nature of Newtonian gravity. (Quoted in Bantz 1980, p. 311)

Here, thirty five years after Mill's death, we find a highly distinguished chemist suggesting that chemical bonding may well involve fundamental chemical forces.

When they were writing in the British Emergentist tradition, Alexander, Morgan, and Broad knew of electrons and protons (albeit not neutrons, which were not discovered until 1932). But they knew of no plausible micro-explanations of chemical bonding. As we will see later, both Alexander and Broad mention the possibility that chemical bonding might be explained by appeal to properties of electrons, but they considered it an open question whether the possibility of such an explanation could be realized. The gap between chemistry and physics seemed vast. And so did the gap between biology and chemistry. It seemed very much an empirical question whether the gaps could be bridged.

Broad's *The Mind and Its Place in Nature* consists of the Tarner Lectures he delivered in Trinity College, Cambridge in 1923. Some months earlier, in June 1922, Niels Bohr gave a series of lectures at the University of Göttingen on quantum theory and atomic structure (Gribbin 1984, p. 71). In these lectures, he presented a revolutionary new application of his solar system model of the atom.[8] Using his model, Bohr proposed to explain the periodicity of the periodic table, and some of the chemical properties of elements in terms of the electrons in their outermost shells. While his solar system model of the atom proved wrong, his qualitative explanation of chemical properties in terms of electrons survives in essentially the same form today. Quantum mechanics deepened the qualitative explanations of chemical bonding in terms of electrons. (You will recall that in the years form around 1926 to 1928, Heisenberg, Schrödinger, and Dirac independently worked out theories of the mechanics of quanta; and the theories were shown to be mathematically equivalent by von Neumann.) The explanation of chemical phenomena is generally touted as one of the greatest achievements of quantum mechanics. It took a little time for the dust to settle after the initial quantum mechanical revolution, of course, and for its influence to spread to biology.[9] But once it did, it was all quickly downhill for the Emergentist tradition. It is not at all surprising that Broad's *The Mind and Its Place in Nature* was the last truly major work in the British Emergentist tradition. Its publication in 1925 was followed soon after by the advent of the quantum mechanical revolution.

British Emergentism flourished in a time before that revolution. Prior to that revolution, Emergentism's main doctrines appeared to many to be exciting empirical hypotheses worthy of serious consideration. Emergentism was mainly inspired by the dramatic advances in chemistry and biology in the nineteenth century that made psychologically salient the conceptual chasms between physics and chemistry and between chemistry and biology, and by the failures of various attempts to build conceptual bridges to cross those chasms. (The conceptual gap between the psychological and the biological had, of course, been a topic of extensive discussion since the seventeenth century.) And the proliferation of special sciences made the world seem to many like a fundamentally diverse place.

Given the epistemic situation during the first third of this century, it is not surprising that there was much talk of emergence. Alfred Whitehead defended a brand of emergentism.[10] There were American Emergentists, William James, Arthur Lovejoy, and Roy Wood Sellars. The French philosopher Henri Bergson developed his own brand of emergentism. And there was, in the nineteen twenties, a raging debate in the Soviet Union between "mechanists" and the emergentists of the Deborin School, headed by A. M. Deborin.[11] The Deborin School spoke of the emergence of new forms in nature, and maintained that the mechanists "neglected the specific character of the definite levels or stages of the development of matter" (Kamenka 1972, p. 164).[12]

I will, however, focus exclusively on the British Emergentists. Their views are particularly close; and later figures in the British Emergentist tradition cite earlier figures. Moreover, the British Emergentist tradition, I believe, inspired many of the others.[13] Let us turn now to the texts of that tradition.

## 1.3

This section is mainly intended as a brief overview of a small piece of intellectual history, presented assembly-line style: from Mill, to Bain, Lewes, Alexander, and then to Morgan. To keep the conveyor-belt moving to meet production deadlines, I will let some philosophical issues pass unaddressed. But I try to address the main philosophical issues raised by this historical discussion in sections 1.4 and 1.5. In section 1.4, I discuss in detail how configurational forces could figure in classical mechanics; and I briefly discuss how such forces might fare in special and general relativity. In section 1.5, I address some key philosophical issues as they arise in the context of Broad's Emergentism. Emergence implies, for instance, a certain kind of irreducibility. And, in section 1.5, I compare and contrast Broad's notion of reduction with those of Ernest Nagel (1961) and Robert Causey (1977). But turn now to some intellectual history.

On the second page of *Emergent Evolution*, Lloyd Morgan says:

> The concept of emergence was dealt with (to go no further back) by J. S. Mill in his *Logic* (Bk. III. ch. vi 2) under the discussion of "heteropathic laws" in causation. (1923, p. 2)

Mill did not, however, use the word "emergent." Morgan tells us that:

> The word "emergent," as contrasted with "resultant," was suggested by G. H. Lewes in his *Problems of Life and Mind* (Vol. II Prob. V. ch. iii, p. 412) (pp. 2–3).

And Alexander says:

> I use the word "emergent" after the example of Mr. Lloyd Morgan...it contrasts with..."resultant"...The word ["emergent"] is used by G. H. Lewes...as Mr. Lloyd Morgan reminds me. (1920, p. 14)

Lewes's notion of a "resultant" is just the notion of a vector sum of state types. His notion of an emergent is, as we will see in due course, just Mill's notion of a heteropathic effect; and his distinction between emergents and resultants is directly owed to Mill. Nagel (1961) cites Mill's "Of the Composition of Causes" in *System of Logic* as the *locus classicus* on the notion of emergence. And, indeed, it is. Mill, it is fair to say, is the father of British Emergentism.[14]

In "Of the Composition of Causes" Mill introduces the notions of a heteropathic law and a heteropathic effect by appeal to a certain distinction. The distinction, he tells us, is "so radical, and of so much importance, as to require a chapter to itself" (1843, p. 427). The distinction is between "two modes of the conjoint action of causes, the mechanical and the chemical" (p. xviii).[15] Since Mill's discussion of this distinction launched the British Emergentist movement, let us examine it detail.

Of the mechanical mode, Mill says:

> In this important class of cases of causation, one cause never, properly speaking, defeats or frustrates another; both have their full effect. If a body is propelled in two directions by two forces, one tending to drive it to the north and the other to the east, it is caused to move in a given time exactly as far in both directions as the two forces would separately have carried it; and is left precisely where it would have arrived if it had been acted upon first by one of the two forces, and afterwards by the other. This law of nature is called, in dynamics, the principle of the Composition of Forces: and in imitation of that well-chosen expression, I shall give the name of the Composition of Causes to the principle which is exemplified in all cases in which the joint effect of several causes is identical with the sum of their separate effects. (1843, p. 428)

Since forces are vectors, the principle of the Composition of Forces employs, of course, vector addition. Mill takes the laws of the vector addition of forces (such as, e.g., the parallelogram law) to be the paradigmatic principles of Composition of Causes. Given the Composition of Forces, forces acting together always exhibit the mechanical mode of conjoint action of causes: For the effect of two or more forces acting together is the vector sum of the effect of each force.[16] More generally, two or more types of causes acting together would produce a certain type of effect in the mechanical mode if and only if the effect type is the sum, the vector sum or the algebraic sum (as the case may be), of the type of effects each of the cause types has according to the laws in which it figures as the sole causal factor (its "laws as a separate agent"). Mill calls a type of effect of two or more types of causes which would produce it in the mechanical mode, a "homopathic effect." Mill calls laws which assert causal relations between causes and homopathic effects of those causes, "homopathic laws." Causal transactions in the mechanical mode are covered by homopathic laws.

Turn now to the chemical mode of the conjoint action of causes. In the chemical mode, in contrast to the mechanical, the effect of the joint action of two or more types of causes is not, Mill says, the sum of the effects each type of cause has according to the laws in which it figures as the sole causal factor. Thus, the chemical mode is just the absence of the mechanical mode. Mill speaks here of the chemical mode of the conjoint action of causes since, he maintains, chemical transactions typically exhibit it. For example, consider the following type of chemical process:

$$CH_4 + 2O_2 \rightarrow CO_2 + 2H_2O$$

(Methane + oxygen produces carbon dioxide + water).

Here the product is not, in any sense, the sum of the effects of each reactant. This is, Mill holds, characteristic of chemical processes. Mill calls an effect type of two or more causes types which would combine in the chemical mode to produce it, a "heteropathic effect." Laws which assert causal relations between causes and heteropathic effects of those causes, he calls "heteropathic laws." A heteropathic law owes its existence, Mill says, to a breach of the Composition of Causes. Heteropathic laws cannot, he holds, be derived from the laws of the causal factors "as separate agents" and any principle of Composition of Causes.

The fact that there are breaches of the Composition of Causes is, according to Mill, part of the explanation of the existence of the various special sciences. He says that: "where the principle of Composition of Causes . . . fails . . . the concurrence of causes is such as to determine a change in the properties of the body generally, and render it subject to new laws, more or less dissimilar to those to which it conformed in its previous state" (1843, p. 435). The new laws are the concern of the various special sciences.

The new laws, Mill says, can "supersede" the old:

> in some instances, at some particular points in the transition from separate to united action, the laws change, and an entirely new set of effects are either added to, or take the place of, those which arise from the separate agency of the same causes: the laws of these new effects being again susceptible of composition, to an indefinite extent, like the laws which they superseded. (1843, pp. 433–434)

The last point is that heteropathic effects of certain causes can combine with each other in accordance with the Composition of Causes. But of that, more shortly. When the laws of separate agency contain "unless counteracted" clauses (1843, p. 517), a "new set of effects takes the place of those which arise from the separate agency of the same causes." But, when they do not contain such clauses, new effects may still be "added to" the effects of the separate agents as separate agents; and such new effects will be heteropathic. While the new laws will supersede the old ones, they will not contravene them. The old laws will continue to hold. Speaking of this situation in the case of vegetable and animal substances, he says:

> Those bodies continue, as before, to obey mechanical and chemical laws, in so far as the operation of those laws is not counteracted by the new laws which govern them as organized beings. (1843, p. 431)

When the "unless counteracted" clause of a law fails to be satisfied, the law is inapplicable, and so is *not* contravened. In cases in which the laws contain no such clauses, as Mill thinks is the case with respect to, for instance, the law of gravity, "new effects" are added to (i.e., occur in addition to) the "old effects." For example, when an unqualified (i.e., nonceteris paribus) heteropathic law and unqualified laws citing the individual agents alone are all instantiated in a situation, both the effects cited in the laws of the separate agents *and* the effect cited in the heteropathic law concerning their joint action occur. So, the laws of the separate agents as separate agents are not contravened.

Let us look briefly at how the distinction between homopathic and heteropathic laws fits into Mill's picture of the scientific enterprise. Mill claims that all sciences should strive to be deductive. There are, he says,

> weighty scientific reasons for giving to every science as much of the character of a Deductive Science as possible; for endeavouring to construct the science from the fewest and the simplest possible inductions, and to make these, by any combinations however complicated, suffice for proving even such truths, relating to complex cases, as could be proved, if we chose, by inductions from specific experience. (1843, p. 251)

A science is deductive in Mill's sense, if it contains a small group of systematically well-integrated laws from which all its other laws can be derived. Chemistry, he says, is far from being a deductive science. Virtually all of its laws must, he says, be arrived at empirically, by induction. The only way to arrive at any law of chemical bonding is to perform experiments to see which elements bond with which others. Certain elements have the power to bond with certain others, and that is all there is to it. Moreover, Mill says "the different actions of a chemical compound will never, undoubtedly, be found to be the sums of the actions of its separate elements" (1843, p. 432).

However, he tells us not to despair over the prospects of chemistry and physiology ever becoming deductive. For:

> Though there are laws which, like those of chemistry and physiology, owe their existence to a breach of the principle of Composition of Causes, it does not follow that these peculiar, or as they might be termed, *heteropathic* laws, are not capable of composition with one another. The causes which by one combination have had their laws altered, may carry their new laws with them unaltered into their ulterior combinations. And hence there is no reason to despair of ultimately raising chemistry and physiology to the condition of deductive sciences; for though it is

impossible to deduce all chemical and physiological truths from the laws or properties of simple substances or elementary agents, they may possibly be deducible from laws which commence when these elementary agents are brought together into some moderate number of not very complex combinations. The Laws of Life will never be deducible from the mere laws of the ingredients, but the prodigiously complex Facts of Life may all be deducible from comparatively simple laws of life; which laws (depending indeed on combinations, but on comparatively simple combinations, of antecedents) may, in more complex circumstances be strictly compounded with one another, and with the physical and chemical laws of the ingredients. The details of the vital phenomena, even now, afford innumerable exemplifications of the Composition of Causes. (1843, pp. 431–432; emphasis Mill's)

His idea seems to be that the laws of life will never be deducible from the laws of chemical elements, for there will be a breach of the Composition of Causes.[17] Vital phenomena are heteropathic effects of chemical agents. But chemistry and physiology themselves may approximate the ideal of being a deductive science. For many laws of chemistry may be deducible from a small group of fundamental laws of chemistry.[18] And many laws of life may be deducible from a small group of fundamental laws of life. One reason is, to use his terminology, that heteropathic effects can combine to produce homopathic effects. This too is part of the explanation of why there are various special sciences. Laws cluster into systematically related groups.[19]

Now the causal agents of the various special sciences have the power to move things in various ways. Chemical processes involve, as Mill recognized, rearrangements of elements; biological processes involve rearrangements of elements and compounds, and so on. Combinations of causal agents can, it seems, influence motion in ways unanticipated by their laws as separate agents and any principle of Composition of Causes. Various "collocations of causal agents" have fundamental powers to influence movement. It is hardly surprising that Bain picked on up this very theme. Bain (1870) explicitly endorses Mill's distinction between homopathic and heteropathic laws. And he discusses how the latter concern types of causal factors which arise only when certain "collocations of agents occur." Bain says that some types of causal transactions follow the rule of Composition of Causes. But he says that combinations of causal factors "that merely make good the collocation for bringing a prime mover into action, or that release a potential force, do not follow any such rule" (1870, ii, p. 31). According to Bain, *certain collocations of agents bring into action new forces of nature*; forces of a sort not exerted by any less complex collocations. Bain quite explicitly espouses the existence of what I called configurational forces.

In the seventh and eighth editions of *System of Logic* Mill frequently speaks approvingly of Bain; though he questions some of Bain's examples of breaches of the Composition of Causes (see Mill 1843, p. 435). He speaks with great approval of Bain's treatment of the new principle of the conservation of energy in connection with the notion of heteropathic effects (1843, Preface, pp. x–xi, see also p. 406). Bain argues that such new forces need not violate the principle of the conservation of energy.[20] Mill applauds Bain's discussion of the principle and joins Bain in speaking of potential forces which are actualized only when certain "collocations of factors" occur.[21]

Moreover, Mill speaks of chemical forces:

> To produce a bonfire, there must not only be fuel, and air, and a spark, which are collocations, but *chemical action* between the air and the materials, *which is a force*. (Mill 1843, p. 407; emphases mine)

This was, at the time, a perfectly reasonable position. Recall that thirty five years after Mill's death, the famous chemist Walter Nernst was still speaking of the need to postulate chemical forces to explain chemical bonding. Mill also seems to hold that there are collocations of agents that possess vital and psychological force-giving properties. And he speaks of the force of will that causally effects bodily movements, arguing that it is a physical force since it is exerted by certain collocations of wholly material agents (1843, p. 410). The forces in question would all be configurational forces.

Following Bain, Mill cautions against reifying forces themselves, however. Certain collocations of agents produce motion. To speak of forces is just to speak of the effects the collocations would have on motion (1843, pp. 406–409). He says:

> To produce a bonfire, there must not only be fuel, and air, and a spark, which are collocations, but chemical action between the air and the materials, which is a force.
>
> To grind corn, there must be a certain collocation of the parts composing a mill, relatively to one another and to the corn; but there must also be the gravitation of water, or the motion of wind, to supply a force. But as the Force in these cases was regarded as a property of the objects in which it is embodied, it seemed a tautology to say that there must be the collocation *and* the force. As the collocation must be a collocation of objects possessing the force-giving property, the collocation, so understood, included the force. (1843, p. 407; emphasis Mill's)

As we will see later, this reluctance to reify forces themselves can be found in most of the other works we will examine. Whether we should countenance forces themselves, however, will not matter for what follows. We could, if we like, recast talk of forces in terms of talk of the properties that influence motion.[22] The important point to note for now is just that Mill holds that collocations of agents can possess fundamental force-giving properties of a sort not possessed by any of the individual agents.

Turn to Lewes (1875). Lewes, as was mentioned earlier, coined the term "emergent." He contrasts emergents with resultants. An emergent, in his sense, is just a heteropathic effect in Mill's sense. A type of effect of two or more types of causes is an emergent, in Lewes's sense, if and only if it is not the sum of the types of effects of each type of cause has according to the laws in which it figures as a separate agent. A resultant, in his sense, is just a type of effect of two or more types of causes that is the sum of the effect of each type of cause as a separate agent. Explicating the notion of a resultant, Lewes says: "Again, in the somewhat more complicated effect of compound motions, say the orbit of a planet—the resultant of its tangential direction and its direction towards the sun—every student learns that the resultant motion of two impressed forces is the diagonal of those directions which the body would take were each force separately applied. Every resultant is either a sum or a difference of the co-operant forces" (1875, p. 413). Lewes's idea, then, is that heteropathic effects *emerge* from the causal factors relative to which they are heteropathic. And heteropathic laws citing such cause and effect relationships (the ones involving the chemical mode) are *emergent* laws: They *emerge* from the laws governing the causal factors as separate agents and any additive principle of composition of causes (such as, e.g., the parallelogram law for force-vectors).

The introduction of the term "emergent" is, I should note, Lewes's main contribution to Emergentism. Emergentism does not figure heavily in his work. His primary aim is to defend British Empiricism, not what I have called British Emergentism. But his contribution to British Emergentism is an important one. "Emergence" captures the imagination in ways that "heteropathic law" and "heteropathic effect" do not.[23] There is something in a name.

Closely related notions of emergence figure prominently in the work of Morgan and Alexander. Let us turn to them.

Alexander speaks of emergent "qualities." He says:

> The emergence of a new quality from any level of existence means that at that level there comes into being a certain constellation or collocation of the motions belonging to that level, and this collocation possesses a new quality distinctive of the higher-complex. (1920, p. 45)

The idea, then, is that a certain collocation of motions possess a new "emergent" quality. Alexander goes on to say:

> The higher-quality emerges from the lower level of existence and has its roots therein, but it emerges therefrom, and it does not belong to that lower level, but constitutes its possessor a new order of existent with its special laws of behavior. The existence of emergent qualities thus described is something to be noted, as some would say, under the compulsion of brute empirical fact, or, as I should prefer to say in less harsh terms, to be accepted with the "natural piety" of the investigator. It admits no explanation.
>
> To adopt the ancient distinction of form and matter, the kind of existent from which the new quality emerges is the "matter" which assumes a certain complexity of configuration and to this pattern or universal corresponds the new emergent quality. (1920, pp. 46–47)

I am hesistant in my interpretation of Alexander, since, to be frank, I find apparently conflicting passages in his texts and I am uncertain how to resolve the apparent conflicts.[24] I will spare the reader a discussion of these conflicts, however, since it would require a detailed examination of Alexander's texts; and they are especially difficult since they are loaded with his technical terminology. A proper treatment of Alexander's views will have to await another occasion. But for now, I want to note a reading of Alexander's texts that I find plausible: A complex configuration of elements of a given level is itself on a higher-level, if it possesses a certain kind of quality not possessed by the elements. The examples Alexander gives of qualities are often examples of powers, dispositions, or capacities. The qualities that elevate the configuration to a new level must be those of having the power, or capacity, or disposition to produce certain sorts of effects, where the laws that connect the configurations with these effects are special laws of behavior specific to those configurations. And the special laws must in no sense be derivative from the laws of behavior of the elements as separate agents and any principle of composition of causes. That is the sense in which the emergent quality admits of no explanation and, so, must be accepted with natural piety.

Thus, Alexander's main idea is, I believe, that a certain complex configuration of elements of a given level may possess capacities and dispositions to produce certain types of effects that are, in Mill's sense, heteropathic relative to the elements in the configuration. Qualities (causal powers, capacities, and behaviorial dispositions) *emerge* from the configuration. And a configuration's possession of the qualities involves its figuring in special laws of behavior not derivative from lower-level laws, principles for combining causes, and boundary conditions.

This interpretation is supported by Alexander's discussion of Driesch's Vitalism (see 1920, pp. 61–73). Alexander rejects the thesis that there are entelechies. He thinks that vital behavior can be fully explained in terms of the internal structures of organisms. These internal structures are structures of physico-chemical processes. These processes are of the vital order because of the complexity of their organizational structure. Vital capacities, dispositions, and powers emerge from the structure. And in virtue of their complex physico-chemical structures, organisms obey special laws of vital behavior. Finally, I should note that Alexander maintains that if the behavior of chemical elements can some day be explained in terms of laws governing electrons, then chemical elements do not possess emergent qualities (see, especially, pp. 52–55).

Turn now to Morgan. The first section of Morgan's *Emergent Evolution* (1923) is entitled "Emergents and Resultants." In it he cites his debt to Mill and Lewes and proceeds to provide some of the same examples they provide of emergents and resultants. His chief example of an emergent is, for instance, a chemical example. He says, "When carbon having certain properties combines with sulphur having other properties there is formed, not a mere mixture but a new compound, some of the properties of which are quite different from those of either component" (1923, p. 3). He contrasts this case with a case involving a resultant: "the weight of the compound is an additive resultant, the sum of the weights of the components" (1923, p. 3). Morgan's chief concern is to argue that, through a process of evolution, new, unpredictable complex phenomena emerge. He rejects Cartesian Dualism and the existence of entelechies, vital elan, or the like. Every substance is or is wholly composed of elementary material particles. However, he combines this idea with the idea of emergence and with an evolutionary cosmology inspired by Darwinian evolution.

Morgan contrasts his emergentist, evolutionary cosmology with a "mechanistic" cosmology, rejecting the latter. And he says:

> The essential feature of a mechanical—or, if it be preferred, a mechanistic—interpretation is that it is in terms of resultant effects only, calculable by algebraic summation. It ignores the something more that must be accepted as emergent... Against such a mechanical interpretation—such a mechanistic dogma—emergent evolution rises in protest. The gist of its contention is that such an interpretation is quite inadequate. Resultants there are; but there is emergence also. Under naturalistic treatment, however, the emergence, in all its ascending grades, is loyally accepted, on the evidence, with natural piety. That it cannot be mechanically interpreted in terms of resultants only, is just that for which it is our aim to contend with reiterated emphasis. (1923, p. 8)

The various emergent levels in the ascending grades of complexity of matter are the subject matter of the various special sciences.

The various grades of complexity have their own laws. Morgan speaks of the laws in which a structure of particles participates as "modes of extrinsic relatedness." And he says that such laws are "effective." By this, he says he means that they make a difference to the "go of events" (to use his expression) at lower levels of complexity. And the difference they make is unanticipated by lower-level laws concerning less complex wholes. The laws of a higher level, he says, "involve" lower-level laws. But, he holds, they are not wholly implemented by them. The course of events at lower-levels "depends," he says, in part, on the higher-level laws. In Morgan, one finds the notion of downward causation clearly and forcefully articulated.

Morgan says that he does not speak explicitly of new forces only because "There is... some ambiguity in the word 'force'" (1923, p. 21). He takes it that in one sense, it suggests a special kind of agency at work. He rejects forces in this sense. He does not, for example, want to be accused of postulating a special agency, such as an entelechy. In the other sense, the "scientific sense," the word "force" functions as a kind of middle term between the general laws of motion and the so-called force-laws. He thinks that talk of forces, in that sense, can be eliminated in favor of direct talk of the factors that causally influence acceleration. (Here he follows Mill and Bain, of course.) In the scientific sense of "force," he is perfectly happy to admit he holds that there are chemical and vital forces (1923, pp. 21–22). The forces in question would, of course, be configurational forces.

That, then, is my brief, assembly-line style, history of British Emergentism from Mill to Morgan. There are, no doubt, a number of unanswered questions. I try to anticipate the main ones in the following two sections. In section 1.4, I examine the notion of a configurational force itself; and, in the process, I help myself to some quotes from Broad's *Scientific Thought* (1923). Then, in section 1.5, I turn to Broad's *The Mind and Its Place in Nature*, where, as I said, the main doctrines of British Emergentism receive their most mature formulation.

## 1.4

Classical mechanics includes, of course, Newton's three general laws of motion: the law of inertia, the law of acceleration, and the law of the conservation of momentum. As Broad noted in *Scientific Thought*:

> [These] laws of motion do not profess to tell us in detail how motions are caused or modified. What they do is to tell us the general conditions which all motions, however produced, must conform to. They take no account of the kind of matter which is moved, or of its physical or chemical state at the time... The special laws of nature, on the other hand, tell us about the various causes of motion. They have to take into account all sorts of properties of bodies beside their inertial masses. (1923, p. 177)

Special laws are, of course, force-laws. They tell us the types of causal factors that, as Broad says, "start and modify motion." Special laws, he thinks, must "take into account all sorts of properties of bodies beside their inertial masses." In this passage, Broad speaks of special laws having to cite chemical states. As we will see in the next section, Broad thinks chemical elements can influence movements in fundamental ways. While he does not specifically mention vital properties in *Scientific Thought*, that is because the work contains no detailed discussion of the kinds of properties that will actually have to figure in special laws. As we will see in the next section, Broad seems committed to maintaining that special laws must take some vital properties into account too.[25] In this section, however, let us focus on the notion of configurational forces itself. Let us see how such forces could figure in the framework of classical mechanics. (I will also make a few remarks about configurational forces in relation to special and general relativity.)

To begin, consider the law of gravity: Two objects will exert a gravitational force on each other that is directly proportional to the product of their masses, and inversely proportional to the square of the distance between them. Broad thinks that this force-law is in some ways atypical. Here is why. The gravitational force-generating property is mass; and the strength of the gravitational force, always an attractive force, is affected by spatial distance. Every object possesses gravitational mass and has spatial location, and so every object participates in gravitational attractions. For this reason, Broad says:

> there seems to be a very much closer connexion between the [general] laws of motion and the law of gravitation than between any of the special laws of nature and the [general] laws of motion. (1923, p. 177)

Most "special laws of nature," Broad thinks, concern properties specific to specific kinds of matter. The special laws of nature involve, he holds, properties "that vary from one bit of nature to another" (1923, p. 167).

There is of course nothing problematic in the classical framework with the idea that there may be force-generating properties that are possessed only by a limited range of kinds of matter. Consider the other inverse-square law of classical mechanics, namely Coulomb's law. It says that the electrostatic force exerted between two objects is directly proportional to the product of the charges of the objects, and inversely proportional to the square of the distance between them. When this law was proposed, it was thought that many kinds of objects had no charge at all. But, nonetheless, Coulomb's law was taken to be a fundamental force-law. Moreover, the magnetic force was thought to be exerted only by things which contain moving electrical charges. Many things, it was thought, exert no magnetic force. (Nowadays, we speak of the strong nuclear force, and that force is not exerted by electrons, for instance. My point is just that there may be force-generating properties not possessed by every kind of matter.)

Let us see how force-laws work together. Consider the inverse-square laws: the law of gravity and Coulomb's law. When two objects have mass and are charged, the resultant of the force due to gravity and the force due to electricity is arrived at by vector addition. Since forces combine by vector addition, we can, in principle, arrive at the resultant, the vector sum, of all the forces exerted on a given object. We can then plug this vector sum into the second general law of motion, $F = ma$, to determine the acceleration of the object in question. Further special laws, in addition to the inverse square laws, are easily accommodated in the framework of classical mechanics. They just specify further basic factors that, when present, causally influence the accelerations of objects.[26] Even with the addition of further special laws, we can still, in principle, use vector addition to arrive at the resultant of all the forces exerted on an object in any situation. And then we can, in principle, use $F = ma$ to determine the acceleration of the object. Force-laws act together without contravening one another.[27] And it should go without saying that the general laws of motion and the inverse-square laws are consistent with there being many fundamental special laws, or force-laws, which cite "all sorts of properties of bodies," including many specific to only certain kinds of substances.

Now the idea of a configurational force is, you will recall, that of a force that can be exerted only by substances with certain types of structures, where the forces are such that they cannot be exerted by any kinds of pairs of elementary particles. Our concern is, of course, not with different forces exerted by different kinds of elementary (non-complex) substances. Emergentism, you will also recall, maintains that all the different kinds of substances there are can be wholly composed of elementary particles in different proportions and different spatial or spatio-temporal arrangements. And it even allows that there may be a single kind of elementary particle. Aggregates of particles are type-individuated just by the differences in their proportions (if there is more than one kind) and by their spatial or spatio-temporal arrangements: No two types of aggregates of particles can be wholly composed of exactly the same kind or kinds of elementary particles in the same proportions and spatial or spatio-temporal arrangements. And Emergentism maintains that certain types of aggregates generate fundamental forces not generated by any pairs of elementary particles. We want to know whether configurational forces can be accommodated in the classical framework. They can, easily.

Configurational forces are not excluded by the general laws of motion. As Broad noted, they just state the general conditions to which motion must conform. They do not tell us the factors that "start or modify" motion. They say nothing about whether forces most be particle-pair forces or whether they can be configurational forces. The law of inertia, you will recall, just states that every object remains in a state of rest or in uniform motion in a straight line unless acted upon by net outside forces. There is no conflict here with the existence of configurational forces. The second general law of motion, F = ma, just states that the net force on an object equals the product of its inertial mass and its acceleration. Configurational forces could be among the forces acting on an object. Configurational forces would be vectors that would combine by vector addition with particle-pair forces, and so could be component vectors of the net force on an object. And thus we could use F = ma to determine the acceleration of either an aggregate or an individual particle acted on by a configurational force or a force that has a configurational force as a component force-vector. Moreover, the force-fields generated by appropriate configurations can originate with particle force-fields. It would just be the case that particles have certain fields which have a value of 0 until the particle figures in an appropriate configuration. Nor are configurational laws excluded by the third law, the law of the conservation of momentum. According to this law, there is for every action an equal and opposite reaction; momentum is conserved; when two bodies exert a force on each other, their momenta (the products of their inertial masses and their velocities) due to the exertion will be equal and in opposite directions. There is no reason to think this cannot be so even when the force exerted is a configurational force.

Let us turn to other conservation principles of classical mechanics. In the classical framework, it was assumed that chemical processes, for example, do not violate the principle of the conservation of mass.[28] But configurational chemical forces could respect that principle. The mass of a configuration of particles exerting a configurational force could be the sum of the masses of the constituents of the configuration. It would just follow that mass is not a configurational force generating property. (No Emergentist claimed it was.) Consider next the principle of the conservation of energy. That principle imposes no constraint that would exclude the existence of configurational forces. Configurations of particles exerting configurational forces will have the capacity to do work that could not be anticipated just by considerations of the particles in isolation and their spatial arrangements. But that does not violate the conservation of energy. It is possible for it to be the case that a particle contains a certain kind of potential energy that can be released only when the particle figures in an appropriate configuration.[29]

Classical mechanism is often portrayed as being deterministic.[30] Many of the Emergentists were causal determinists. Mill, for example, is well-known as a champion of causal determinism. Consider the following often quoted remark:

> the state of the whole universe at any instant ... [is] the consequent of its state at the present instant; inasmuch that one who knew all the agents which exist at the present moment, *their collocation in space, and all their properties*, in other words, the laws of their agency, could predict the whole subsequent history of the universe. (1843, p. 247; emphasis mine)

Causal determinism can be squared with the existence of configurational forces. Certain collocations of agents will have properties in virtue of which they exert configurational forces. A Laplacean Demon would just have to take into his calculations the configurational forces as well as the particle-pair forces.[31]

This completes my discussion of how configurational forces can be accommodated in classical mechanics.[32] Before turning to Broad, however, a few remarks about relativity theory are in order.

The theories of special and general relativity have had, you will recall, some striking consequences for our views about what the forces of nature are. General relativity treats gravity not as a force, but as a geometrical property of spacetime. And it was relativity theory that enabled us to see that electricity and magnetism were manifestations of the same (frame-invariant) force, the electro-magnetic force. Let us briefly pause to ask whether configurational forces are compatible with special and general relativity.[33]

Relativity theory introduced the principle of the conservation of mass-energy.[34] Configurational forces need not involve any violation of this principle. When chemicals bond there is, of course, a slight increase in their masses; and when compounds break apart there is a slight decrease in the masses of the elements involved. Configurational forces could involve various compensating shifts in mass and energy that maintained conformance to the principle of mass-energy. There is no logical incompatibility in maintaining that there are configurational forces and in maintaining the principle of the conservation of mass-energy. It is also worthwhile noting here that Einstein's field equations are nonlinear. According to general relativity, the gravitational field of two or more objects will not be, in any sense, the sum of the fields of each object. Indeed, it will not even be a linear function of the fields of each object. Gravity is not a force, and so not a configurational force. However, the gravitational effect of two or more objects on another will be, in Mill's terminology, a *heteropathic* effect; in Lewes's, Alexander's, and Morgan's terminology, it will be emergent.[35]

Does this mean the Emergentists were right about there being emergents? In a word, yes. There are emergent or heteropathic effects, and emergent powers. In fact, they were right in maintaining that where chemical processes are concerned, the effects of two or more chemical agents will typically be a heteropathic effect of the agents. It will not be the sum, in any sense, of the effects of the individual reactants. (Just think of our earlier example: Methane + oxygen produces carbon dioxide + water.)

Nonetheless, as I said earlier, many of the central doctrines of Emergentism fly in the face of many of the major scientific achievements of this century. The Emergentists were wrong in thinking that there are configurational forces. They were wrong in thinking that there is downward causation from either the biological or the chemical level. They were wrong in thinking that the types of structures that compose special science kinds endow them with fundamental causal powers to influence motion. They were wrong in thinking that the bonding of chemical elements could not be given micro-explanations in terms of subatomic particles; and they were wrong about much else. But let us give credit where credit is due. As Morgan (1923) said: "Resultants there are; but there is emergence also" (p. 8). Strictly speaking, there is emergence also. I will have occasion to recur to these points in the next two sections.

## 1.5

Turn now, at long last, to Broad's *The Mind and Its Place in Nature* (1925). In this work, he introduces Emergentism by way of contrast with a view he labels "Mechanism." (In this section, unless I indicate otherwise, all references to are to Broad 1925.) So, let us begin with his characterization of Mechanism.

Broad characterizes what he calls the "ideal of Pure Mechanism" this way:

> On a purely mechanical theory all the apparently different kinds of matter would be made of the same stuff. They would differ only in the number, arrangement and movements of their constituent particles. And their apparently different kinds of behaviour would not be ultimately different. For they would all be deducible by a single simple principle of composition from the mutual influences of the particles taken by pairs; and these mutual influences would all obey a single law which is quite independent of the configuration and surroundings in which the particles happen to find themselves. The ideal which we have been describing may be called "Pure Mechanism." (pp. 45–46)

He illustrates the ideal thus:

> A set of gravitating particles, on the classical theory of gravitation, is an almost perfect example of the ideal of Pure Mechanism. The single elementary law is the inverse-square law for any pair of particles. The single and simple principle of composition is the rule that the influence of any set of particles on a single particle is the vector sum of the influences that each would exert taken by itself. (p. 45)

The single elementary law is, of course, the classical law of gravity. It connects pairs of particles in a way that is "quite independent of the configurations and surroundings in which the particles happen to find themselves." The principle of composition is, of course, vector addition.

Broad says that according to the ideal of Pure Mechanism,

> there is one and only one kind of material. Each particle of this obeys one elementary law of behaviour, and continues to do so no matter how complex may be the collection of particles of which it is a constituent. There is one uniform law of composition, connecting the behaviour of groups of these particles as wholes with the behaviour which each would show in isolation and with the structure of the group. All the apparently different kinds of stuff are just differently arranged groups of different numbers of the one kind of elementary particle; and all the apparently peculiar laws of behaviour are simply special cases which could be deduced in theory from the structure of the whole under consideration, the one elementary law of behaviour for isolated particles, and the one universal law of composition. On such a view the external world has the greatest amount of unity which is conceivable. There is really only one science and the various "special sciences" are just particular cases of it. (p. 76)

Mechanism, however, does not require Pure Mechanism. Broad goes on to say:

> An electronic theory of matter departs to some extent from this ideal. In the first place, it has to assume at present that there are two ultimately different kinds of particle, viz., protons and electrons. Secondly, the laws of electro-magnetics cannot, so far as we know, be reduced to central forces. Thirdly, gravitational phenomena do not at present fall within the scheme; and so it is necessary to ascribe masses as well as charges to the ultimate particles, and to introduce other elementary forces beside those of electro-magnetics. (p. 45)

Broad maintains, however, that these departures from Pure Mechanism are consistent with Mechanism itself.

Mechanism can thus tolerate certain departures from the ideal of Pure Mechanism. It can accommodate further kinds of elementary particles in addition to (say) electrons and protons, further basic force-generating properties in addition to (say) mass and charge, and further fundamental forces in addition to (say) gravity and electromagneticism. Such departures from the ideal of Pure Mechanism are themselves consistent with Mechanism.

What we may call "Comprehensive Mechanism" holds if and only if the following four conditions are met. First, every object is or is entirely made up of elementary material particles. Second, the force-generating properties are possessed by (at least) some kinds of elementary material particles. Third, the value of any force-generating property of a whole is determined, in accordance with a *compositional principle,* by the values of that sort of property for at least some of its parts. Fourth, forces combine by a principle of vector addition; such principles are themselves compositional principles: the value of the force exerted by a whole is determined in accordance with such a principle by the value of that force as it is exerted by the components of the whole.

Three comments about compositional principles or laws are in order so as to avert misunderstanding. First, the paradigms Broad offers are all principles of addition (of vectors, or of scalars). He counts the parallelogram law of vector addition as a compositional principle. Principles of additivity for scalars such as mass and charge are compositional principles. And relativistic addition principles for velocities would, I think, also count as compositional principle on his view (cf. Feigl 1958). Now Broad nowhere says that compositional principles (or laws[36]) must be additive. I am morally certain, however, that they must not employ nonlinear functions.[37] In any case, Broad offers no general characterization of what counts as a compositional principle. And I will not venture one here. That leaves a gap in the discussion. But I do not see how to fill it without literally putting words in Broad's mouth.

Second, Broad emphasizes that compositional principles are essential in mechanistic explanations of the behavior of wholes. A mechanistic explanation of the force a whole exerts on a particle in terms of the forces its constituent particles exert on it must appeal, for instance, to a principle of vector addition. And there must be a compositional principle linking the force-generating property of the whole with the force-generating properties of its parts. Thus suppose, for instance, that the force in question is the gravitational force. Then, the mechanistic explanation must appeal to the principle of the additivity of mass. For suppose that the mass of a whole were not the sum of the masses of its parts. Then, while both whole and parts would obey the law of gravity, the gravitational influence of a whole on the particle in question would not be derivative from the gravitational influences of its constituents. Broad points out that whenever we seem to explain the behavior of a whole mechanistically without compositional principles that

> is because we are using a suppressed premise which is so familiar that it has escaped our notice. The suppressed premise is the fact that we have examined other complexes in the past and have noted their behaviour; that we have found a general law connecting the behaviour of these wholes with that which their constituents would show in isolation; and that we are assuming that this law of composition will hold also of the particular complex whole at present under consideration. (p. 63)

And he cites the parallelogram law of vector addition as an example.

Third, Broad knows full-well that compositional principles are logically contingent and a posteriori. Of the parallelogram law, he says, "There is not the least possibility of deducing this law of composition from the laws of each force taken separately" (p. 62). But, he says, the law has been "verified" (p. 63). And he talks about why certain other additivity principles are logically contingent.[38] So much, then, by way of discussing compositional principles.

Consider once again the four individually necessary and jointly sufficient conditions for Comprehensive Mechanism. First, every object is or is entirely made up of elementary material particles. Second, the force-generating properties are possessed by (at least) some kinds of elementary material particles. Third, the value of any force-generating property of a whole is determined, in accordance with a compositional principle, by the values of that sort of property for at least some of its parts. And, fourth, forces combine by a principle of vector addition.

Emergentism accepts that the first and fourth conditions are satisfied. But it denies that the second and third conditions are. And it denies the third is because it denies the second is. Wholes can possess force-generating properties of a sort not possessed by any of their parts. The properties in question will be the properties of being composed of certain sorts of constituents in certain spatial or spatio-temporal relations. Such properties will endow a whole with the power to exert a fundamental force not exerted by any less complex wholes. These structural properties will generate configurational forces.

Now Broad does not actually speak of configurational forces. He tells us in Broad (1923) that "force" just acts like a middle term between the general laws of motion and the special laws (the force-laws). He thus cautions against reifying forces themselves. (Recall that this view is shared by Mill, Bain, and Morgan.) However, he talks of aggregates affecting the movements of other aggregates or of individual particles in fundamental ways. But of this, more later.

Turn to Broad's discussion of Emergentism. He says that on the Emergentist view, in contrast to the Mechanist view,

> we have to reconcile ourselves to much less unity in the external world and a much less intimate connexion between the various sciences. At best the external world and the various sciences that deal with it will form a hierarchy. (p. 77)

Physics will be at the base of the hierarchy, which will include in ascending order, chemistry, biology, and psychology. The kinds of substances specific to each level of organizational complexity are wholly made up of kinds of lower-orders. But there will be properties specific to kinds of a given order. Broad calls these the "ultimate characteristics" of the order. He cites "the power of reproduction" as an ultimate characteristic of the vital order. And he would count bonding as an ultimate characteristic of the chemical order. Broad contrasts ultimate characteristics of an order with "ordinally neutral characteristics" and "reducible characteristics." Reducible characteristics of an order, as the name suggests, reduce to characteristics of lower-orders. He says that some properties involved in the beating of the heart may be reducible. Ordinally neutral characteristics, as the name suggests, are characteristics that are possessed by aggregates at all levels of complexity. Broad cites inertial and gravitational mass as examples (p. 79). Physics resides at the base of the hierarchy in that it concerns itself with the most general characteristics of matter, the ordinally neutral properties that are possessed by (at least some) elementary particles and that are "unchanged in living bodies, chemical compounds, etc." (p. 79). Physics studies the organizational relationships objects participate in in virtue of their ordinally neutral properties. In classical physics, for example, all objects participate in organizational relationships due to gravitational attractions.

Broad says that Emergentism can "keep the view that there is only one fundamental kind of stuff" (p. 77). But, he says, if Emergentism is correct, then

> we should have to recognise aggregates of various orders. And there would be two fundamentally different types of law, which might be called "intra-ordinal" and "trans-ordinal" respectively. A trans-ordinal law would be one which connects the properties of aggregates of adjacent orders. A and B would be adjacent, and in ascending order, if every aggregate of order B is composed of aggregates of order A, and if it has certain properties which no aggregate of order A possess and which cannot be deduced from the A-properties and the structure of the B-complex by any law of composition which has manifested itself at lower-levels. An intra-ordinal law would be one which connects the properties of aggregates of the same order. A trans-ordinal law would be a statement of the irreducible fact that an aggregate composed of aggregates of the next lower order in such and such proportions and arrangements has such and such characteristic and non-deducible properties. (pp. 77–78)

And to illustrate the notion of a trans-ordinal law, Broad says:

> The law which asserts that all aggregates composed of such and such chemical substances in such and such proportions and relations have the power of reproduction would be an instance of a Trans-ordinal law. (pp. 78–79)

Some comments are in order about these passages.

First, as we saw, Broad remarks that: "A trans-ordinal law would be a statement of the irreducible fact that an aggregate composed of aggregates of the next lower order in such and such proportions and arrangements has such and such characteristic and non-deducible properties" (p. 78). This remark suggests that trans-ordinal laws are, by definition, emergent. But he sometimes uses "trans-ordinal law" just to mean a law that asserts that all aggregates of a certain sort have a certain property. Thus, he sometimes insists that it is an empirical question whether a trans-ordinal law is emergent, that is, a question which cannot be settled on a priori grounds.[39] Second, trans-ordinal laws are, however, the candidates for being emergent laws. Intra-ordinal laws are laws of the special sciences. Broad *never* speaks of intra-ordinal laws themselves as emergent. So far as I can tell, he does not think that special science laws are emergent. I believe that he thinks they will be deducible from lower-level laws and conditions together with trans-ordinal laws. However, he thinks that at least some of the trans-ordinal laws in question will be emergent. Third, a trans-ordinal law is emergent if and only if it is not deducible from the laws of lower orders, lower-level conditions, and any compositional principles manifested at lower-levels. (A compositional law is manifested at lower levels if it connects properties that are possessed by substances at some lower level with properties of their constituents.) Emergent trans-ordinal laws are, Broad holds, brute nomological facts that "cannot be explained" (p. 55). They must, he says, "simply be swallowed whole with that philosophic jam which Professor Alexander calls 'natural piety'" (p. 55). They are fundamental, nonderivative laws.

Broad's talk of the nondeducibility of emergent trans-ordinal laws invites a certain rejoinder that attempts to trivialize the doctrine of emergent laws. The rejoinder is that of course trans-ordinal laws may not be deducible from lower-level laws, conditions, and compositional principles. The reason is that the trans-ordinal law may well contain terms not contained in any lower-level laws, compositional principles or statements of lower-level conditions. For example, the term "power of reproduction," will not occur in chemical laws or compositional principles manifested at the chemical level. So, the fact that a trans-ordinal law cannot be deduced from lower-order laws and compositional principles can be due to the trivial fact that it contains different terms from the would-be deduction base. Emergent trans-ordinal laws thus do not reflect any radical "disunity in the external world."[40]

This would-be trivialization rests on misunderstandings of Broad's position. Let me draw out some relevant features of that position. First, I will discuss kinds of substances. Then, I will discuss properties (or attributes, or characteristics): ways things might be.

First, Broad would allow any necessary truth to count as an implicit premise in a would-be deduction of a trans-ordinal law. He has in mind a semantic notion of deduction: B is deducible from A iff when A is true, B is true. He does not, of course, have in mind deducibility in virtue of logical form. Second, he would allow appeal to a posteriori statements of kind identities in any would-be deduction of a trans-ordinal law. He speaks interchangeably of water and $H_2O$ and of salt and NaCl (sodium chloride). He moves from the "macro-vocabulary" to the "micro-vocabulary" without the slightest hint of concern. It is perfectly plain that he thinks that water = $H_2O$ and that salt = NaCl.[41] And he fully recognizes that we know that water is $H_2O$, for instance, only a posteriori, through chemical investigations. He plainly does not require that statements of kind identities (even those flanked by the sorts of terms Kripke (1971) would count as rigid designators) be a priori. While I lack the space to pursue this point here, the concern with a prioricity in such cases arose, I believe, only when philosophers began holding "use theories of meaning." But, in any case, the concern is entirely missing in Broad's work. As I said, he would be happy to include any kind identities, even a posteriori ones, in the would-be deduction base for a trans-ordinal law. Third, Broad seems to take laws to be singular relations between universals, not to be linguistic entities. He would, I believe, maintain that (the law that NaCl dissolves in $H_2O$) = (the law that salt dissolves in water). Fourth, trans-ordinal laws cite lower-level aggregates, rather than the special science kinds the aggregates compose.[42] If one thinks of the laws as linguistic entities, they specify the aggregates in the lower-level vocabulary. Where aggregates are concerned, at least, the issue of a shift in vocabulary simply need not arise at all. Indeed, while Broad speaks of trans-ordinal laws as citing aggregates of the adjacent lower level, he holds that substances of all levels may be wholly composed of elementary particles. He says, as I noted, that it is consistent with Emergentism that there is even a single kind of elementary particle, and that all kinds differ only in the numbers and spatial or spatio-temporal arrangements of the single kind of particle. There will be trans-ordinal laws attributing the ultimate characteristics of every level with aggregates of elementary particles (however, the laws may well not be stateable). These trans-ordinal laws could be emergent. Here the issue of a shift from micro-kind terms to macro-kind terms would not arise.

What, then, about the *properties* cited in trans-ordinal laws? There the issue of vocabulary shifts arises. It should be noted first of all, however, that Broad did not think two predicates have to be synonymous to express the same property. And he thinks it is often an a posteriori issue whether a property of aggregates of a certain order reduces to a property possessed by aggregates of lower orders. He says, for instance, that when the property in question is that of having the power to produce a "kind of behaviour which can be completely described in terms of changes of position, size, shape, and arrangement of parts, etc." (p. 72), it is an a posteriori issue whether the property is reducible or emergent (pp. 80–81). He mentions, in this connection, the property of having the capacity to breathe. That property, he says, appears to be an ultimate characteristic of the vital order; and it appears that the power to breathe emerges from certain complex aggregates of chemical compounds. But he allows that it could turn out that the characteristic power is a reducible characteristic. The question is, he insists, an a posteriori one. Broad allows that an intertheoretical property identity claim might be justifiable only a posteriori. There is no doubt that he would allow appeal to such claims in trying to deduce a trans-ordinal law. If a trans-ordinal law could be deduced, in principle, from lower-level laws, compositional principles, and a posteriori property identity statements, Broad would count the trans-ordinal law as *non*emergent.

Broad would not, however, take the fact that some nonemergent property Q is, as a matter of law, coextensive with a property P to suffice to show that P is nonemergent. For that law will itself be emergent if it is not implied by the sorts of factors we mentioned above. Broad would deny, for example, that "bridge laws" in Ernest Nagel's sense suffice for reduction (Nagel 1961, chapter eleven). I will elaborate.

Subtleties aside, Nagel (1961) maintains that to reduce one theory to another, the laws of the first must be deduced from the laws of the second. When the laws of the theories employ different vocabularies, principles are required to connect terms from the different vocabularies. According to Nagel, when different theoretical vocabularies are involved, in order to deduce the laws of a theory $T_1$ from the laws of a theory $T_2$, $T_2$ must be supplemented with a set of bridge laws (1961, pp. 336–366). Nagel allows that bridge laws can be universal biconditionals or universal conditionals (which contain only terms of $T_2$, the reducing theory, in the antecedent). Now Broad too seems to think of theory reduction as a matter of deducing the laws of the reduced theory from the laws of the reducing theory. If we think of trans-ordinal laws in the formal rather than the material mode, *trans-ordinal laws would count as Nagelian bridge laws*. Broad, however, would regard appeal to unexplainable trans-ordinal laws as insufficient for theory reduction. *Unexplainable bridge laws of the Nagelian sort would count as emergent trans-ordinal laws in Broad's sense.*[43]

The objection to Nagelian reduction that appeals to unexplained bridge laws does not count as reduction was, by the way, forcefully argued by Robert Causey (1967; 1972; 1977, chapter 4) and others (e.g., Schaffner 1967) in the late nineteen sixties and well into the nineteen seventies. Causey et al. made the straightforward point that if appeal must be made to unexplained bridge laws in order to deduce the laws of a theory $T_1$ from a theory $T_2$, then the laws of $T_1$ do not count as reducible to the laws of $T_2$. Causey went on to argue that law reduction requires kind and property identities (see, especially, 1977, chapter 4). These identities, he pointed out, require *justifications*. And he claimed that typically the justifications will only be a posteriori.[44] However, identity claims, he maintained, do not require *explanations*. So, he said, law reductions that appeal to them do not invoke laws that themselves require explanation. As I said above, Broad would allow a posteriori statements of property identity. And he would allow appeal to them in a would-be deduction of a trans-ordinal law. Let us, then, briefly compare and contrast Broad's and Causey's notions of reduction. This will, I hope, make Broad's notion clearer by contrast.

On Causey's view (1977, chapter 3), a theory T will consist of a set of sentences. The set will contain the set of fundamental law sentences of T, call it F, a set of true kind and property identity statements I, and a set D of derivative laws of T. (As I mentioned, he holds that the identity statements will typically be justifiable only a posteriori.) If the laws of T concern structured wholes, then the domain of T will contain basic and compound elements, according to Causey. The basic elements will satisfy open sentences of T that contain only basic attribute predicates, while the compound elements will satisfy only open sentences containing compound attribute predicates. With the help of truth functions, quantifiers, and set theoretic terminology, the compound attribute predicates will be definable in terms of basic attribute predicates. Now Causey says, "a basic element is *free* when it is not combined with other basic elements in a compound element; when it is so combined, we say that it is *bound*" (1977, p. 67). And he says: "We would then normally expect $T_1$ to contain some laws of the following type: a free basic element of a certain kind possesses a certain attribute under certain environmental conditions. A law of this kind would seem to be a natural candidate for a fundamental law. Likewise, laws expressing nonstructural relations holding between free basic elements would seem to be fundamental laws" (1977, p. 67). And he goes on to say that we want laws that contain compound attribute predicates to count as derivative, rather than fundamental laws. The reason is that "we want to understand the behavior of the wholes (compound elements) in terms of the behavior of their parts (basic elements)" (1977, p. 67). We thus want to be able to deduce the laws of T concerning compounds from laws of T concerning the basic elements in its domain. So, laws of T that contain compound attribute predicates are to be *derivative* laws of T. They are to be laws that can be deduced from the laws of T that concern only free basic elements of T or nonstructural relations between free basic elements of T and a boundary condition that is stated using a structural predicate of T.[45]

Turn now to Causey's account of what he calls "micro-reduction." The laws of a theory $T_2$ are micro-reducible to the laws of a theory $T_1$ if and only if they are deducible from the *fundamental* laws of $T_1$ and certain other statements. The other allowable statements in the would-be deduction base are (a) any property or kind identity statements, (b) any statement of boundary conditions, and (c) any analytical statements.

Broad would, I think, maintain that Causey's conditions for micro-reduction are *too strong*. They do not allow appeal to compositional principles. Perhaps, Causey thinks such principles are analytic. It is uncertain from his texts since he does not mention such principles. But, in any case, as Broad notes, and as we discussed earlier, they are not analytic. (The principle of the additivity of mass, for instance, is contingent.) Broad insists that, nonetheless, appeal to such principles is legitimate in a would-be micro-reduction of a law. Indeed, as we saw earlier, he insists that laws concerning (what Causey would call) compound elements can be deduced from laws concerning (what Causey would call) free elements only if appeal is made to compositional principles.

Broad could, however, concede that Causey has provided a sufficient condition for micro-reduction. And he would maintain that *emergent* trans-ordinal laws will not be micro-reducible in Causey's sense. They will, of course, not be identity statements of any sort, and they won't be analytic. They will attribute a certain property to a certain kind of aggregate. Thus, they will make reference to compound elements of the lower-level theory. And an emergent trans-ordinal law will not be deducible, even in principle, from fundamental laws of the lower-level theory (that do not cite the aggregates in question), kind or property identities, compositional principles applicable at the lower level, and any analytical principles. So, emergent trans-ordinal laws would not be micro-reducible in Causey's sense. Moreover, when there are such emergent trans-ordinal laws linking theories, at least some of the intra-ordinal laws of the (relatively) macro-theory will not be micro-reducible to the laws of the micro-theory either. So much, then, by way of comparing and contrasting Broad's and Causey's notions of reduction.

Earlier I defined a Broadian notion of Comprehensive Mechanism. I called it Comprehensive Mechanism since, as Broad points out, various forms of limited Mechanism are possible. He points out that biology might be emergent and chemistry mechanistic; or chemistry might be emergent and biology mechanistic (biological laws being deducible from chemical laws, compositional principles, etc.). Let us turn again to his texts.

Concerning mechanistic chemistry and biology, Broad says:

> I do not see any *a priori* impossibility in a mechanistic biology or chemistry, so long as it confines itself to that kind of behaviour which can be completely described in terms of changes of position, size, shape, and arrangement of parts, etc. (p. 72)

He goes on to say that we can see a priori that it is impossible if the biological or chemical theory concerns itself with secondary properties such as, for example, the odor of a chemical compound. But even if chemistry and biology confine themselves to laws concerning "the kind of behaviour which can be completely described in terms of changes of position, size, shape, and arrangements of parts, etc.," he thinks it very likely that chemistry and biology will not be mechanistic.

In the case of biology, he cites with approval a body of biological data gathered by Driesch (p. 58, p. 69). Driesch postulates an entelechy; and Broad calls this view "Substantial Vitalism." Broad rejects Substantial Vitalism: He eschews the existence of entelechies, holding, instead, that everything either is or is wholly composed of material particles. He maintains, however, that Driesch's data support Emergent Vitalism over what he calls Mechanistic Vitalism. And he maintains that the data support Emergent Vitalism over Substantial Vitalism, in part, since the data do not warrant the postulation of an entelechy. Emergent Vitalism maintains that certain types of structures of chemical compounds are responsible for vital behavior, not the possession of a special kind of component. But the power to produce such behavior in certain circumstances is, he holds, an emergent property of the structures. And the trans-ordinal laws that ascribe the powers to the types of structures of chemical compounds in question are emergent laws. He holds that this seems so, on the evidence, even when the powers are powers to produce "the kind of behaviour which can be completely described in terms of changes of position, size, shape, and arrangements of parts, etc." It is clear that he maintains that *certain structures of chemical compounds can influence motion in fundamental ways.*

It is particular interesting to note that Broad says:

> The situation with which we are faced in chemistry . . . seems to offer the most plausible example of emergent behaviour. (p. 65)

And Broad asks us to "consider the mechanistic alternative about chemical . . . behaviour, so as to make the emergent theory still clearer by contrast" (p. 69). Indeed, the contrasts he draws are quite instructive and support my reading of him. So, let us examine them in detail.

Emergent and mechanistic chemistry agree:

> that all the different chemical elements are composed of positive and negative electrified particles in different numbers and arrangements; and that these differences of number and arrangement are the only ultimate difference between them. (p. 69)

However, if mechanistic chemistry were true:

> it would be *theoretically* possible to deduce the characteristic behaviour of any element from an adequate knowledge of the number and arrangement of the particles in its atom, without needing to observe a sample of that substance. We could, *in theory*, deduce what other elements it would combine with and in what proportions; which of these compounds would react in the presence of each other under given conditions of temperature, pressure, etc. And all this should be *theoretically* possible without needing to observe samples of these compounds. (p. 70; emphases Broad's)

Broad emphasizes that Mechanistic Chemistry need only require that this be *theoretically* possible, since, as he points out, "the mathematical difficulties might be overwhelming in practice." So, we want to grant Mechanistic Chemistry that, perhaps, only a "Mathematical Archangel" with greater computational powers than even Sir Ernest Rutherford (p. 70) would be able to overcome those computational difficulties. Chemical Emergentism, however, denies that such deductions are possible, even in theory by a Mathematical Archangel whose computational powers are not constrained by matter. The problem is not that the vocabulary we use to talk about chemical elements and compounds is different from the vocabulary we use to talk about electrons and protons. Broad insists that Chemical Mechanism's

> difference from the emergent theory is . . . profound, even when we allow for our mathematical and *perceptual limitations*. If the emergent theory of chemical compounds be true, a mathematical archangel, *gifted with the further power of perceiving the microscopic structure of atoms as easily as we can perceive hay-stacks*, could no more predict the behaviour of silver or of chlorine or the properties of silver-chloride without having observed samples of those substances than we can at present. And he could no more deduce the rest of the properties of a chemical element or compound from a selection of its properties than we can. (pp. 70–71; emphases mine)

Broad's points is that even a being who could literally *see* the subatomic constituents of two chemical elements would be in no better position than we are to know how the elements would affect each other chemically. As this illustrates, Broad would allow appeal to principles linking aggregates of chemical elements with the chemical elements they compose in a would-be deduction of a chemical law. But he maintains that chemical laws could not be deduced from laws of subatomic particles even in conjunction with such principles, kind and property identities, compositional principles, and analytical truths. The reason is that the laws concerning which aggregates of subatomic particles have the power to bond with which others are fundamental, nonderivative laws. Allowing talk of forces, chemical forces emerge at the level of chemical elements that are as fundamental as the electro-magnetic force. The nomological fact that three aggregates of electrons and protons two of which compose hydrogen atoms and one of which composes an oxygen atom have the power to bond is a brute fact that admits of no explanation. The nomological fact that an aggregate of $H_2O$ molecules has the power to dissolve an aggregate of NaCl will likewise be a brute fact that admits of no explanation. Such laws must be accepted, on the evidence, with natural piety.

Scruples about reifying forces themselves aside, Broad's Emergentism commits him to configurational forces. Chemical processes, for example, involve the rearrangements of chemical elements; and these involve the motions of chemical elements. They must move about in rearranging themselves. The fundamental forces that influence such motions include chemical forces: fundamental forces exerted only by aggregates of subatomic particles themselves. Similarly, biological processes will involve motions influenced by fundamental vital forces. Now Emergentism concedes that nothing moves unless some elementary particle moves. And it maintains that aggregates of particles that can wholly compose biological kinds will influence the accelerations of particles in ways that are in no sense derivative from the influences of particle-pair forces or configurational chemical forces. At the chemical and biological levels of organizational complexity of elementary material particles, wholes (aggregates of particles composing chemical or biological kinds) influence each other's movements and, so, the movements of (at least some of) their constituent particles in ways that are in no sense derivative from "the influences of their constituents particles taken as pairs." If Chemical and Vital Emergentism are correct, there are configurational chemical and vital forces.

As I keep saying, quantum mechanics and various scientific advances it has made possible have provided us with compelling evidence that there are no such forces. As I said in section 1.1, it is not that quantum mechanics itself is logically incompatible with the existence of configurational chemical and vital forces. It is not. Schrödinger's equation, as I noted in section 1.1, does not itself tell us what forces or energies there are. The Hamiltonian must be determined independently. But quantum mechanical explanations of chemical bonding in terms of the electro-magnetic force, and the advances this led to in molecular biology and genetics render the doctrines of configurational chemical and vital forces enormously implausible.

Broad did not anticipate the type of micro-explanation of chemical bonding which quantum mechanics provides. Mechanistic chemistry would attempt to explain chemical bonding in terms of additive properties of individual electrons. Broad was exactly right in thinking that chemical bonding could not be explained in that way. Quantum mechanical explanations of chemical bonding are not "mechanical" in this sense.[46] But quantum mechanical explanations of chemical bonding suffice to refute central aspects of Broad's Chemical Emergentism: Chemical bonding can be explained by properties of electrons, and there are no fundamental chemical forces. The moral is that both Emergentism and Mechanism proved wrong in certain important respects. The quantum mechanical revolution left both Emergentism and Mechanism behind.

### 1.6

I will close with some general observations. In a span of roughly one hundred years, British Emergentism enjoyed a great rise and suffered a great fall. It is one of my central contentions that its rise was not due to "philosophical mistakes," nor its fall to the uncovering of such mistakes. So far as I can tell, British Emergentism does not rest on any "philosophical mistakes."[47] It is one of my main contentions that advances in science, not philosophical criticism, led to the fall of British Emergentism. Here, in a nutshell, is what happened. In their quest to discover "the connexion or lack of connexion of the various sciences" (Broad 1923, pp. 41–42), the Emergentists left the dry land of the a priori to brave the sea of empirical fortune. (The only route is by sea, of course.) They set off in a certain direction, and for awhile winds of evidence were in their sails; but the winds gradually diminished, and eventually ceased altogether to blow their way. Without these winds in its sails, the British Emergentist movement has come to an almost complete halt.

Only a few are left manning the oars. Witness the following remarks by Roger Sperry:

> If the micro-determinists are correct . . . then we live in a cosmos driven entirely by physical forces of the most elemental kind, that is, quantum mechanics, the four fundamental forces of physics, or some even more elemental unifying field force yet to be discovered. These are the forces that are in control, the forces that made and move the universe and created man . . .
>
> On the other hand, if we are correct about macro-determination, or "emergent determination," these lower-level physical forces, though still active, are successfully enveloped, overwhelmed, and superseded by the emergent forces of higher and higher levels that in our own biosphere include vital, mental, political, religious, and other social forces of civilization . . . The result is a vastly transformed scientific view of human and nonhuman nature. Organic evolution becomes a gradual emergence of increased directedness . . . among the forces and properties that move and govern living things. (1986, p. 269)

Sperry also says that "the properties, forces and laws of micro-events are . . . encompassed and superseded, not disrupted, by the properties, forces and laws at the macro-levels" (1986, pp. 267–268). These views place him squarely in the British Emergentist tradition: Indeed, he himself considers Lloyd Morgan to be a predecessor.

Sperry is, obviously, committed to the existence of configurational vital, mental, and social forces. His doctrine of emergent determination is, I contend, perfectly internally coherent. I do not object to it on a priori grounds. I defend his *logical* right to express it. Unfortunately, however, he offers not a scintilla of evidence that there are such configurational forces.[48] That there are such forces is, on the evidence, enormously implausible. The lattice forces that hold together organic molecules are electro-magnetic in origin. And there are no vital or psychological forces. The doctrine of emergent determination due to configurational vital, psychological, or social forces is, I contend, simply false. As truly remarkable as it is, it seems to be a fact about our world that the fundamental forces which influence acceleration (the electro-magnetic-weak force and the strong force) are all exerted at the subatomic level. (Who would have thought it on a priori grounds?!) On the evidence, this remarkable fact must, it seems, be accepted with natural piety.

I wrote this chapter because I thought British Emergentism, once a fairly widely held doctrine, deserved another look; not because it was right, but because its mistakes were fewer and far more interesting than had been generally recognized. It went wrong for deep empirical reasons. It took many of the greatest scientific achievements of the twentieth century to refute it.[49]

## Notes

This chapter originally appeared in Ansgar Beckerman, Hans Flohr, and Jaegwon Kim (eds.), *Emergence or Reduction?: Essays on the Prospects of Nonreductive Physicalism*, pp. 49–93, Berlin: Walter de Gruyter, 1992.

1. Some British Emergentists, however, try to analyze matter in phenomenalist terms. For example, as is well-known, Mill characterizes matter as "the permanent possibility of sensation." I will not be concerned here at all with the phenomenalist themes in the literature in question. The British Emergentists were, for the most part, British Empiricists. But I focus exclusively on their Emergentism.

2. The Emergentists were well-aware that most special science kinds are such that many different kinds of *aggregates* of particles can wholly compose them. The Emergentists recognize that there is a great deal of "configurational plasticity." Functionalist themes abound in Lewes (1875), Alexander (1920), and Broad (1925). But I lack the space here to discuss those themes. Suffice it to note that the Emergentists do *not* argue against the reducibility of special science laws on functional grounds.

3. The term "downward causation" was introduced by Donald Campbell (1974). But my use of the term here is, however, really due to Klee (1984). Jaegwon Kim (this volume) and Achim Stephan (this volume) discuss downward causation in the sense in question here. I should also mention that Karl Popper (1977) and Roger Sperry (1986), both of whom regard themselves as emergentists, explicitly claim that there is downward causation. I briefly discuss Sperry's views in the concluding section of this paper since his notion of emergence is especially close to the British Emergentist notion.

4. Most of the Emergentists would emphatically deny that laws concerning conscious experiences and so-called secondary qualities (e.g., color, etc.) are derivative from laws of mechanics. I will, however, *ignore* the Emergentist's views about conscious experiences and secondary qualities in this paper. It is, I claim, consistent with their views that laws concerning emergent powers to

produce certain kinds of movements are derivative from laws of mechanics. This is discussed in great detail in section 1.5.

5. If there were configurational forces, then downward causation would be a nomological possibility. I will leave open here whether downward causation requires configurational forces, however. And I will also leave open whether downward causation is nomologically possible. David Bohm's (1989) interpretation of the formalism of quantum mechanics understands the quantum potential in a way that seems to involve downward causation. (This observation about Bohm was made to me independently by Barry Loewer, Tim Maudlin, and Robert Weingard.) And Einstein's field equations of general relativity may count as involving downward causation. (See the discussion of general relativity in section 1.4.) Of course, downward causation from the psychological, biological, or chemical level is another matter. That is enormously implausible. There is overwhelming reason to reject that idea and the existence of configurational chemical, vital, or psychological forces. Or so I argue below.

6. After essentially completing this chapter, I was pleased to discover that the relevance of quantum mechanics as a serious challenge to the main theses of Emergentism had been noted by Herbert Feigl (1958) (see pp. 411–413) and Ernest Nagel (1963). It is hardly surprising that the relevance of quantum mechanics was noticed by philosophers. It was quantum mechanics that made possible many of the scientific advances that now make Emergentism seem so very implausible. The epistemic situation for Emergentism has, I believe, only worsened since Feigl (1958) and Nagel (1963).

7. I should note here that I will not discuss the psychological in what follows. I will hereafter restrict my discussion of British Emergentism to its views about the chemical and the biological. As we will see, the primary examples of emergence offered by the Emergentists were chemical and biological. It is worth repeating, however, that the Emergentists rejected Cartesian Dualism. Lewes (1875) and Alexander (1920), for example, both insist that every particular mental process is identical with a neurophysiological process; but that mental qualities or properties are *emergent*. And Broad (1925) defends a doctrine he calls "Emergent Materialism." This doctrine implies that everything is wholly made of matter, and that all particular mental processes are processes in the central nervous system. But it also implies that irreducible mental powers *emerge* from the minute internal structures of the central nervous system (1925, p. 436).

8. He had proposed that model some years before, in 1913. It is important to appreciate how revolutionary that model itself was. Consider the following remarks by Norman Campbell, quoted in the obituary by L. Hartshorn:

> Some algebraic formulae caught my eye... It was part of a paper by Mr. N. Bohr of whom I had never heard... I sat down and began to read. In half an hour I was in a state of excitement and ecstasy, such as I have never experienced before or since in my scientific career. I had just finished a year's work revising my book on *Modern Electrical Theory*. These few pages made everything I had written entirely obsolete. That was a little annoying, no doubt; but the annoyance was nothing to the thrill of a new revelation, such as must have inspired Keat's most famous sonnet. (Quote taken from French and Kennedy 1985, p. 77)

9. It was, I believe, Professor Wimsatt who informed me that, for a while after having proposed his wave mechanics, Schrödinger continued to encourage Driesch to develop his theory of vitalism.

10. While British and an emergentist, I have not counted him as a British Emergentist since his views differ in certain ways I cannot discuss here from the views of the authors mentioned above.

11. I have been unable to obtain the texts of the Deborin School. But I speculate there is a line of influence from Mill to Marx and Engels and from them to the Deborin School.

12. Emergence won out for a time in Soviet philosophy. But its success was short-lived. The debate between the mechanists and the emergentists came to an abrupt political end. On January 25, 1931, after a speech by Stalin in which he condemned the disputants on both sides, the Central Committee of the Communist Party called an end to the debate (Kamenka 1972, p. 164).

13. For references to emergentist traditions which I have not mentioned, see Stephan [1992].

14. I am no Mill scholar, but from my research on the secondary literature on Mill, it seems that his influence on British Emergentism is a badly neglected area of scholarship.

15. Mill tells us in his *Autobiography* (p. 113) that he first encountered the distinction in Thomson's book on chemistry which, he says, he loved as a boy. I presume that he was referring to Thomas Thomson's *System of Chemistry* (1807). That work, by the way, contained a presentation of Dalton's new atomic theory.

16. Nancy Cartwright (1980) has attacked Mill's notion of the Composition of Causes. While I cannot pause to discuss her attack here, I should note that it fails. As Creary (1981) has pointed out, she fails to appreciate that force-laws are "laws of causal influence." (See the discussion of force-laws in section 1.4 and note 20 for more details.) Cartwright (1983) has conceded this point about force-laws to Creary and has conceded also that it defeats her attack on the Composition of Causes in mechanics. Mill, I should mention, takes the full effect of a force exerted on one object by another to be a certain *pressure* on the former object in virtue of which it will have a certain *tendency* or *disposition* to accelerate (pp. 427–428). How an object will in fact accelerate will depend on the *net* force on the object, the vector sum of all the forces exerted on it. Mill anticipates Creary's idea that force-laws are "laws of causal influence." (Creary does not cite Mill.) Moreover, Mill *explicitly* raises the very concerns expressed by Cartwright (one hundred and thirty nine years later) and, I believe, correctly responds to them. (I pursue these matters in "Mill's Response to Cartwright" (in preparation).)

17. Mill's talk of deducing truths will, no doubt, invoke a certain knee-jerk response in anyone trained in the analytical tradition. Of course physiological truths cannot be deduced from chemical truths, one will say, the statements expressing such truths employ different vocabularies. I cannot pause here to address this sort of would-be trivialization of Mill's thesis. But in section 1.5, I discuss the same issue as it arises for Broad.

18. Mill says that while "the different actions of a chemical compound will never, undoubtedly, be found to be the sums of the actions of its separate elements... there may exist, between the properties of a compound and those of its elements, some constant relation" (1843, p. 432). To provide an example of what he has in mind, he says "The law of definite proportions, first discovered in its full generality by Dalton, is a complete solution of this problem in one, though but a secondary aspect, that of quantity" (1843, p. 433). Mill thus holds that some laws of chemistry may be derivative.

19. While I cannot pursue this here, it is interesting to note that Mill held that political theory is deductive (see, for example, Mill 1924, pp. 110–133). One of his ideas is that the effects of group actions are homopathic effects of the actions of the individuals who are acting together.

20. In this, he was right. I briefly discuss the principle of the conservation of energy in the next section.

21. It is worthwhile noting here a remark from Mill's *Autobiography*: "My father's Analysis of the Mind, my own Logic, and Professor Bain's great treatise, had attempted to re-introduce a better mode of philosophizing" (1924, p. 193). Here we see his deep respect for Bain's work.

22. See Bigelow et al. (1988), however, for some, to my mind, convincing reasons why we should countenance forces themselves. But, as I said, nothing in what follows turns on this.

23. I want to mention that Lewes lived for many years with George Elliot; and they frequently talked to each other about their work. I cannot help but wonder whether George Elliot had anything to do with coining the term "emergent."

24. I suspect the apparent conflicts arise from the fact that Alexander thinks it is a priori that secondary qualities are emergent but a posteriori whether other qualities are and that he is not always careful to express this. (Broad (1925) is.)

25. *Scientific Thought*, unfortunately, contains no mention of Emergentism.

26. Creary (1981) aptly calls force-laws "laws of causal influence." Force-laws express fundamental factors that, when present, will causally influence the acceleration of an object in a certain way. But they do not tell us how any object will in fact accelerate. How an object will in fact accelerate will depend on the *net* force on the object, the vector sum of all the forces exerted on the object. As Mill (1843) noted, the component forces (that correspond to actual agents) of the net force exerted on an object *dispose* the object to accelerate in a certain way.

27. This, in a nutshell, is the point Cartwright (1980) failed to appreciate. This led her to claim that the fundamental laws of mechanics are false. (See "Mill's Response to Cartwright" (in preparation).)

28. That assumption proved wrong, of course. I briefly discuss the principle of the conservation of mass-energy below.

29. As I mentioned above, Bain (1870) argues, on essentially the same grounds, that the fact that certain fundamental forces arise only with certain collocations of agents is consistent with the principle of the conservation of energy. And Mill wholeheartedly states his agreement with Bain about this. Indeed, to put the point this way is somewhat misleading. It suggests that they thought the principle of the conservation of energy posed a serious challenge to their views. But, in fact, the idea of potential energy, which received great attention when the conservation of energy was first proposed, seems to have been a source of inspiration for them. It led Bain to speak of potential forces that can be exerted only when circumstances are right. Bain and Mill were actually encouraged in their views by contemplation of the (then newly proposed) principle of the conservation of energy.

For Broad's views on the principle of the conservation of energy, see Broad (1925), pp. 97–103. He argues there that even dualistic psychophysical causal interaction is consistent with the principle of the conservation of energy. Nothing here, however, turns on whether he is right on that point. But, as should be apparent, he would not be concerned that the principle of the conservation of energy would be violated by configurational forces. And rightly so.

30. Actually, however, as John Earman (1986) has pointed out, it is not. (For details see Earman (1986), especially chapters IV and X, and, in particular, his discussions of "invaders from infinite space"). As Earman also points out (1986) we can formulate causal determinism, if we eschew absolute space and time and help ourselves to an idea from relativity theory. In four-dimensional spacetime, a Cauchy Surface is a three-dimensional hypersurface without a temporal dimension. It is everywhere spacelike and intersects every lightlike and timelike worldline in its past and future. So, it intersects every lightcone. Causal determinism can be formulated this way: the laws of mechanics and the arrangement of particles and fields on any Cauchy Surface determine the arrangement of particles and fields on any Cauchy Surface in its future. I won't pursue this, however, since whether causal determinism holds is orthogonal to the issue of whether Emergentism is correct.

31. The reader will no doubt have noticed that Mill speaks of predictability in his statement of causal determinism. So, by the way, does Laplace. Predictability is, of course, an epistemological notion, while causal determinism is an ontological one. These authors are not always as careful as they might have been in distinguishing epistemological from ontological theses. But charitable commentators have rightly taken them to be stating the ontological thesis of causal determinism. I should note in this connection that Emergentists often speak of emergent properties and laws as unpredictable from what they emerge from. But, contra what some commentators have thought, the Emergentists do not maintain that something is an emergent because it is unpredictable. Rather, they maintain that something can be unpredictable because it is an emergent. Emergence implies a kind of unpredictability. But it is a mistake to conflate emergence with this consequence of emergence. The British Emergentists do not.

32. I mentioned at the outset, you will recall, that they are compatible with Schrödinger's equation, the fundamental equation of nonrelativistic quantum mechanics.

33. Broad's *Scientific Thought*, from which I quoted earlier, is devoted in large part to explaining special and general relativity to the nonspecialist. Alexander and Morgan were also aware of special and general relativity. Alexander (1920, Vol. I) thought that Einstein's theories of relativity vindicated his own view that spacetime is ontologically fundamental (see, especially, pp. vii, 58, 87). And Morgan (1923) briefly discusses views of "the young Mr. Einstein." (Mill, Bain, and Lewes, of course, knew nothing of relativity. Einstein, you will recall, proposed special relativity in 1905 and general relativity in 1916.) Alexander, Morgan, and Broad do not mention Emergentism in connection with relativity theory. They apparently thought that the issues raised by relativity theory are orthogonal to the issue of whether Emergentism is correct.

34. Broad briefly discusses how "the Special Theory of Relativity leads to a connexion between the three principles of the Conservation of Momentum, of Mass, and of Energy, which was not obvious on the traditional view" in (1923, pp. 182–183).

35. And as we will be apparent later, Einstein's field equations do not fit Broad's model of "Mechanism."

36. My occasional shifts from talk of compositional principles to talk of compositional laws is merely stylistic.

37. As I indicated earlier, the gravitational field of a whole will not be a linear function of the gravitational fields of its constituent particles. If compositional principles must not employ non-linear functions, Mechanism is false. That, of course, would not disturb Broad. He thinks that Mechanism is false. Indeed, he thinks that we can see that it is false a priori, given the existence of secondary qualities. But, as I said earlier, I will not discuss secondary qualities here.

38. I belabor this point only because some of Broad's most distinguished commentators have missed it. Hempel and Oppenheim (1948) challenge the interest of Broad's Emergentism on the grounds that it would imply that the weight of a whole is an emergent property of the whole. Why? Because, they say, the principle of the additivity of weight is logically contingent. But Broad has no objection whatsoever to Mechanism appealing to logically contingent compositional principles. Indeed, he insists that they must. Moreover, in a recent article, Van Cleve (1990) misinterprets Broad in the same way. He describes the behavior of a certain three body system and says Broad must hold that the behavior of the system is emergent on the grounds that it can be explained only by appeal to the parallelogram law. And that law, Van Cleve says, is logically contingent. But Broad invokes the parallelogram law in his description of Pure Mechanism. And, as we have just seen, he emphasizes that the principle is a posteriori. As a good Empiricist, he would taken that to imply logical contingency. I mention these misreadings as an excuse for my lengthy quotes.

39. Actually, he thinks the question can be settled on a priori grounds when the trans-ordinal law connects properties of aggregates with conscious properties or secondary properties (e.g., color). He calls such trans-ordinal laws "transphysical laws." And he holds that they are by necessity emergent. But, as I have said, I will ignore here the Emergentist's discussions of consciousness and secondary properties. I flag this point just so as not to mislead the reader.

40. This is the main line of objection to Emergentism in Nagel (1961, ch. 11). Nagel (1963) does better.

41. By the way, Lewes (1875) insisted that water = $H_2O$; and he criticized the idea that $H_2O$ causes water.

42. Broad is, by the way, well-aware of the completely obvious point that there will typically be no single kind of aggregate of lower-level entities that composes a special science kind. He thinks this is particularly clear in the case of biological organisms (see 1925, chapter x). He recognizes that there is plenty of configurational plasticity.

43. I owe this point to Jaegwon Kim. Also see Beckermann [1992] for a fine discussion of related points.

44. Causey (1977, pp. 96–97) resists Kripke's idea, however, that the sort of identity claims in question will be metaphysically necessary. Causey maintained they would be contingent. Nothing in what follows turns on this issue.

45. Now it is uncertain what, exactly, Causey counts as a "nonstructural relation" among free basic elements. He may have had in mind merely spatial or spatio-temporal relations. But, as we

have noted, it is logically possible for spatial or spatio-temporal relations among particles to result in the exertion of a fundamental force not exerted by any pairs of particles. Causey fails to appreciate this logical possibility. But I won't press this point here.

46. I cannot pursue this point here. It is enough to say where Emergentist goes wrong. I cannot attempt to say in detail how Mechanism goes wrong. To attempt to do so would take me into some thorny questions concerning the interpretation of quantum mechanics. However, I want to mention that Feigl (1958) correctly notes that quantum mechanics invokes "holistic principles." He cites Pauli's Exclusion Principle as an example. (Also, see note 5.)

47. I speak here only of British Emergentism. There are, as I noted earlier, other emergentist traditions. I do not speak for them.

48. Sperry defends his thesis of emergent determination mainly by appeal to examples. But suffice it to say that the examples are easily accommodated by the "micro-determinist" (Cf. Klee, 1985). He also cites, however, the fall of behaviorism and the advent of cognitive science. Cognitive science, he points out, invokes mental processes. Indeed, it does. But cognitive science attempts to understand cognition in computational terms. It does not invoke new forces! Sperry also raises the problem of consciousness, however; and cognitive science has been mostly silent about consciousness. As I said, I won't try to tackle the issue of consciousness here. Consciousness, it seems, is the last refuge of an Emergentist. Suffice it to say that there is, I believe, no evidence whatsoever that consciousness involves a new force in nature or that it exerts downward causation.

49. I wish to thanks audiences from the Bielefeld Conference on Emergence, Supervenience, and Nonreductive Materialism, Duke University, North Carolina State University, and Davidson College, where sections of this chapter were read. I wish to thank members of my graduate seminar in Metaphysics in the Spring of 1991. I wish also to thank Martha Bolton, Barry Loewer, Van McGee, Colin McGinn, Renee Weber, and Robert Weingard for helpful discussions which improved the paper. I owe a special debt to Jaegwon Kim for getting me interested in the topic of emergence and for his philosophical guidance. Finally, I want to thank Tim Maudlin for dozens of hours of private lectures on quantum mechanics and matters philosophical.

## References

Alexander, S. (1920) *Space, Time, and Deity. 2 Vols*. London: Macmillan.

Bain, A. (1870) *Logic*, Book II & III. London.

Bantz, D. A. (1980) "The Structure of Discovery: Evolution of Structural Accounts of Chemical Bonding," in: Nickles, T. (ed.), *Scientific Discovery: Case Studies*. Dordrecht: Reidel.

Beckermann, A. (1992) "Supervenience, Emergence, and Reduction," in: Beckermann, A., Flohr, H., and Kim, J. (eds.), *Emergence or Reduction? Essays on the Prospects of Nonreductive Physicalism*. Berlin: Walter de Gruyter, pp. 94–118.

Bigelow, J., Ellis, B., and Pargetter, R. (1988) "Forces." *Philosophy of Science* 55, pp. 614–630.

Bohm, D. (1989) *Quantum Theory*. New York: Dover.

Broad, C. D. (1923) *Scientific Thought*. New York: Humanities Press.

——— (1925) *The Mind and Its Place In Nature.* London: Routledge and Kegan Paul.

Campbell, D. T. (1974) "'Downward Causation' in Hierarchically Organised Biological Systems," in: Ayala, F. J., and Dobzhansky, T. (eds.), *Studies in the Philosophy of Biology.* Berkeley and Los Angeles: University of California Press, pp. 179–186.

Cartwright, N. (1980) "Do the Laws of Physics State the Facts?" *Pacific Philosophical Quarterly* 61, pp. 75–84.

——— (1983) *How the Laws of Physics Lie.* Oxford: Clarendon Press.

Causey, R. L. (1969) "Polanyi on Structure and Reduction." *Synthese* 25, pp. 230–237.

——— (1972) "Attribute-Identities and Microreductions." *The Journal of Philosophy* 69, pp. 407–422.

——— (1977) *Unity of Science.* Dordrecht: Reidel.

Creary, L. G. (1981) "Causal Explanation and the Reality of Natural Component Forces." *Pacific Philosophical Quarterly* 62, pp. 148–157.

Earman, J. (1986) *A Primer on Determinism.* Dordrecht: Reidel.

Edwards, P. (ed.) (1972) *Encyclopedia of Philosophy. Vol. 2.* Repr. ed. New York: Macmillan.

Feigl, H. (1958) "The 'Mental' and the 'Physical,'" in: Feigl, H., Scriven, M., and Maxwell, G. (eds.), *Concepts, Theories and the Mind-Body Problem. Minnesota Studies in the Philosophy of Science. Vol. 2.* Minneapolis: University of Minnesota Press, pp. 320–492.

French, A. P., and Kennedy, P. J. (eds.) (1985) *Niels Bohr: A Centenary Volume.* Cambridge: Harvard University Press.

Gribbin, J. (1984) *In Search of Schrödinger's Cat.* New York: Bantam Books.

Hempel, C. G., and Oppenheim, P. (1948) "Studies in the Logic of Explanation." *Philosophy of Science* 15, pp. 135–175. Reprinted in: Hempel, C. G., *Aspects of Scientific Explanation and Other Essays in the Philosophy of Science.* New York 1965, pp. 245–290.

Kamenka, E. (1972) "Communism, Philosophy Under," in: Edwards, P. (1972), pp. 163–168.

Klee, R. (1984) "Micro-Determinism and Concepts of Emergence." *Philosophy of Science* 51, pp. 44–63.

Kripke, S. (1971) "Identity and Necessity," in: Munitz, M. K. (ed.), *Identity and Individuation.* New York: University Press, pp. 135–164.

Lewes, G. H. (1875) *Problems of Life and Mind. Vol. 2.* London: Kegan Paul, Trench, Turbner, & Co.

Mill, J. S. (1843) *System of Logic.* London: Longmans, Green, Reader, and Dyer. (Eighth edition, 1872).

——— (1924) *Autobiography.* New York.

Morgan, C. Lloyd (1923) *Emergent Evolution.* London: Williams & Norgate.

Nagel, E. (1961) *The Structure of Science.* New York: Harcourt, Brace and World.

——— (1963) "Wholes, Sums, and Organic Unities," in: Lerner, D. (ed.), *Parts and Wholes*. New York: The Free Press.

Pepper, S. (1926) "Emergence." *Journal of Philosophy* 23, pp. 241–245.

Popper, K. R., and Eccles, J. C. (1977) *The Self and its Brain*. Berlin/Heidelberg/London/New York: Springer.

Schaffner, K. F. (1967) "Approaches to Reduction." *Philosophy of Science* 34, pp. 137–147.

Sperry, R. W. (1986) "Discussion: Macro- Versus Micro-Determinism." *Philosophy of Science* 53, 265–270.

Stephan, A. (1992) "Emergence—A Systematic View on its Historical Facts," in: Beckermann, A., Flohr, H., and Kim, J. (eds.), *Emergence or Reduction? Essays on the Prospects of Nonreductive Physicalism*. Berlin: Walter de Gruyter, pp. 25–48.

Thomson, T. (1807) *System of Chemistry*.

Van Cleve, J. (1990) "Emergence vs. Panpsychism: Magic or Mind Dust?" in: Tomberlin, J. E. (ed.), *Philosophical Perspectives. Vol. 4*. Atascadero, Cal.: Ridgeview Publishing Company, pp. 215–226.

# 2 On the Idea of Emergence

**Carl Hempel and Paul Oppenheim**

## Levels of Explanation. Analysis of Emergence

A phenomenon may be explainable by sets of laws of different degrees of generality.[1] The changing positions of a planet, for example, may be explained by subsumption under Kepler's laws, or by derivation from the far more comprehensive general law of gravitation in combination with the laws of motion, or finally by deduction from the general theory of relativity, which explains—and slightly modifies—the preceding set of laws. Similarly, the expansion of a gas with rising temperature at constant pressure may be explained by means of the Gas Law or by the more comprehensive kinetic theory of heat. The latter explains the Gas Law, and thus indirectly the phenomenon just mentioned, by means of (1) certain assumptions concerning the micro-behavior of gases (more specifically, the distributions of locations and speeds of the gas molecules) and (2) certain macro-micro principles, which connect such macro-characteristics of a gas as its temperature, pressure and volume with the micro-characteristics just mentioned.

In the sense of these illustrations, a distinction is frequently made between various *levels of explanation*.[2] Subsumption of a phenomenon under general laws directly connecting observable characteristics represents the first level; higher levels require the use of more or less abstract theoretical constructs which function in the context of some comprehensive theory. As the preceding illustrations show, the concept of higher-level explanation covers procedures of rather different character; one of the most important among them consists in explaining a class of phenomena by means of a theory concerning their micro-structure. The kinetic theory of heat, the atomic theory of matter, the electromagnetic as well as the quantum theory of light, and the gene theory of heredity are examples of this method. It is often felt that only the discovery of a micro-theory affords real scientific understanding of any type of phenomenon, because only it gives us insight into the inner mechanism of the phenomenon, so to speak. Consequently, classes of events for which no micro-theory was available have frequently been viewed as not actually understood; and concern with the theoretical status of phenomena which are unexplained in this sense may be considered as one of the roots of the doctrine of emergence.

Generally speaking, the concept of *emergence* has been used to characterize certain phenomena as "novel," and this not merely in the psychological sense of being unexpected,[3] but in the theoretical sense of being unexplainable, or unpredictable, on the basis of information concerning the spatial parts or other constituents of the systems in which the phenomena occur, and which in this context are often referred to as "wholes." Thus, e.g., such characteristics of water as its transparence and liquidity at room temperature and atmospheric pressure, or its ability to quench thirst have been considered as emergent on the ground that they could not possibly have been predicted from a knowledge of the properties of its chemical constituents, hydrogen and oxygen. The weight of the compound, on the contrary, has been said not to be emergent because it is a mere "resultant" of its components and could have been predicted by simple addition even before the compound had been formed. The conceptions of explanation and prediction which underly this idea of emergence call for various critical observations, and for corresponding changes in the concept of emergence.

1. First, the question whether a given characteristic of a "whole," *w*, is emergent or not cannot be significantly raised until it has been stated what is to be understood by the parts or constituents of *w*. The volume of a brick wall, for example, may be inferable by addition from the volumes of its parts if the latter are understood to be the component bricks, but it is not so inferable from the volumes of the molecular components of the wall. Before we can significantly ask whether a characteristic *W* of an object *w* is emergent, we shall therefore have to state the intended meaning of the term "part of." This can be done by defining a specific relation *Pt* and stipulating that those and only those objects which stand in *Pt* to *w* count as parts of constituents of *w*. "*Pt*" might be defined as meaning "constituent brick of" (with respect to buildings), or "molecule contained in" (for any physical object), or "chemical element contained in" (with respect to chemical compounds, or with respect to any material object), or "cell of" (with respect to organisms), etc. The term "whole" will be used here without any of its various connotations, merely as referring to any object *w* to which others stand in the specified relation *Pt*. In order to emphasize the dependence of the concept of part upon the definition of the relation *Pt* in each case, we shall sometimes speak of *Pt*-parts, to refer to parts as determined by the particular relation *Pt* under consideration.

2. We turn to a second point of criticism. If a characteristic of a whole is counted as emergent simply if its occurrence cannot be inferred from a knowledge of all the properties of its parts, then, as Grelling has pointed out, no whole can have any emergent characteristics. Thus, to illustrate by reference to our earlier example, the properties of hydrogen include that of forming, if suitably combined with oxygen, a compound which is liquid, transparent, etc. Hence the liquidity, transparence, etc., of water *can*

be inferred from certain properties of its chemical constituents. If the concept of emergence is not to be vacuous, therefore, it will be necessary to specify in every case a class *G* of attributes and to call a characteristic *W* of an object *w* emergent relatively to *G* and *Pt* if the occurrence of *W* in *w* cannot be inferred from a complete characterization of all the *Pt*-parts with respect to the attributes contained in *G*, i.e., from a statement which indicates, for every attribute in *G*, to which of the parts of *w* it applies. Evidently, the occurrence of a characteristic may be emergent with respect to one class of attributes and not emergent with respect to another. The classes of attributes which the emergentists have in mind, and which are usually not explicitly indicated, will have to be construed as nontrivial, i.e., as not logically entailing the property of each constituent of forming, together with the other constituents, a whole with the characteristics under investigation. Some fairly simple cases of emergence in the sense so far specified arise when the class *G* is restricted to certain simple properties of the parts, to the exclusion of spatial or other relations among them. Thus, the electromotive force of a system of several electric batteries cannot be inferred from the electromotive forces of its constituents alone without a description, in terms of relational concepts, of the way in which the batteries are connected with each other.[4]

3. Finally, the predictability of a given characteristic of an object on the basis of specified information concerning its parts will obviously depend on what general laws or theories are available.[5] Thus, the flow of an electric current in a wire connecting a piece of copper and a piece of zinc which are partly immersed in sulfuric acid is unexplainable, on the basis of information concerning any nontrivial set of attributes of copper, zinc and sulphuric acid, and the particular structure of the system under consideration, unless the theory available contains certain general laws concerning the functioning of batteries, or even more comprehensive principles of physical chemistry. If the theory includes such laws, on the other hand, then the occurrence of the current is predictable. Another illustration, which at the same time provides a good example for the point made under (2) above, is afforded by the optical activity of certain substances. The optical activity of sarco-lactic acid, for example, i.e., the fact that in solution it rotates the plane of polarization of plane-polarized light, cannot be predicted on the basis of the chemical characteristics of its constituent elements; rather, certain facts about the relations of the atoms constituting a molecule of sarco-lactic acid have to be known. The essential point is that the molecule in question contains an asymmetric carbon atom, i.e., one that holds four different atoms or groups, and if this piece of relational information is provided, the optical activity of the solution can be predicted provided that furthermore the theory available for the purpose embodies the law that the presence of one asymmetric carbon atom in a molecule implies optical activity of the solution; if the theory does not include this micro-macro law, then the phenomenon is emergent with respect to that theory.

An argument is sometimes advanced to the effect that phenomena such as the flow of the current, or the optical activity, in our last examples, are absolutely emergent at least in the sense that they could not possibly have been predicted before they had been observed for the first time; in other words, that the laws requisite for their prediction could not have been arrived at on the basis of information available before their first occurrence.[6] This view is untenable, however. On the strength of data available at a given time, science often establishes generalizations by means of which it can forecast the occurrence of events the like of which have never before been encountered. Thus, generalizations based upon periodicities exhibited by the characteristics of chemical elements then known enabled Mendeleev in 1871 to predict the existence of a certain new element and to state correctly various properties of that element as well as of several of its compounds; the element in question, germanium, was not discovered until 1886. A more recent illustration of the same point is provided by the development of the atomic bomb and the prediction, based on theoretical principles established prior to the event, of its explosion under specified conditions, and of its devastating release of energy.

As Grelling has stressed, the observation that the predictability of the occurrence of any characteristic depends upon the theoretical knowledge available, applies even to those cases in which, in the language of some emergentists, the characteristic of the whole is a mere resultant of the corresponding characteristics of the parts and can be obtained from the latter by addition. Thus, even the weight of a water molecule cannot be derived from the weights of its atomic constituents without the aid of a law which expresses the former as some specific mathematical function of the latter. That this function should be the sum is by no means self-evident; it is an empirical generalization, and at that not a strictly correct one, as relativistic physics has shown.

Failure to realize that the question of the predictability of a phenomenon cannot be significantly raised unless the theories available for the prediction have been specified has encouraged the misconception that certain phenomena have a mysterious quality of absolute unexplainability, and that their emergent status has to be accepted with "natural piety," as C. L. Morgan put it. The observations presented in the preceding discussion strip the idea of emergence of these unfounded connotations: emergence of a characteristic is not an ontological trait inherent in some phenomena; rather it is indicative of the scope of our knowledge at a given time; thus it has no absolute, but a relative character; and what is emergent with respect to the theories available today may lose its emergent status tomorrow.

The preceding considerations suggest the following *redefinition of emergence*: The occurrence of a characteristic $W$ in an object $w$ is emergent relative to a theory $T$, a part relation $Pt$, and a class $G$ of attributes if that occurrence cannot be deduced by means of $T$ from a characterization of the $Pt$-parts of $w$ with respect to all the attributes in $G$.

This formulation explicates the meaning of emergence with respect to *events* of a certain kind, namely the occurrence of some characteristic $W$ in an object $w$. Frequently, emergence is attributed to *characteristics* rather than to events; this use of the concept of emergence may be interpreted as follows: A characteristic $W$ is emergent relatively to $T$, $Pt$, and $G$ if its occurrence in *any* object is emergent in the sense just indicated.

As far as its cognitive content is concerned, the emergentist assertion that the phenomena of life are emergent may now be construed, roughly, as an elliptic formulation of the following statement: Certain specifiable biological phenomena cannot be explained, by means of contemporary physico-chemical theories, on the basis of data concerning the physical and chemical characteristics of the atomic and molecular constituents of organisms. Similarly, the thesis of an emergent status of mind might be taken to assert that present-day physical, chemical, and biological theories do not suffice to explain all psychological phenomena on the basis of data concerning the physical, chemical, and biological characteristics of the cells or of the molecules or atoms constituting the organisms in question. But in this interpretation, the emergent character of biological and psychological phenomena becomes trivial; for the description of various biological phenomena requires terms which are not contained in the vocabulary of present-day physics and chemistry; hence we cannot expect that all specifically biological phenomena are explainable, i.e., deductively inferable, by means of present-day physico-chemical theories on the basis of initial conditions which themselves are described in exclusively physico-chemical terms. In order to obtain a less trivial interpretation of the assertion that the phenomena of life are emergent, we have therefore to include in the explanatory theory all those presumptive laws presently accepted which connect the physico-chemical with the biological "level," i.e., which contain, on the one hand, certain physical and chemical terms, including those required for the description of molecular structures, and on the other hand, certain concepts of biology. An analogous observation applies to the case of psychology. If the assertion that life and mind have an emergent status is interpreted in this sense, then its import can be summarized approximately by the statement that no explanation, in terms of micro-structure theories, is available at present for large classes of phenomena studied in biology and psychology.[7]

Assertions of this type, then, appear to represent the rational core of the doctrine of emergence. In its revised form, the idea of emergence no longer carries with it the connotation of absolute unpredictability—a notion which is objectionable not only because it involves and perpetuates certain logical misunderstandings, but also because, not unlike the ideas of neovitalism, it encourages an attitude of resignation which is stifling for scientific research. No doubt it is this characteristic, together with its theoretical sterility, which accounts for the rejection, by the majority of contemporary scientists, of the classical absolutistic doctrine of emergence.[8]

## Notes

This chapter originally appeared in Carl G. Hempel, *Aspects of Scientific Explanation and other Essays in the Philosophy of Science*, pp. 258–264, New York: Free Press, 1965.

1. This selection is part II of the authors' original article, "Studies in the Logic of Explanation." The omitted material contains the first detailed exposition of the deductive-nomological model of explanation. The notes have been renumbered to reflect the text reprinted here.

2. For a lucid brief exposition of this idea, see Feigl (1945), pp. 284–88.

3. Concerning the concept of novelty in its logical and psychological meanings, see also Stace (1939).

4. This observation connects the present discussion with a basic issue in Gestalt theory. Thus, e.g., the insistence that "a whole is more than the sum of its parts" may be construed as referring to characteristics of wholes whose prediction requires knowledge of certain structural relations among the parts. For a further examination of this point, see Grelling and Oppenheim (1937–1938) and (1939).

5. Logical analyses of emergence which make reference to the theories available have been propounded by Grelling and recently by Henle (1942). In effect, Henle's definition characterizes a phenomenon as emergent if it cannot be predicted, by means of the theories accepted at the time, on the basis of the data available before its occurrence. In this interpretation of emergence, no reference is made to characteristics of parts or constituents. Henle's concept of predictability differs from the one implicit in our discussion (and made explicit in Part III of this article) in that it implies derivability from the "simplest" hypothesis which can be formed on the basis of the data and theories available at the time. A number of suggestive observations on the idea of emergence and on Henle's analysis of it are presented in Bergmann's article (1944). The idea that the concept of emergence, at least in some of its applications, is meant to refer to unpredictability by means of "simple" laws was advanced also by Grelling in the correspondence mentioned in note (1). Reliance on the notion of simplicity of hypotheses, however, involves considerable difficulties; in fact, no satisfactory definition of that concept is available at present.

6. C. D. Broad, who in chapter 2 of his book (1925) gives a clear account and critical discussion of the essentials of emergentism, emphasizes the importance of "laws" of composition in predicting the characteristics of a whole on the basis of those of its parts (*op. cit.*, pp. 61ff.); but he subscribes to the view characterized above and illustrates it specifically by the assertion that "if we want to know the chemical (and many of the physical) properties of a chemical compound, such as silver-chloride, it is absolutely necessary to study samples of *that particular compound*.... The essential point is that it would also be useless to study chemical compounds in general and to compare their properties with those of their elements in the hope of discovering a *general* law of composition by which the properties of *any* chemical compound could be foretold when the properties of its separate elements were known." (p. 64) That an achievement of precisely this sort has been possible on the basis of the periodic system of the elements is noted above.

7. The following passage from Tolman (1932) may serve to support this interpretation: "...'behavior-acts,' though no doubt in complete one-to-one correspondence with the underly-

ing molecular facts of physics and physiology, have, as 'molar' wholes, certain emergent properties of their own.... Further, these molar properties of behavior-acts cannot in the present state of our knowledge, i.e., prior to the working-out of many empirical correlations between behavior and its physiological correlates, be known even inferentially from a mere knowledge of the underlying, molecular, facts of physics and physiology" (*op. cit.*, pp. 7–8). In a similar manner, Hull uses the distinction between molar and molecular theories and points out that theories of the latter type are not at present available in psychology. Cf. (1943a), pp. 19ff.; (1943), p. 275.

8. This attitude of the scientist is voiced, for example, by Hull in (1943a), pp. 24–28.

## Bibliography

Bergmann, Gustav (1994). "Holism, Historicism, and Emergence," *Philosophy of Science* 11, pp. 209–221.

Broad, C. D. (1925). *The Mind and Its Place in Nature*. London: Routledge and Kegan Paul.

Feigl, Herbert (1945). "Operation and Scientific Method," *Psychological Review* 52, pp. 250–259, 284–288.

Grelling, Kurt, and Paul Oppenheim (1937–1938). "Logical Analysis of Gestalt as 'Functional Whole,'" Reprinted for distribution at the Fifth International Congress for the Unity of Science, Cambridge, MA.

Henle, Paul (1942). "The Status of Emergence," *The Journal of Philosophy* 39, pp. 486–493.

Hull, Clark L. (1943a). *Principles of Behavior*. New York: Appelton-Century.

Stace, W. T. (1939). "Novelty, Indeterminism, and Emergence," *Philosophical Review* 48, pp. 296–310.

Tolman, Edward Chace (1932). *Purposive Behavior in Animals and Men*. New York: The Century Company.

Bibliography

# 3 Reductionism and the Irreducibility of Consciousness

John Searle

The view of the relation between mind and body that I have been putting forward is sometimes called "reductionist," sometimes "antireductionist." It is often called "emergentism," and is generally regarded as a form of "supervenience." I am not sure that any one of these attributions is at all clear, but a number of issues surround these mysterious terms, and in this chapter I will explore some of them.

## 3.1 Emergent Properties

Suppose we have a system, *S*, made up of elements *a*, *b*, *c*. . . . For example, *S* might be a stone and the elements might be molecules. In general, there will be features of *S* that are not, or not necessarily, features of *a*, *b*, *c*. . . . For example, *S* might weigh ten pounds, but the molecules individually do not weigh ten pounds. Let us call such features "system features." The shape and the weight of the stone are system features. Some system features can be deduced or figured out or calculated from the features of *a*, *b*, *c*. . . . just from the way these are composed and arranged (and sometimes from their relations to the rest of the environment). Examples of these would be shape, weight, and velocity. But some other system features cannot be figured out just from the composition of the elements and environmental relations; they have to be explained in terms of the causal interactions among the elements. Let's call these "causally emergent system features." Solidity, liquidity, and transparency are examples of causally emergent system features.

On these definitions, consciousness is a causally emergent property of systems. It is an emergent feature of certain systems of neurons in the same way that solidity and liquidity are emergent features of systems of molecules. The existence of consciousness can be explained by the causal interactions between elements of the brain at the micro level, but consciousness cannot itself be deduced or calculated from the sheer physical structure of the neurons without some additional account of the causal relations between them.

This conception of causal emergence, call it "emergent1," has to be distinguished from a much more adventurous conception, call it "emergent2." A feature *F* is emergent2 iff *F* is emergent1 and *F* has causal powers that cannot be explained by the causal interactions of *a, b, c*.... If consciousness were emergent2, then consciousness could cause things that could not be explained by the causal behavior of the neurons. The naive idea here is that consciousness gets squirted out by the behavior of the neurons in the brain, but once it has been squirted out, it then has a life of its own.

It should be obvious from the previous chapter[1] that on my view consciousness is emergent1, but not emergent2. In fact, I cannot think of anything that is emergent2, and it seems unlikely that we will be able to find any features that are emergent2, because the existence of any such features would seem to violate even the weakest principle of the transitivity of causation.

## 3.2 Reductionism

Most discussions of reductionism are extremely confusing. Reductionism as an ideal seems to have been a feature of positivist philosophy of science, a philosophy now in many respects discredited. However, discussions of reductionism still survive, and the basic intuition that underlies the concept of reductionism seems to be the idea that certain things might be shown to be *nothing but* certain other sorts of things. Reductionism, then, leads to a peculiar form of the identity relation that we might as well call the "nothing-but" relation: in general, *A*'s can be reduced to *B*'s, iff *A*'s are nothing but *B*'s.

However, even within the nothing-but relation, people mean so many different things by the notion of "reduction" that we need to begin by making several distinctions. At the very outset it is important to be clear about what the relata of the relation are. What is its domain supposed to be: objects, properties, theories, or what? I find at least five different senses of "reduction"—or perhaps I should say five different kinds of reduction—in the theoretical literature, and I want to mention each of them so that we can see which are relevant to our discussion of the mind-body problem.

1. Ontological Reduction The most important form of reduction is ontological reduction. It is the form in which objects of certain types can be shown to consist in nothing but objects of other types. For example, chairs are shown to be nothing but collections of molecules. This form is clearly important in the history of science. For example, material objects in general can be shown to be nothing but collections of molecules, genes can be shown to consist in nothing but DNA molecules. It seems to me this form of reduction is what the other forms are aiming at.

2. Property Ontological Reduction This is a form of ontological reduction, but it concerns properties. For example, heat (of a gas) is nothing but the mean kinetic energy

of molecule movements. Property reductions for properties corresponding to theoretical terms, such as "heat," "light," etc., are often a result of theoretical reductions.

3. Theoretical Reduction Theoretical reductions are the favorite of theorists in the literature, but they seem to me rather rare in the actual practice of science, and it is perhaps not surprising that the same half dozen examples are given over and over in the standard textbooks. From the point of view of scientific explanation, theoretical reductions are mostly interesting if they enable us to carry out ontological reductions. In any case, theoretical reduction is primarily a relation between theories, where the laws of the reduced theory can (more or less) be deduced from the laws of the reducing theory. This demonstrates that the reduced theory is nothing but a special case of the reducing theory. The classical example that is usually given in textbooks is the reduction of the gas laws to the laws of statistical thermodynamics.

4. Logical or Definitional Reduction This form of reduction used to be a great favorite among philosophers, but in recent decades it has fallen out of fashion. It is a relation between words and sentences, where words and sentences referring to one type of entity can be translated without any residue into those referring to another type of entity. For example, sentences about the average plumber in Berkeley are reducible to sentences about specific individual plumbers in Berkeley; sentences about numbers, according to one theory, can be translated into, and hence are reducible to, sentences about sets. Since the words and sentences are *logically* or *definitionally* reducible, the corresponding entities referred to by the words and sentences are *ontologically* reducible. For example, numbers are nothing but sets of sets.

5. Causal Reduction This is a relation between any two types of things that can have causal powers, where the existence and a fortiori the causal powers of the reduced entity are shown to be entirely explainable in terms of the causal powers of the reducing phenomena. Thus, for example, some objects are solid and this has causal consequences: solid objects are impenetrable by other objects, they are resistant to pressure, etc. But these causal powers can be causally explained by the causal powers of vibratory movements of molecules in lattice structures.

Now when the views I have urged are accused of being reductionist—or sometimes insufficiently reductionist—which of these various senses do the accusers have in mind? I think that theoretical reduction and logical reduction are not intended. Apparently the question is whether the causal reductionism of my view leads—or fails to lead—to ontological reduction. I hold a view of mind/brain relations that is a form of causal reduction, as I have defined the notion: Mental features are caused by neurobiological processes. Does this imply ontological reduction?

In general in the history of science, successful causal reductions tend to lead to ontological reductions. Because where we have a successful causal reduction, we simply redefine the expression that denotes the reduced phenomena in such a way that the phenomena in question can now be identified with their causes. Thus, for example, color terms were once (tacitly) defined in terms of the subjective experience of color perceivers; for example, "red" was defined ostensively by pointing to examples, and then real red was defined as whatever seemed red to "normal" observers under "normal" conditions. But once we have a causal reduction of color phenomena to light reflectances, then, according to many thinkers, it becomes possible to redefine color expressions in terms of light reflectances. We thus carve off and eliminate the subjective experience of color from the "real" color. Real color has undergone a property ontological reduction to light reflectances. Similar remarks could be made about the reduction of heat to molecular motion, the reduction of solidity to molecular movements in lattice structures, and the reduction of sound to air waves. In each case, the causal reduction leads naturally to an ontological reduction by way of a redefinition of the expression that names the reduced phenomenon. Thus, to continue with the example of "red," once we know that the color experiences are caused by a certain sort of photon emission, we then redefine the word in terms of the specific features of the photon emission. "Red," according to some theorists, now refers to photon emissions of 600 nanometers. It thus follows trivially that the color red is nothing but photon emissions of 600 namometers.

The general principle in such cases appears to be this: Once a property is seen to be *emergent1*, we automatically get a causal reduction, and that leads to an ontological reduction, by redefinition if necessary. The general trend in ontological reductions that have a scientific basis is toward greater generality, objectivity, and redefinition in terms of underlying causation.

So far so good. But now we come to an apparently shocking asymmetry. When we come to consciousness, we cannot perform the ontological reduction. Consciousness is a causally emergent property of the behavior of neurons, and so consciousness is causally reducible to the brain processes. But—and this is what seems so shocking—a perfect science of the brain would still not lead to an ontological reduction of consciousness in the way that our present science can reduce heat, solidity, color, or sound. It seems to many people whose opinions I respect that the irreducibility of consciousness is a primary reason why the mind-body problem continues to seem so intractable. Dualists treat the irreducibility of consciousness as incontrovertible proof of the truth of dualism. Materialists insist that consciousness must be reducible to material reality, and that the price of denying the reducibility of consciousness would be the abandonment of our overall scientific world view.

I will briefly discuss two questions: First, I want to show why consciousness is irreducible, and second, I want to show why it does not make any difference at all to our scientific world view that it should be irreducible. It does not force us to property dualism or anything of the sort. It is a trivial consequence of certain more general phenomena.

## 3.3 Why Consciousness Is an Irreducible Feature of Physical Reality

There is a standard argument to show that consciousness is not reducible in the way that heat, etc., are. In different ways the argument occurs in the work of Thomas Nagel (1974), Saul Kripke (1971), and Frank Jackson (1982). I think the argument is decisive, though it is frequently misunderstood in ways that treat it as merely epistemic and not ontological. It is sometimes treated as an epistemic argument to the effect that, for example, the sort of third-person, objective knowledge we might possibly have of a bat's neurophysiology would still not include the first-person, subjective experience of what it feels like to be a bat. But for our present purposes, the point of the argument is ontological and not epistemic. It is a point about what real features exist in the world and not, except derivatively, about how we know about those features.

Here is how it goes: Consider what facts in the world make it the case that you are now in a certain conscious state such as pain. What fact in the world corresponds to your true statement, "I am now in pain"? Naively, there seem to be at least two sorts of facts. First and most important, there is the fact that you are now having certain unpleasant conscious sensations, and you are experiencing these sensations from your subjective, first-person point of view. It is these sensations that are constitutive of your present pain. But the pain is also caused by certain underlying neurophysiological processes consisting in large part of patterns of neuron firing in your thalamus and other regions of your brain. Now suppose we tried to reduce the subjective, conscious, first-person sensation of pain to the objective, third-person patterns of neuron firings. Suppose we tried to say the pain is really "nothing but" the patterns of neuron firings. Well, if we tried such an ontological reduction, the essential features of the pain would be left out. No description of the third-person, objective, physiological facts would convey the subjective, first-person character of the pain, simply because the first-person features are different from the third-person features. Nagel states this point by contrasting the objectivity of the third-person features with the what-it-is-like features of the subjective states of consciousness. Jackson states the same point by calling attention to the fact that someone who had a complete knowledge of the neurophysiology of a mental phenomenon such as pain would still not know what a pain was if he or she did not know what it felt like. Kripke makes the same point when he says that pains could not be identical with neurophysiological states such as neuron firings in the thalamus and elsewhere, because any such identity would have to be necessary, because both sides of the identity statement are rigid designators, and yet we know that the identity could not be necessary.[2] This fact has obvious epistemic consequences: my knowledge that I am in pain has a different sort of basis than my knowledge that you are in pain. But the antireductionist point of the argument is ontological and not epistemic.

So much for the antireductionist argument. It is ludicrously simple and quite decisive. An enormous amount of ink has been shed trying to answer it, but the answers are all so much wasted ink. But to many people it seems that such an argument paints us into a corner. To them it seems that if we accept that argument, we have abandoned our scientific world view and adopted property dualism. Indeed, they would ask, what is property dualism but the view that there are irreducible mental properties? In fact, doesn't Nagel accept property dualism and Jackson reject physicalism precisely because of this argument? And what is the point of scientific reductionism if it stops at the very door of the mind? So I now turn to the main point of this discussion.

## 3.4 Why the Irreducibility of Consciousness Has No Deep Consequences

To understand fully why consciousness is irreducible, we have to consider in a little more detail the pattern of reduction that we found for perceivable properties such as heat, sound, color, solidity, liquidity, etc., and we have to show how the attempt to reduce consciousness differs from the other cases. In every case the ontological reduction was based on a prior causal reduction. We discovered that a surface feature of a phenomenon was caused by the behavior of the elements of an underlying microstructure. This is true both in the cases in which the reduced phenomenon was a matter of subjective appearances, such as the "secondary qualities" of heat or color; and in the cases of the "primary qualities" such as solidity, in which there was both an element of subjective appearance (solid things feel solid), and also many features independent of subjective appearances (solid things, e.g., are resistant to pressure and impenetrable by other solid objects). But in each case, for both the primary and secondary qualities, the point of the reduction was to carve off the surface features and redefine the original notion in terms of the causes that produce those surface features.

Thus, where the surface feature is a subjective appearance, we redefine the original notion in such a way as to exclude the appearance from its definition. For example, pretheoretically our notion of heat has something to do with perceived temperatures: Other things being equal, hot is what feels hot to us, cold is what feels cold. Similarly with colors: Red is what looks red to normal observers under normal conditions. But when we have a theory of what causes these and other phenomena, we discover that it is molecular movements causing sensations of heat and cold (as well as other phenomena such as increases in pressure), and light reflectances causing visual experiences of certain sorts (as well as other phenomena such as movements of light meters). We then *redefine* heat and color in terms of the underlying causes of both the subjective experiences and the other surface phenomena. And in the redefinition we eliminate any reference to the subjective appearances and other surface effects of the underlying causes. "Real" heat is now defined in terms of the kinetic energy of the molecular movements, and the subjective feel of heat that we get when we touch a hot object is now treated as just a subjective appearance caused by heat, as an effect of heat. It is no longer part of real heat. A similar distinction is made between real color and the subjective experience of color. The same pattern works for the primary qualities: Solidity is defined in terms of the vibratory movements of molecules in lattice structures, and objective, observer-independent features, such as impenetrability by other objects, are now seen as surface effects of the underlying reality. Such redefinitions are achieved by way of carving off all of the surface features of the phenomenon, whether subjective or objective, and treating them as effects of the real thing.

But now notice: The actual pattern of the facts in the world that correspond to statements about particular forms of heat such as specific temperatures are quite similar to the pattern of facts in the world that correspond to statements about particular forms of consciousness, such as pain. If I now say, "It's hot in this room," what are the facts? Well, first there is a set of "physical" facts involving the movement of molecules, and second there is a set of "mental" facts involving my subjective experience of heat, as caused by the impact of the moving air molecules on my nervous system. But similarly with pain. If I now say, "I am in pain," what are the facts? Well, first there is a set of "physical" facts involving my thalamus and other regions of the brain, and second there is a set of "mental" facts involving my subjective experience of pain. So why do we regard heat as reducible and pain as irreducible? The answer is that what interests us about heat is not the subjective appearance but the underlying physical causes. Once we get a causal reduction, we simply redefine the notion to enable us to get an ontological reduction. Once you know all the facts about heat—facts about molecule movements, impact on sensory nerve endings, subjective feelings, etc.—the reduction of heat to molecule movements involves no new *fact* whatever. It is simply a trivial consequence of the redefinition. We don't first discover all the facts and then discover a new fact, the fact that heat is reducible; rather, we simply redefine heat so that the reduction follows from the definition. But this redefinition does not eliminate, and was not intended to eliminate, the subjective experiences of heat (or color, etc.) from the world. They exist the same as ever.

We might not have made the redefinition. Bishop Berkeley, for example, refused to accept such redefinitions. But it is easy to see why it is rational to make such redefinitions and accept their consequences: To get a greater understanding and control of reality, we want to know how it works causally, and we want our concepts to fit nature at its causal joints. We simply redefine phenomena with surface features in terms of the underlying causes. It then looks like a new discovery that heat is *nothing but* mean kinetic energy of molecule movement, and that if all subjective experiences disappeared from the world, real heat would still remain. But this is not a new discovery, it is a trivial consequence of a new definition. Such reductions do not show that heat, solidity, etc., do not really exist in the way that, for example, new knowledge showed that mermaids and unicorns do not exist.

Couldn't we say the same thing about consciousness? In the case of consciousness, we do have the distinction between the "physical" processes and the subjective "mental" experiences, so why can't consciousness be redefined in terms of the neurophysiological processes in the way that we redefined heat in terms of underlying physical processes? Well, of course, if we insisted on making the redefinition, we could. We could simply define, for example, "pain" as patterns of neuronal activity that cause subjective sensations of pain. And if such a redefinition took place, we would have achieved the same sort of reduction for pain that we have for heat. But of course, the reduction of pain to its physical reality still leaves the subjective experience of pain unreduced, just as the reduction of heat left the subjective experience of heat unreduced. Part of the point of the reductions was to carve off the subjective experiences and exclude them from the definition of the real phenomena, which are now defined in terms of those features that interest us most. But where the phenomena that interest us most are the subjective experiences themselves, there is no way to carve anything off. Part of the point of the reduction in the case of heat was to distinguish between the subjective appearance on the one hand and the underlying physical reality on the other. Indeed, it is a general feature of such reductions that the phenomenon is defined in terms of the "reality" and not in terms of the "appearance." But we can't make that sort of appearance-reality distinction for consciousness because consciousness consists in the appearances themselves. *Where appearance is concerned we cannot make the appearance-reality distinction because the appearance is the reality.*

For our present purposes, we can summarize this point by saying that consciousness is not reducible in the way that other phenomena are reducible, not because the pattern of facts in the real world involves anything special, but because the reduction of other phenomena depended in part on distinguishing between "objective physical reality," on the one hand, and mere "subjective appearance," on the other; and eliminating the appearance from the phenomena that have been reduced. But in the case of consciousness, its reality is the appearance; hence, the point of the reduction would be lost if we tried to carve off the appearance and simply defined consciousness in terms of the underlying physical reality. In general, the pattern of our reductions rests on rejecting the subjective epistemic basis for the presence of a property as part of the ultimate constituent of that property. We find out about heat or light by feeling and seeing, but we then define the phenomenon in a way that is independent of the epistemology. Consciousness is an exception to this pattern for a trivial reason. The reason, to repeat, is that the reductions that leave out the epistemic bases, the appearances, cannot work for the epistemic bases themselves. In such cases, the appearance is the reality.

But this shows that the irreducibility of consciousness is a trivial consequence of the pragmatics of our definitional practices. A trivial result such as this has only trivial consequences. It has no deep metaphysical consequences for the unity of our overall scientific world view. It does not show that consciousness is not part of the ultimate furniture of reality or cannot be a subject of scientific investigation or cannot be brought into our overall physical conception of the universe; it merely shows that in the way that we have decided to carry out reductions, consciousness, by definition, is excluded from a certain pattern of reduction. Consciousness fails to be reducible, not because of some mysterious feature, but simply because by definition it falls outside the pattern of reduction that we have chosen to use for pragmatic reasons. Pretheoretically, consciousness, like solidity, is a surface feature of certain physical systems. But unlike solidity, consciousness cannot be redefined in terms of an underlying microstructure, and the surface features then treated as mere effects of real consciousness, without losing the point of having the concept of consciousness in the first place.

So far, the argument of this chapter has been conducted, so to speak, from the point of view of the materialist. We can summarize the point I have been making as follows: The contrast between the reducibility of heat, color, solidity, etc., on the one hand, and the irreducibility of conscious states, on the other hand, does not reflect any distinction in the structure of reality, but a distinction in our definitional practices. We could put the same point from the point of view of the property dualist as follows: The apparent contrast between the irreducibility of consciousness and the reducibility of color, heat, solidity, etc., really was *only* apparent. We did not really eliminate the subjectivity of red, for example, when we reduced red to light reflectances; we simply stopped calling the subjective part "red." We did not eliminate any subjective phenomena whatever with these "reductions"; we simply stopped calling them by their old names. Whether we treat the irreducibility from the materialist or from the dualist point of view, we are still left with a universe that contains an irreducibly subjective physical component as a component of physical reality.

To conclude this part of the discussion, I want to make clear what I am saying and what I am not saying. I am not saying that consciousness is not a strange and wonderful phenomenon. I think, on the contrary, that we ought to be amazed by the fact that evolutionary processes produced nervous systems capable of causing and sustaining subjective conscious states. As I remarked in chapter 4,[3] consciousness is as empirically mysterious to us now as electromagnetism was previously, when people thought the universe must operate entirely on Newtonian principles. But I am saying that once the existence of (subjective, qualitative) consciousness is granted (and no sane person can deny its existence, though many pretend to do so), then there is nothing strange, wonderful, or mysterious about its *irreducibility*. Given its existence, its irreducibility is a trivial consequence of our definitional practices. Its irreducibility has no untoward scientific consequences whatever. Furthermore, when I speak of the irreducibility of consciousness, I am speaking of its *irreducibility according to standard patterns of reduction.* No one can rule out a priori the possibility of a major intellectual revolution that would give us a new—and at present unimaginable—conception of reduction, according to which consciousness would be reducible.

## 3.5 Supervenience

In recent years there has been a lot of heavy going about a relationship between properties called "supervenience" (e.g., Kim 1979, 1982; Haugeland 1982). It is frequently said in discussions in the philosophy of mind that the mental is supervenient on the physical. Intuitively, what is meant by this claim is that mental states are totally dependent on corresponding neurophysiological states in the sense that a difference in mental states would necessarily involve a corresponding difference in neurophysiological states. If, for example, I go from a state of being thirsty to a state of no longer being thirsty, then there must have been some change in my brain states corresponding to the change in my mental states.

On the account that I have been proposing, mental states are supervenient on neurophysiological states in the following respect: Type-identical neurophysiological causes would have type-identical mentalistic effects. Thus, to take the famous brain-in-the-vat example, if you had two brains that were type-identical down to the last molecule, then the causal basis of the mental would guarantee that they would have the same mental phenomena. On this characterization of the supervenience relation, the supervenience of the mental on the physical is marked by the fact that physical states are causally sufficient, though not necessarily causally necessary, for the corresponding mental states. That is just another way of saying that as far as this definition of supervenience is concerned, sameness of neurophysiology guarantees sameness of mentality; but sameness of mentality does not guarantee sameness of neurophysiology.

It is worth emphasizing that this sort of supervenience is *causal* supervenience. Discussions of supervenience were originally introduced in connection with ethics, and the notion in question was not a causal notion. In the early writings of Moore (1922) and Hare (1952), the idea was that moral properties are supervenient on natural properties, that two objects cannot differ solely with respect to, for example, their goodness. If one object is better than another, there must be some other feature in virtue of which the former is better than the latter. But this notion of moral supervenience is not a causal notion. That is, the features of an object that make it good do not *cause* it to be good, they rather *constitute* its goodness. But in the case of mind/brain supervenience, the neural phenomena cause the mental phenomena.

So there are at least two notions of supervenience: a constitutive notion and a causal notion. I believe that only the causal notion is important for discussions of the mind-body problem. In this respect my account differs from the usual accounts of the supervenience of the mental on the physical. Thus Kim (1979, especially p. 45ff.) claims that we should not think of the relation of neural events to their supervening mental events as causal, and indeed he claims that supervening mental events have no causal status apart from their supervenience on neurophysiological events that have "a more direct causal role." "If this be epiphenomenalism, let us make the most of it," he says cheerfully (p. 47).

I disagree with both of these claims. It seems to me obvious from everything we know about the brain that macro mental phenomena are all caused by lower-level micro phenomena. There is nothing mysterious about such bottom-up causation; it is quite common in the physical world. Furthermore, the fact that the mental features are supervenient on neuronal features in no way diminishes their causal efficacy. The solidity of the piston is causally supervenient on its molecular structure, but this does not make solidity epiphenomenal; and similarly, the causal supervenience of my present back pain on micro events in my brain does not make the pain epiphenomenal.

My conclusion is that once you recognize the existence of bottom-up, micro to macro forms of causation, the notion of supervenience no longer does any work in philosophy. The formal features of the relation are already present in the causal sufficiency of the micro-macro forms of causation. And the analogy with ethics is just a source of confusion. The relation of macro mental features of the brain to its micro neuronal features is totally unlike the relation of goodness to good-making features, and it is confusing to lump them together. As Wittgenstein says somewhere, "If you wrap up different kinds of furniture in enough wrapping paper, you can make them all look the same shape."

### Notes

This chapter originally appeared as chapter 5 of John Searle's *The Rediscovery of the Mind*, pp. 111–126, Cambridge: The MIT Press, 1992.

1. Not included in this collection.

2. For further discussion of this point, see chapter 2 in the original book (not included in this collection).

3. Not included in this collection.

### Bibliography

Hare, R. M. (1952). *The Language of Morals*. Oxford: Oxford University Press.

Haugeland, John (1982). "Weak Supervenience," *American Philosophical Quarterly* 19, pp. 93–104.

Jackson, Frank (1982). "Epiphenomenal Qualia," *Philosophical Quarterly* 32, pp. 127–136.

Kim, Jaegwon (1979). "Causality, Identity, and Supervenience in the Mind-Body Problem," *Midwest Studies in Philosophy* 4, pp. 31–49.

Kim, Jaegwon (1982). "Psychophysical Supervenience," *Philosophical Studies* 41, pp. 51–70.

Kripke, Saul (1971). "Naming and Necessity," pp. 253–355, 763–769 in *Semantics of Natural Language,* D. Davidson and G. Harman (eds). Dordrecht: D. Reidel Publishing Company.

Moore, G. E. (1922). *Philosophical Studies.* London: Routledge and Kegan Paul.

Nagel, Thomas (1974). "What Is It Like to Be a Bat?" *Philosophical Review* 83, pp. 435–450.

# 4 Emergence and Supervenience

**Brian P. McLaughlin**

The notion of emergence played a prominent role in philosophy in the first half of the twentieth century. In this last decade of the century, the notion has once again become a focus of attention. Jaegwon Kim (1992) has claimed that some of the varieties of nonreductive materialism—currently the most popular brand of materialism—appear to be versions of emergent materialism: the doctrine that mental properties emerge from physical properties. One issue that remains unclear is what it is exactly for one sort of property to emerge from properties of another sort. It is generally acknowledged that when there is such emergence, the emerging property is a macro property relative to its emergence base properties (and thus they are micro properties relative to it), and that emergence precludes reducibility. But beyond that, there is little agreement. The aim of this paper is to formulate a notion of an emergent property such that if certain sorts of properties (e.g., mental properties) emerge from physical properties, then no version of reductive materialism is true. Whether any sort of property indeed emerges from physical properties will be left an open question. My aim is to develop a notion of an emergent property that is useful for formulating the dispute between reductive and nonreductive materialism, not to attempt to adjudicate that dispute.

In what follows, I shall first present a short history of the modern emergentist tradition, a tradition that begins with John Stuart Mill's *System of Logic* (1843), and traces through Alexander Bain's *Logic* (1870), George Henry Lewes's *Problems of Life and Mind* (1875), Sammuel Alexander's two-volume *Space, Time, and Deity* (1920), Lloyd Morgan's *Emergent Evolution* (1923), and C. D. Broad's *The Mind and Its Place in Nature* (1925). Then, I shall present some twentieth century results, both philosophical and scientific, that bear on the conclusions drawn by members of that tradition. After that, I shall examine an attempt by James van Cleve (1990) to define the notion of an emergent property by appeal to supervenience. Finally, I shall offer my own definition of an emergent property appealing to supervenience.

## 4.1 A Short History

Ernest Nagel (1961) aptly cites John Stuart Mill's chapter "Of the Composition of Causes" in Mill's *System of Logic* as the locus classicus of the notion of emergence. Indeed, Mill is the father of a philosophical tradition that I have labeled "British Emergentism" (McLaughlin 1992).

In "Of the Composition of Causes," Mill distinguishes "two modes of the conjoint action of causes, the mechanical and the chemical" (p. xviii). Of the mechanical mode, he says:

> In this important class of cases of causation, one cause never, properly speaking, defeats or frustrates another; both have their full effect. If a body is propelled in two directions by two forces, one tending to drive it to the north and the other to the east, it is caused to move in a given time exactly as far in both directions as the two forces would separately have carried it; and is left precisely where it would have arrived if it had been acted upon first by one of the two forces, and afterwards by the other. This law of nature is called, in dynamics, the principle of the Composition of Forces: and in imitation of that well-chosen expression, I shall give the name of the Composition of Causes to the principle which is exemplified in all cases in which the joint effect of several causes is identical with the sum of their separate effects. (1843, p. 428)

The principle of the Composition of Forces is of course a principle of vector addition. Forces acting together exhibit the mechanical mode of the conjoint action of causes: the effect of two or more forces acting together is the vector sum of the effect each force would have had if it had acted alone. The Composition of Forces is Mill's paradigm of the Composition of Causes. Mill calls a type of effect of two or more types of causes which would produce it in the mechanical mode "a homopathic effect," and laws which assert causal relations between causes and their homopathic effects, "homopathic laws." Homopathic laws thus subsume causal transactions in the mechanical mode.

According to Mill, in the chemical mode of the conjoint action of causes, the type of effect of the action of two or more types of causes is not the sum of the effects each of the causes would have had had it been acting alone. Thus, a causal transaction involving two or more causes is in the chemical mode if and only if it is not in the mechanical mode. Mill calls this mode of the conjoint action of causes the chemical mode precisely because chemical transactions typically exhibit it. Thus, consider the following type of chemical process:

$$CH_4 + 2O_2 \rightarrow CO_2 + 2H_2O$$

(Methane + oxygen produces carbon dioxide + water).

The product of this chemical process is not, in any sense, the sum of the effects of each reactant. Mill labels an effect of two or more types of causes which would combine in the chemical mode to produce it "a heteropathic effect," and laws subsuming such causal transactions, "heteropathic laws." Mill says that a heteropathic law owes its existence to a breach of the Composition of Causes.

Mill maintains that one finds heteropathic laws not only in chemistry, but throughout the special sciences—the sciences concerned with special properties of special kinds of things: biology, psychology, etc. Indeed, he explains the existence of the various special sciences partly in terms of breaches of the Composition of Causes. He says:

> Where the principle of Composition of Causes . . . fails . . . the concurrence of causes is such as to determine a change in the properties of the body generally, and render it subject to new laws, more or less dissimilar to those to which it conformed in its previous state. (1843, p. 435)

Moreover, Mill tells us that in some instances, at some particular points in the transition from separate to united action, the laws change, and an entirely new set of effects are either added to, or take the place of, those which arise from the separate agency of the same causes: the laws of these new effects being again susceptible of composition, to an indefinite extent, like the laws which they superseded (1984, pp. 433–434).

Heteropathic effects can themselves combine with each other in accordance with the Composition of Causes. Thus, a special science might well admit of a small group of laws and compositional principles from which other laws of the science can be deduced. He says:

> Though there are laws which, like those of chemistry and physiology, owe their existence to a breach of the principle of the Composition of Causes, it does not follow that these peculiar, or as they might be termed, heteropathic laws, are not capable of composition with one another. The causes which by one combination have had their laws altered, may carry their new laws with them unaltered into their ulterior combinations. And hence there is no reason to despair of ultimately raising chemistry and physiology to the condition of deductive sciences; for though it is impossible to deduce all chemical and physiological truths from the laws or properties of simple substances or elementary agents, they may possibly be deducible from laws which commence when these elementary agents are brought together into some moderate number of very complex combinations. The Laws of Life will never be deducible from the mere laws of ingredients, but the prodigiously complex Facts of Life may all be deducible from comparatively simple laws of life; which laws (depending indeed on combinations, but on comparatively simple combinations, of antecedents) may, in more complex circumstances be strictly compounded with one another, and with the physical and chemical laws of the ingredients. The details of vital phenomena, even now, afford innumerable exemplifications of the Composition of Causes. (1843, pp. 431–432)

Special sciences can thus aspire to be what Mill called "deductive sciences": sciences that have a small group of laws from which all other laws of the science can be deduced. However, the fundamental laws of the special science will not themselves be deducible from laws of sciences concerned with more general, more pervasive properties of substances (such as, for example, charge or mass).

While the new laws will supersede the old ones, they will not contravene them. The old laws will continue to hold. Speaking of vegetable and animal substances, Mill says:

> Those bodies continue, as before, to obey mechanical and chemical laws, in so far as the operation of those laws is not counteracted by the new laws which govern them as organized beings. (1843, p. 431)

Sometimes the old laws (e.g., chemical laws) will contain "ceteris paribus" clauses, and thus will not be contravened. Moreover, the old mechanical or dynamical laws will not be contravened even when, as Mill seems to hold, new forces are exerted as a result of "simple substances" or "elementary agents" being configured in certain ways. When "simple substances" become so configured as to make up a living organism, for instance, the exertion of a vital force may come into play. The vital force will be a fundamental force that must be taken into account when calculating the net force acting on a body. The force would have a value of 0 until the "simple substances" or "elementary agents" becomes configured in such a way as to constitute a living organism. But when they become so configured, a vital force would come into play in affecting dynamic behavior. Since forces are additive, other force laws will not be contravened. It is just that the fundamental force laws of mechanics will have to include in addition to, say, the inverse square laws, force laws for configurational forces, that is, forces that come into play only when elementary agents are organized in a certain way.

Alexander Bain (1870) embraced Mill's distinctions between heteropathic and homopathic effects and laws, and argued that certain collocations of causal agents bring into action new forces of nature (1870, ii, p. 31). Another contemporary of Mill's, George Henry Lewes embraced Mill's distinction and coined the term "emergent." By an emergent, Lewes meant what Mill called a heteropathic effect: an effect that is not the sum of what would have been the effects of each of its causes had they acted separately. Lewes contrasted emergents with resultants: effects that are the sum of what would have been the effects of each of their causes had those causes acted alone. Lewes says:

> In the somewhat more complicated effect of compound motions, say the orbit of a planet—the resultant of its tangential direction and its direction towards the sun—every student learns that the resultant motion of two impressed forces is the diagonal of those directions which the body would take were each force separately applied. Every resultant is either a sum or a difference of the co-operant forces. (1875, p. 413)

In Lewes's terminology, heteropathic effects emerge from the causal factors that produce them.

Very closely related notions of emergence figure in the work of the metaphysician and theologian Samuel Alexander (1920) and the biologist Llyod Morgan (1923). Embracing Lewes term "emergent," Alexander speaks of emergent qualities thus:

> The emergence of a new quality from any level of existence means that at that level there comes into being a certain constellation or collocation of the motions belonging to that level, and this collocation possesses a new quality distinctive of the higher-complex.... The higher-quality emerges from the lower level of existence and has is roots therein, but it emerges therefrom, and it does not belong to that lower level, but constitutes its possessor a new order of existent with its special laws of behavior. The existence of emergent qualities thus described is something to be noted, as some would say, under the compulsion of brute empirical fact, or, as I should prefer to say in less harsh terms, to be accepted with the "natural piety" of the investigator. It admits of no explanation. (1920, p. 45–47)

Alexander's main idea is that a certain complex configuration of elements of a given level may possess capacities to produce certain types of effects that are, in Mill's sense, heteropathic relative to the elements in the configuration. Qualities emerge from the configuration, and the configuration is governed by special laws of behavior not derivative from the laws that govern behavior at lower levels of organizational complexity.

The first section of Morgan's *Emergent Evolution* (1923) is entitled "Emergents and Resultants." In it, he cites his debt to Mill and Lewes. Morgan's principal example of an emergent is a chemical one. He says:

> When carbon having certain properties combines with sulphur having other properties there is formed, not a mere mixture but a new compound, some of the properties of which are quite different from those of either component. (1923, p. 3)

Here is one of his paradigms of a resultant: "the weight of the compound is an additive resultant, the sum of the weights of the components" (1923, p. 3).

Morgan's principal purpose in his book is to argue that through the process of evolution, new, unpredictable complex phenomena emerge. He thus combines the idea of emergence with a cosmology inspired by Darwinian evolution. He contrasts his evolutionary cosmology with a mechanistic cosmology, which he rejects saying:

> The essential feature of a mechanical—or, if it be preferred, a mechanistic—interpretation is that it is in terms of resultant effects only, calculable by algebraic summation. It ignores the something more that must be accepted as emergent.... Against such a mechanical interpretation—such a mechanistic dogma—emergent evolution rises in protest. The gist of its contention is that such an interpretation is quite inadequate. Resultants there are; but there is emergence also. Under naturalistic treatment, however, the emergence, in all its ascending grades, is loyally accepted, on the evidence, with natural piety. That it cannot be mechanically interpreted in terms of resultants only, is just that for which it is our aim to contend with reiterated emphasis. (1923, p. 8)

The various emergent levels in the ascending grades of complexity of matter are the subjects of the various special sciences.

The last major work in the British Emergentist tradition is C. D. Broad's *The Mind and Its Place in Nature* (1925). In this important work, Broad contrasts emergentism with mechanism. Of what he calls the "ideal of Pure Mechanism." Broad says:

> On a purely mechanical theory all the apparently different kinds of matter would be made of the same stuff. They would differ only in the number, arrangement and movements of their constituent particles. And their apparently different kinds of behaviour would not be ultimately different. For they would all be deducible by a single simple principle of composition from the mutual influences of the particles taken by pairs; and these mutual influences would all obey a single law which is quite independent of the configuration and surroundings in which the particles happen to find themselves. The ideal which we have been describing may be called "Pure Mechanism." (1925, pp. 45–46)

He offers the following illustration:

> A set of gravitating particles, on the classical theory of gravitation, is an almost perfect example of the ideal of Pure Mechanism. The single elementary law is the inverse-square law for any pair of particles. The single and simple principle of composition is the rule that the influence of any set of particles on a single particle is the vector sum of the influence that each would exert taken by itself. (1925, p. 45)

The single elementary law is of course the law of gravity, the principle of composition, vector addition. (Broad cites the parallelogram law.) Broad tells us that on a such view:

> There is one and only one kind of material. Each particle of this obeys some elementary law of behaviour, and continues to do so no matter how complex may be the collection of particles of which it is a constituent. There is one uniform law of composition, connecting the behaviour of groups of these particles as wholes with the behaviour which each would show in isolation and with the structure of the group. All the apparently different kinds of stuff are just differently arranged groups of different numbers of the one kind of elementary particle; and all the apparently peculiar laws of behaviour are simple special cases which could be deduced in theory from the structure of the whole under consideration, the one elementary law of behaviour for isolated particles, and the one universal law of composition. On such a view the external world has the greatest amount of unity which is conceivable. There is really only one science and the various "special sciences" are just particular cases of it. (1925, p. 76)

The electronic theory of matter, he notes, departs to some extent from the ideal of Pure Mechanism in that it postulates more than one kind of elementary particle, and in that "the laws of electro-magnetics cannot, so far as we know, be reduced to central forces" (1925, p. 45). He maintains, however, that such departures are compatible with Mechanism itself.

Broad tells us that on the Emergentist view, in contrast to the Mechanist view,

> We have to reconcile ourselves to much less unity in the external world and a much less intimate connexion between the various sciences. At best the external world and the various sciences that deal with it form a hierarchy. (p. 77)

At the base of the hierarchy will be physics, for it concerns itself with the most general characteristics of matter. The hierarchy includes in ascending order: chemistry, biology, psychology, and the social sciences. Eschewing Cartesian souls, entelechies, or indeed substance dualism of any sort, Broad maintains that the kinds of substances specific to any level will be wholly made up of kinds of substances of lower-orders. Every substance either is or is wholly made up of elementary particles. There are, however, properties or qualities or characteristics that are specific to kinds of a given order. He cites "the power of reproduction" as a property specific to the vital order. Broad calls the properties that are specific to a given order "the ultimate characteristics" of that order. He calls characteristics of an order that are reducible to characteristics of lower-orders, "reducible characteristics." And he calls characteristics that are possessed by aggregates at all levels of complexity, "ordinally neutral characteristics." Thus, inertial and gravitational mass are examples of ordinally neutral characteristics. Physics studies the organizational relationships objects participate in in virtue of their ordinally neutral properties. The various special sciences study the properties distinctive of certain kinds of complex substances; these features are the ultimate characteristics of the orders of complexity in question. Some of these properties are reducible; but the ultimate properties of the order that are irreducible are emergent properties.

Broad notes that emergentism can "keep the view that there is only one fundamental kind of stuff" (1925, p. 77). It is consistent with emergentism that every complex object is wholly made of one kind of elementary particle. However, there are irreducible ultimate characteristics or properties of the various orders of complexity. These properties emerge from properties exhibited at lower orders of complexity; and these various layers of reality are the subject matter of the various special sciences.

Broad tells us that if Emergentism is correct, then:

> We should have to recognize aggregates of various orders. And there would be two fundamentally different types of law, which might be called "intra-ordinal" and "trans-ordinal" respectively. A trans-ordinal law would be one which connects the properties of aggregates of adjacent orders. A and B would be adjacent, and in ascending order, if every aggregate of order B is composed of aggregates of order A, and if it has certain properties which no aggregate of order A possesses and which cannot be deduced from the A-properties and the structure of the B-complex by any law of composition which has manifested itself at lower-levels. An intra-ordinal law would be one which connects the properties of aggregates of the same order. A trans-ordinal law would be a statement of the irreducible fact that an aggregate composed of aggregates of the next lower order in such and such proportions and arrangements has such and such characteristic and non-deducible properties. (1925, pp. 77–78)

Broad illustrates the notion of a trans-ordinal law as follows:

> The law which asserts that all aggregates composed of such and such chemical substances in such and such proportions and relations have the power of reproduction would be an instance of a Trans-ordinal law. (1925, pp. 78–79)

Trans-ordinal laws, according to Broad, are emergent laws: they are not deducible from the laws of lower orders, lower-level conditions, and any compositional principles instantiated at lower-levels. Emergent trans-ordinal laws are "unique and ultimate" (1925, p. 65). They are, so to speak, brute nomological facts that "cannot be explained" (1925, p. 55). They are fundamental, nonderivative laws that must be "simply swallowed whole with that philosophic jam which Professor Alexander calls 'natural piety'" (1925, p. 55). Emergent trans-ordinal laws are themselves fundamental compositional principles.

The British Emergentist notion of an emergent property is thus explicated in terms of the notion of an emergent (trans-ordinal) law. Such a law, Broad tells us: "would be a statement of the irreducible fact that an aggregate composed of aggregates of the next lower order in such and such proportions and arrangements has such and such characteristic and non-deducible properties" (1925, p. 78).

## 4.2 Quantum Mechanics, Functionalism, and A Posteriori Necessity

As I mentioned, Broad's *The Mind and Its Place in Nature* is the last major work in the British Emergentist tradition. The reason, I have speculated elsewhere, is that the quantum mechanical revolution occurred shortly after its publication. One of the crowning achievements of this scientific revolution was the reductive explanation of chemical bonding.

The members of the British Emergentist tradition were perfectly correct in claiming that the product of two chemical reactants is in no sense the sum of what would have been the effect of each reactant had it acted alone. Chemical processes indeed produce heteropathic or emergent effects; and chemical laws are indeed heteropathic or emergent. To take our earlier example, carbon dioxide + water is indeed a heteropathic effect of combining methane and oxygen. But that chemistry is emergent in the sense in question poses no problem for reductive materialism. The quantum mechanical reduction of chemistry is held as the leading paradigm of reductive materialism. The British Emergentists all worked with a Newtonian conception of mechanism. Quantum mechanics has broadened our conception of mechanism—introducing a holistic notion of mechanism—and thereby of reductive explanation. Quantum mechanics reductively explains chemistry, but without appeal to additive or even linear compositional principles, and without the postulation of new irreducible higher-level forces (General relativity too invokes nonlinearity). Moreover, quantum mechanics has led to the development of molecular biology, and the successes of this discipline (e.g., the discovery of the structure of DNA) have virtually eradicated any sort of vitalism from biology. On the current evidence, it appears that all fundamental forces are exerted below the level of the atom.

While chemical properties are reducible and biological properties seem to be as well, the question still persists whether all mental properties are reducible. Broad articulates a doctrine he calls "Emergent Materialism," according to which everything is wholly made of matter, all particular mental processes are processes in the central nervous system, but mental properties emerge from the minute internal structures of the central nervous system (1925, p. 436). In this, he follows Lewes (1875) and Alexander (1920), who both insist that every particular mental process is identical with a neurophysiological process, but that mental qualities or properties emerge from neurophysiological properties. The British Emergentists were mistaken in taking reductive materialism to require linear compositional principles. But is it possible to salvage a notion of an emergent property from this tradition that will allow us to formulate a version of emergent materialism that is a competitor with reductive materialism—at least where certain mental properties are concerned? I believe that the answer is "yes." But before pursuing this issue, I want to mention two relevant philosophical results.

One philosophical result is that dispositions and capacities can be functionally analyzed. To functionally analyze a disposition or capacity is to analyze it as a second-order state of being in a state that plays a certain causal role. Functional analysis reveals that the disposition of water-solubility, for instance, is the state of being in a state that disposes its occupant to dissolve in water; and that fragility is the state of being in a state that disposes its occupant to shatter when struck. Dispositions and capacities are thus second-order states. The first-order state that disposes the substance in question to dissolve or shatter is "the base" for the respective disposition (a disposition or capacity need not have a unique base; it can have multiple bases). When the base property is a microstructural property, and the manifestation of the disposition or capacity (dissolving in water is the manifestation of water-solubility) is expressible in physical and/or topic-neutral terms (see White 1991, ch. 3), the dispositional property or capacity is physicalistically reducible. For the higher-level laws concerning the dispositions or capacities in question will be directly deducible from the lower-level laws governing their bases and whatever lower-level factors make them their bases. (Keep in mind that the notion of deducibility here is a semantic one, not a syntactic one. P is deducible from Q if and only if whenever Q is true, P is.)

To see how this philosophical result bears on British Emergentist doctrines, recall that Broad speaks of the power of reproduction as an ultimate characteristic of the vital order. The power to reproduce is, however, a capacity that is susceptible to functional analysis: it is the property of having a property that enables the organism in question to produce a duplicate or near duplicate. As we noted, properties that are susceptible to functional analysis in physical and topic-neutral terms are reducible. Indeed, they are (semantically) deducible from physical laws and physical conditions. For the functional analyses will yield necessary (definitional) truths. The notion of functional analysis is, it should be noted, not entirely foreign to the British Emergentists. Lewes (1875) claimed that while all mental processes are neural processes, not all neural processes are mental processes. What makes a neural process a mental process, he claimed, is its role in the organism. Moreover, Broad had extensive discussions of dispositions and their bases. The members of the British Emergentist tradition apparently failed to appreciate, however, that dispositions and capacities that are functionally analyzable are ipso facto reducible. Their failure to appreciate this was, perhaps, due to their focus on the Newtonian conception of mechanism, rather than on the broader notion of reductive explanation.

A further philosophical result is also relevant: identities, even *a posteriori* as opposed to *a priori* knowable identities, are necessary. The British Emergentists held that water = $H_2O$, that salt = NaCl, and so on. (Indeed, Lewes (1875) argued that it was a mistake to think that water was caused by $H_2O$ since, in fact, water = $H_2O$.) It was Saul Kripke (1971), however, who demonstrated that if A = B, then necessarily A = B. Identities are metaphysically necessary (they hold under all possible circumstances), even when they are knowable only *a posteriori* via empirical investigation. Given that, and given a semantic notion of deduction, we need not even appeal to such identities to deduce truths about water, salt, and the like from truths about $H_2O$, NaCl, and the like.

The question we shall now turn to is whether we can extract from the British Emergentist tradition a notion of emergent properties that is such that (a) it remains an open question whether certain mental properties are emergent, and (b) if some properties are emergent, then no brand of reductive materialism is true. I want to pursue this question for the remainder of this chapter.

## 4.3 van Cleve's Notion of an Emergent Property

James van Cleve (1990) has attempted to define just such a notion of an emergent property. He has attempted to define a notion of an emergent property based on Broad's and Alexander's notion that is such that it is a genuinely open question whether conscious properties, in particular, are emergent.

van Cleve's definition of an emergent property invokes the notion of supervenience; so some brief remarks about supervenience are in order. There are two core ideas of supervenience that one finds in today's literature. One is the idea that there cannot be a difference of one sort without a difference of another sort: for example, that there cannot be a mental difference without a physical difference, or that there cannot be a moral difference without a descriptive difference. The second core idea is that of a required-sufficiency relationship: the idea that having a certain sort of property requires having a property of another sort that is sufficient for it, for example, that having a mental property requires having some physical property that suffices for its possession. van Cleve employs this second idea. More specifically, he employs a technical definition intended to capture one version of this second idea.

van Cleve (1990, p. 220) employs the following technical definition of supervenience:

A-properties supervene on B-properties = df. Necessarily, for any object x and A-property a, if x has a, then there is a B-property b such that (i) x has B, and (ii) necessarily, if anything has b, it also has a.

Notice that there are two occurrences of "necessarily" in the definition. In van Cleve's definition of emergence (to be stated below), when he says "supervenes with nomological necessity" he means that the second occurrence of "necessarily" is that of nomological necessity; and we shall understand the first occurrence of "necessarily" in the same way.

Armed with this notion of supervenience, van Cleve (1990, p. 222) defines the notion of an emergent property as follows:

If P is a property of w, then P is emergent if and only if P supervenes with nomological necessity, but not with logical necessity, on the properties of the parts of w.

As van Cleve points out, "this is a variety of multiple-domain supervenience, in which the supervening properties are possessed by wholes and the subvening properties by their parts" (1990, p. 220). This definition implies that a property P of a whole w is emergent if and only if it is nomologically necessary that some properties of the parts of w are nomologically sufficient, but not logically sufficient for w's having P.

van Cleve understandably worries that it may be that no property counts as emergent as he defines the notion. The reason is that "for any property P of any whole w, there will always be properties of the parts from which P may be deduced" (1990, p. 223). To illustrate the worry: A part x of a whole w will have the property of being part of a whole with property P.

To avoid this trivialization, van Cleve (1990, p. 223) suggests we might try to adopt a proposal of Broad's, viz. that we include among relevant properties of parts only properties the parts have "taken separately and in other combinations." As van Cleve notes, one can plausibly refuse to regard the property "forming a whole with such and such features" as one the part has taken separately or in other combinations. So revised, then, the definition of emergence is this:

If P is a property of w, then P is emergent if and only if P supervenes with nomological necessity, but not with logical necessity, on properties the parts of w have taken separately or in other combinations.

Let us examine this notion of an emergent property in detail.

This notion of an emergent property so defined is, I believe, too inclusive to be of interest: it would count certain reducible properties as emergent. For the weight or mass of a whole will count as an emergent property by this definition. The reason is that—as Broad well knew—the principles of the additivity of weight and the additivity of mass are logically contingent. The mass of a part, for instance, is a property the part has taken separately and in other combinations. However, given that the principle of the additivity of mass is logically contingent, the mass of a whole will supervene with only nomological necessity on the masses of its parts; and there are no other properties the parts have taken separately or in other combinations on which the mass of the whole supervenes with logical necessity. Thus, the mass of a whole will, by the above definition, count as an emergent property of the whole. The definition is thus too inclusive.

van Cleve appears to recognize this problem. He says:

> There may be a problem with Broad's qualification [that the properties of the parts be ones they have taken separately and in other combinations], however. Consider what Newtonian physics would say about a body A and two more massive bodies B and C. If A and B were the only bodies around, A would gravitate toward B; if A and C were the only bodies around, A would gravitate toward C; and if all three bodies were there, S would gravitate toward a point between B and C. This last fact, however, is not deducible from the laws governing the A-B and A-C systems in isolation. (The parallelogram law for the composition of forces is logically contingent.) It seems therefore to follow from Broad's definition that the behavior of the three-body system is emergent. Yet it also seems that this behavior follows logically, but not in an objectively trivial way, from properties of the parts: B is here, C is there, and A is moving in a certain direction. Broad's account seems therefore to be too liberal in what it counts as emergent. Can it be made less liberal without making anti-emergence trivially true? There must be a way, but at the moment, I do not have a satisfactory proposal. (1990, p. 224)

The above definition is indeed too liberal for just the sort of reason van Cleve gives.

However, van Cleve misunderstands Broad's position. For Broad explicitly says that the parallelogram law must be invoked in deducing the behavior of such systems. Broad himself pointed out that resultants typically have to be deduced using compositional principles, and his paradigm of a compositional principle was the parallelogram law. Moreover, he regarded this and other compositional principles as contingent. van Cleve's definition of an emergent property fails to incorporate the role of contingent compositional principles.

Taking into account that logically contingent compositional principles are required even in the case of resultant properties to deduce properties of wholes from properties of parts, we should revise the definition of emergence as follows:

> If P is a property of w, then P is emergent if and only if P supervenes with nomological necessity, but not with logical necessity, on properties the parts of w have taken separately or in other combinations together with compositional principles that apply to the parts in other combinations.

Now, the weight and mass of wholes are not emergents, but rather resultants. And likewise for the gravitational behavior of the systems van Cleve describes.

However, just as Mill and Broad claimed, chemical properties are emergent on this notion, at least if compositional principles must be linear. As we noted, there is of course nothing sacrosanct about linearity. To require linearity would be to render the notion of an emergent property uninteresting. Linearity is not the issue. How, then, should the notion of a compositional principle be understood so as to yield a theoretically interesting notion of an emergent property?

## 4.4 Emergent Properties

The (modal operator-strong) supervenience thesis in question will imply supervenience principles or laws stating that if the parts of some whole have such and such (subvenient) properties, then the whole will have such and such (supervenient) property. These supervenience principles will be what Broad called trans-ordinal laws. (They may or may not be finitely statable.) Trans-ordinal laws, you will recall, are themselves compositional principles. The key issue is whether the trans-ordinal (supervenience) laws in question are fundamental, irreducible laws, that must simply be accepted with "natural piety," or whether, instead, they are derivative laws.

Let us define the notion of a fundamental law as follows:

A law L is a fundamental law if and only if it is not metaphysically necessitated by any other laws, even together with initial conditions.

Notice that this notion of a fundamental law is like Broad's notion of a law that is "unique and ultimate" in that it is not deducible from other laws and conditions. On this notion of a fundamental law, the laws of thermodynamics count as nonfundamental: for while they are not necessitated by other laws alone, they are necessitated by other laws together with initial conditions. In contrast, Schroedinger's equation, for instance, is a candidate for being a fundamental law.

Here, then, is a two-part definition of an emergent property:

If P is a property of w, then P is emergent if and only if (1) P supervenes with nomological necessity, but not with logical necessity, on properties the parts of w have taken separately or in other combinations; and (2) some of the supervenience principles linking properties of the parts of w with w's having P are fundamental laws.

In the case of weight and mass, the supervenience principles will not be fundamental laws because they will be instances of the general compositional laws of the additivity of weight and the additivity of mass. Chemical properties are not emergent in this sense since the relevant supervenience principles are not fundamental laws: they are in principle derivable from quantum mechanical laws. Dispositional properties susceptible to functional analysis will likewise not be emergent. Since it will be a contingent fact that a given microstructure is a base for a given disposition, condition (1) will be met. However, condition (2) will fail to be met since the manifestations of the dispositional property will be specifiable in physical and/or topic netural terms, and the totality of lower-level laws and conditions will imply that the microstructure in question is a base for the disposition in question.

Chemical properties are not emergent in our sense. Neither, on the evidence, are vital properties. Any mental properties that admit of functional analysis are likewise nonemergent. One group of mental properties, however, appears nonsusceptible to functional analysis: conscious properties. On the current evidence, conscious properties remain the only plausible candidates for emergent properties in the sense defined above.

Whether conscious properties are emergent is a genuinely open question. The issue is this. Suppose that conscious properties of an individual supervene with only nomological necessity on properties the parts of the individual exhibit in isolation and other combinations. Suppose further that at least some of the supervenience principles are fundamental laws. Then, conscious properties count as emergent. If, however, the supervenience principles are nonfundamental, then conscious properties are resultants and pose no threat to reductive materialism.

I am sympathetic to the view that conscious properties are not emergent, even though they do not admit of functional analysis. For I believe that conscious properties are *a posteriori* identical with physical properties (most likely, very abstract neurophysiological properties). If they are, then the supervenience principles (the trans-ordinal laws) connecting such physical properties with conscious properties will be nonfundamental. For they will be deducible from laws governing the physical properties in question. (Identities, you will recall, are necessary truths.) However, whether that is so is a question beyond the scope of this essay. I here simply affirm my faith in reductive materialism. Hopefully, the notion of an emergent property defined above can help to sharpen what is at issue in the debate between emergent materialism and reductive materialism.

## Appendix: "Emergence" and "Supervenience"

We have seen how the notion of supervenience can be employed to explicate the notion of emergence. I want in this appendix to address some issues raised by van Cleve about the terms "emergence" and "supervenience."

At the end of his article, van Cleve says (1990, p. 224):

> In closing, I would like to set down a group of definitions I came upon in the second edition of Webster's Unabridged, published in 1960. Supervene 2. Philos. To occur otherwise than as an additive resultant; to occur in a manner not antecedently predicable, to accrue in the manner of

what is evolutionally emergent. ("Not antecedently predictable": I assume that this means not predictable except with the help of autonomous bridge principles, principles that come to be known only by instantial induction after the advent of the new quality.) supervenient Coming or occurring as something additional, extraneous, or unexpected; also, emergent (sense 4). emergent 4. Philos. and Biol. Appearing as something novel in a process of evolution. Cf. emergent evolution. emergent evolution Philos. and Biol. Evolution conceived of as characterized by the appearance, at different levels, of new and antecedently unpredictable qualities of being or modes of relatedness, such as life and consciousness.

van Cleve goes on to remark:

I was surprised to learn that as recently as three decades ago, "supervenient" was used in some quarters as a synonym of "emergent." I can only suppose it is a coincidence that today's technical sense of "supervenience" permits a definition of emergence in terms of supervenience. (1990, p. 225)

I shall now proceed to argue that it is indeed merely a coincidence.

Llyod Morgan introduced the term "supervenience" into discussions of emergent evolution. He did not, however, use the term in anything like its current philosophical sense. Rather, he used the term in its vernacular sense. The term has a long history in the English language. Dr. Samuel's Johnson's *A Dictionary of the English Language* (1775), Vol. 2 informs us that "supervene" derives from the Latin super, meaning "on," "above," or "additional," and from the Latin verb venire meaning "to come." And Dr. Johnson's dictionary defines "supervene" as "to come as an extraneous addition," and "supervenient" as "added, additional." More recently, *Webster's New International Dictionary*, 3rd edition (1986), defines "supervene" as "coming or occurring as something additional, extraneous, or unexpected." This same definition appears in the early edition of Webster's from which van Cleve quotes. To repeat: when Morgan used "supervenience" in discussing emergents, he used the word in this vernacular sense. He used it to mean that emergent properties are additional to and come unexpectedly or unpredictably from their base properties. This vernacular use of "supervenience" is of course irrelevant to the current philosophical use of "supervenience."

"Supervenience" is a term of art in philosophy. There has been much speculation about when the term "supervenience" entered philosophical discussions in roughly its current philosophical sense. As I have noted, it did not enter via the literature on emergentism. Donald Davidson (1970) introduced the term into contemporary discussions of philosophy of mind with the following often quoted words:

Mental characteristics are in some sense dependent, or supervenient, on physical characteristics. Such supervenience might be taken to mean that there cannot be two events exactly alike in all physical respects but differing in some mental respect. (1970, p. 214)

While Davidson introduced the term into current philosophy of mind, he apparently got the term (used in a similar way) from R. M. Hare (1952). However, Hare

(1984) tells us that while he used the term in Hare (1952), he did not himself introduce it into philosophy. He claimed that the term was being used at Oxford in the 1940s.

My research into the introduction of the term "supervenience" in (roughly) its current philosophical sense confirms a claim made by Peter Geach as reported by Harry Lewis. Lewis (1985, p. 159n) reports that Geach suggested to him "that the term 'supervenient' entered our philosophical vocabulary by way of Latin translations of Aristotle's Nicomachean Ethics 1174B31-3." The Greek at 1174B31-3 reads: *"hos epiginomenon ti telos, hoion toise akmaiois he hora."* Robert Grosseteste's Latin translation of this passage translated *"epiginomenon"* as *"supervenire"* (Gauthier 1973). Sir David Ross used "supervenient" to translate *"epiginomenon."* In Ross's English, 1174B31-3 becomes "as an end which supervenes as the bloom of youth does on those in the flower of their age." This passage occurs in the context of Aristotle's talking of certain properties "naturally following" from other properties. This use of the term is similar to Hare's, which, in turn, is similar to Davidson's. Morgan's vernacular use, in contrast, is altogether different. It is thus indeed a coincidence that today's technical sense of "supervenience" permits a definition of emergence in terms of supervenience.

## Note

This chapter originally appeared in *Intellectica* 25 (1997), 33–43. This reprinting contains historical material also covered in McLaughlin's other contribution to this collection.

## References

Alexander, S. (1920) *Space, Time, and Deity*. 2 vols. London: Macmillan.

Bain, A. (1870) *Logic, Book II & III*. London.

Bantz, D. A. (1980) "The Structure of Discovery: Evolution of Structural Accounts of Chemical Bonding," in: Nickles, T. (ed.) *Scientific Discovery: Case Studies*. Dordrecht: Reidel.

Beckermann, A. (1992) "Supervenience, Emergence, and Reduction" in: Beckermann, A., Flohr, H., and Kim, J. *Emergence or Reduction?* Berlin: Walter de Gruyter, pp. 94–118.

Broad, C. D. (1925) *The Mind and Its Place in Nature*. London: Routledge and Kegan Paul.

Chalmers, D. J. (1996) *The Conscious Mind*. New York: Oxford University Press.

Davidson, D. (1970) "Mental Events," in: Foster, L. and Swanson, J. W. (eds.), *Experience and Theory*. Amherst: University of Massachusetts Press, pp. 79–101.

Hare, R. M. (1952) *The Language of Morals*. Oxford: Oxford University Press.

Hare, R. M. (1984) "Supervenience," Aristotelian Society Supplementary Volume.

Gauthier, R. A. (1973) Aristotleles Latinus XXVI. Leiden.

Johnson, S. (1775) *A Dictionary of the English Language*. vol. 2. Rpt. New York: AMS Press, 1967.

Kim, J. (1984) "Concepts of Supervenience," *Philosophy and Phenomenological Research* 45: 315–326.

Kim, J. (1988) "Supervenience for Multiple Domains," *Philosophical Topics* 16: 129–150.

Kim, J. (1990) "Supervenience as a Philosophical Concept," *Metaphilosophy* 21: 1–27.

Kim, J. (1992) "'Downward Causation' in Emergentism and Nonreductive physicalism" in: Beckermann, A., Flohr, H., and Kim, J. *Emergence or Reduction?* Berlin: Walter de Gruyter, pp. 119–139.

Klee, R. (1984) "Micro-Determinism and Concepts of Emergence," *Philosophy of Science* 51: 44–63.

Kripke, S. (1971) "Identity and Necessity," in: Muntiz, M. K. (ed.), *Identity and Individuation*. New York: University Press, pp. 135–164.

Lewis, H. A. (1985) "Is the Mental Supervenient on the Physical?" in: Vermazen, B. and Hintikka, M. (eds.) *Essays on Davidson: Actions and Events*. Oxford: Clarendon Press, pp. 159–72.

Lewes, G. H. (1875) *Problems of Life and Mind*. vol. 2. London: Kegan Paul, Trench, Turbner, & Co.

McLaughlin, B. P. (1992) "The Rise and Fall of British Emergentism" in: Beckermann, A., Flohr, H., and Kim, J. *Emergence or Reduction?* Berlin: Walter de Gruyter, pp. 49–93.

McLaughlin, B. P. (1995) "Varieties of Supervenience" in: Savellos, E. and Yalcin, U. D. *Supervenience: New Essays*. Cambridge: Cambridge University Press, pp. 16–59.

McLaughlin, B. P. (1995) "Dispositions" in: Kim, J. and Sosa, E. *Companion to Metaphysics*. Oxford: Basil Blackwell.

Mill, J. S. (1843) *System of Logic*. London: Longmans, Green, Reader, and Dyer. (8th edition, 1872).

Mill, J. S. (1924) *Autobiography*. New York.

Morgan, C. Lloyd (1923) *Emergent Evolution*. London: Williams & Norgate.

Nagel, E. (1961) *The Structure of Science*. New York: Harcourt Brace and World.

Prior, E. W. (1985) *Dispositions*. Aberdeen: Aberdeen University Press.

Ross, W. D. (1923) *Aristotle*. London: Methuen.

Stephan, A. (1992) "Emergence—A Systematic View of its Historical Facets" in: Beckermann, A., Flohr, H., and Kim, J. *Emergence or Reduction?* Berlin: Walter de Gruyter, pp. 25–48.

Teller, P. (1992) "Subjectivity and Knowing What It's Like" in: Beckermann, A., Flohr, H., and Kim, J. *Emergence or Reduction?* Berlin: Walter de Gruyter, pp. 180–200.

Thomson, T. (1807) *System of Chemistry*.

Van Cleve, J. (1990) "Emergence vs. Panpsychism: Magic or Mind Dust?" in: Tomberlin, J. E. (ed.) *Philosophical Perspectives*. vol. 4. Atascadero, Cal.: Ridgeview Publishing Company, pp. 215–226.

White, S. (1991) *The Unity of the Self*. Cambridge: MIT Press.

# 5 Aggregativity: Reductive Heuristics for Finding Emergence

William C. Wimsatt

## 5.1 Reduction and Emergence

A traditional philosophical reductionist might suppose that emergence will be a thing of the past—reductionistic explanations for phenomena demonstrate that they are not emergent. As science progresses, emergence claims will be seen as nothing more than temporary confessions of ignorance (e.g., Nagel 1961). Of course, one need not be a reductionist: philosophers of the special sciences who *also* anchor emergence in failures of reduction do not see it as a disease to be cured by reductionistic progress.

Both sides here conflict with most scientists' intuitions about when something is emergent. Discussions of emergent properties in nonlinear dynamics, connectionist modeling, chaos, artificial life, and elsewhere give no support for traditional antireductionism or woolly-headed antiscientism. Emergent phenomena like those discussed here are often subject to surprising and revealing reductionistic explanations.[1] But reductionists often misunderstand the consequences of their explanatory successes. Giving such explanations does not deny their importance or make them any less emergent—quite the contrary: it explains why and how they are important, and ineliminably so.

A reductive analysis of emergence will not satisfy some. Philosophers of mind often want something more—but adopt a very limiting deductivist notion of reduction—which of course makes it easier to reject. This is the wrong notion of reduction for the compositional sciences (Wimsatt 1976, 1998). Many use multiple-realizability as a criterion for emergence, supposing it to banish reduction. But multiple-realizability follows from the existence of compositional levels of organization, is found much more widely than supposed, and is neither mysterious nor contrary to reductionism (Wimsatt 1981, 1994). Philosophers who defend "qualia" type accounts of subjective experience seem to suppose a notion of emergence which—if coherent—*would* be antireductionistic, but it is hard to find a clear analysis of what is required. I do not try to consider it here. Most scientists in the complex sciences have compatible views of reduction and emergence: *a reductive explanation of a behavior or a property of a system is one showing it to be mechanistically explicable in terms of the properties of and interactions among the parts of the system* (see also Kauffman 1971).[2] The explanations are causal, but need not be deductive or involve laws (Wimsatt 1976, Cartwright 1983).

*An emergent property is—roughly—a system property which is dependent upon the mode of organization of the system's parts.* This is compatible with reductionism, and also common. Too weak for most antireductionists, it is still a powerful tool: characterizing what it is for something to depend on the organization of a system's parts provides heuristic ways of evaluating decompositions which can help us to understand why some decompositions are preferred over others, and why we may nonetheless overestimate their powers. Antireductionists should welcome it—as an effective means for clearing away the many cases often called emergent to see what is left, allowing a more focused discussion. And when they see what it can do, some may feel that this kind of emergence is enough.

If emergent system properties depend on the arrangement of the parts, this indicates *context-sensitivity* of relational parts' properties *to intra-systemic conditions*. Some emergentists suppose *extra-systemic context-sensitivity* of system properties. Either can be reductionistic. The latter would require a larger embedding system—moving the boundaries outwards to include the external properties which engage the broader context-sensitivities. If this yields an adequate mechanistic account, the reduction would have failed with the original system boundaries, but a more inclusive one (including variables from the context of the first) would have succeeded.

Such level and scope switching—well-hidden secrets of reductionistic analyses—can also be a useful strategy to remove model building biases resulting from "perceptual focus" on objects at a preferred level. We must work back and forth between the ontologies of different levels to check that features crucial to upper level phenomena are not simplified out of existence when modelling it at the lower level. (See clear cases of this in the group selection controversy described in Wimsatt 1980: one extremely influential simplification found in most of the models of group selection is shown to be equivalent to saying that there are no groups. Not surprisingly, it was hard in such models to demonstrate the efficacy of group selection!) The criteria for aggregativity provide powerful tools for detecting such reductionistic errors.

Classical cases of emergence are those motivating the claim that "the whole is more than the sum of the parts"—like Huygen's ([1673] 1986) discovery that pendulum clocks hung together on a beam became synchronized and kept better time than either did alone. The theory of coupled oscillators predicts that weakly coupled oscillators starting out of phase and with slightly different periods will gradually entrain each other through mutually transmitted effects (small displacements propagated through a connecting beam) until they all oscillate synchronously, in phase, and with a period which is an (appropriately weighted) average of their separate periods. None of them might have this exact period when isolated and oscillating by themselves. If their periods are normally distributed around the "true" mean, each clock whose period differs from that mean will perform better together than by itself. This system has a "virtual metronome" regulating each clock but not located in any of them. Its properties are a product of how the clocks are hooked together, and changes in the way the clocks are connected change its properties. There is nothing antireductionistic here. A mathematical model of the phenomena would satisfy both the formal model of reduction and the weaker characterization given here. It is also an intuitive case of emergence.[3]

There is a long list of well-understood examples of emergence (many are discussed further in Wimsatt 1986, 1997): A classic case at the turn of the century (when emergence was last a common topic) was how the properties of water supervened on the properties of hydrogen and oxygen. In more modern discussions, all of the following would qualify: the variety of fitness interactions among genes, cooperative binding and unloading of oxygen in hemoglobin molecules, traffic jams, critical mass for fissionable materials, and a variety of other threshold phenomena, solvent-solute interactions in chemistry, and even some properties of that paradigm aggregate—a heap of stones (such as its shape or stability!). In each case we have a reductionistic account which depends in some way on the mode of organization of the parts. There are just too many counterexamples to ignore. We need an analysis of emergence which is consistent with reductionism.

## 5.2 Conditions of Aggregativity

Emergence involves some kind of organizational interdependence of diverse parts, but there are many possible forms of such interaction, and no clear way to classify them. It is easier to discuss *failures* of emergence (Wimsatt 1986), by figuring what conditions should be met for the system property *not* to be emergent—for it to be a "mere aggregate" of its parts' properties. Being an aggregate has a straightforward revealing and compact analysis. Forms of emergence can then be classified and analyzed systematically by looking at how *these* conditions may *fail*.

Four conditions (listed below) seem central for *aggregativity*[4]—the non-emergence of a system property relative to properties of its parts (Wimsatt 1986). For each condition, the system property must remain *invariant* under modifications to the system in the specified way. If so, the system property is independent of variations in that kind of relationship among the parts and their properties. Aggregative properties depend on the parts' properties in a very strongly atomistic manner, under all physically possible decompositions. *It is rare indeed that all of these conditions are met.* This is the complete antithesis of functional organization. Post-Newtonian science has focused disproportionately upon such properties, or properties meeting some of these conditions, approximately, or some of the time—and studying them under conditions where they are "well-behaved." I will talk more about studying these "pseudo-aggregative" properties in the last section. Their import in descriptions of the natural world has been substantially exaggerated.

Intuitively, these conditions deal with how the system property is affected by: (1) the intersubstitution or rearrangement of parts; (2) addition or subtraction of parts; (3) decomposition and reaggregation of parts; and (4) a linearity condition: no cooperative or inhibitory interactions among the parts in the production or realization of the system property:

*For a system property to be an aggregate with respect to a decomposition of the system into parts and their properties, the following conditions must be met:*

*Suppose* $\mathbf{P}(\mathbf{S}_i) = \mathbf{F}\{[\mathbf{p}_1, \mathbf{p}_2, \ldots, \mathbf{p}_n(\mathbf{s}_1)], [\mathbf{p}_1, \mathbf{p}_2, \ldots, \mathbf{p}_n(\mathbf{s}_2)], \ldots, [\mathbf{p}_1, \mathbf{p}_2, \ldots, \mathbf{p}_n(\mathbf{s}_m)]\}$ is a composition function for system property $\mathbf{P}(\mathbf{S}_i)$ in terms of parts' properties $\mathbf{p}_1, \mathbf{p}_2, \ldots, \mathbf{p}_n$, of parts $\mathbf{s}_1, \mathbf{s}_2, \ldots, \mathbf{s}_m$. The composition function is an equation—an inter-level synthetic identity, with the lower level specification a realization or instantiation of the system property.[5]

1. IS (InterSubstitution) Invariance of the system property under operations rearranging the parts in the system or interchanging any number of parts with a corresponding numbers of parts from a relevant equivalence class of parts. (This would seem to be satisfied if the composition function is *commutative*).

2. QS (Size scaling) Qualitative Similarity of the system property (identity, or if it is a quantitative property, differing only in value) under addition or subtraction of parts. (This suggests inductive definition of a class of composition functions).

3. RA (Decomposition and ReAggregation) Invariance of the system property under operations involving decomposition and reaggregation of parts. (This suggest an *associative* composition function).

4. CI (Linearity) There are no Cooperative or Inhibitory interactions among the parts of the system for this property. (See discussion of the linear amplifier.)

Conditions IS and RA are implicitly relative to given parts decompositions, as are (less obviously) QS and CI. *For a system property to be truly aggregative, it must meet these conditions for all possible decompositions of the system into parts.* This is an extremely strong condition, and rarely met. More interesting are cases where conditions are met for some decompositions and not for others, or under some special conditions. In many cases we take certain decompositions or constraints on how decompositions are performed so much for granted that we fall into assuming an aggregativity that is not there. This arises often in debates in the units of selection controversy (see Wimsatt 1997).

Figure 5.1 illustrates the first three conditions for the system's amplification ratio in an (idealized) multi-stage linear amplifier. The system property considered is the total amplification ratio, $\Sigma A$. For amplifiers arranged in series, $\Sigma A$ is the product of the

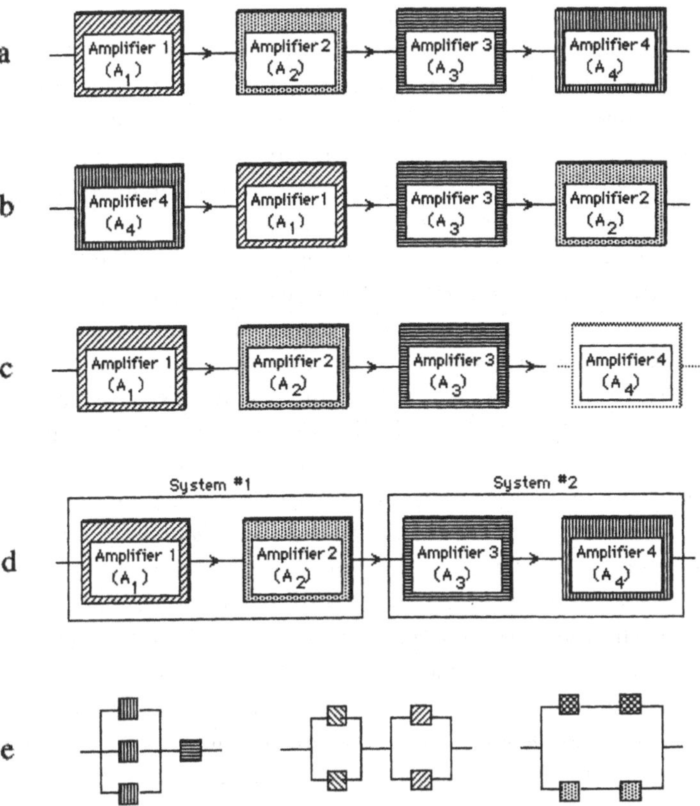


**Figure 5.1**

Conditions of Aggregativity illustrated with idealized linear unbounded amplifiers. 5.1a: Total Amplification Ratio, $\Sigma A$, is the product of the amplification ratios of the individual amplifiers: $\sigma A = A1 \times A2 \times A3 \times A4$. 5.1b: Total Amplification Ratio, $\sigma A = A4 \times A1 \times A3 \times A2$ remains unchanged over intersubstitutions changing the order of the amplifiers (or commutation of the A's in the composition function). 5.1c: Total Amplification Ratio, $\Sigma A(n) = \sigma A(n-1) \times A(n)$, remains qualitatively similar when adding or subtracting parts. 5.1d: Total Amplification Ratio is invariant under subsystem aggregation—it is associative: $A1 \times A2 \times A3 \times A4 = (A1 \times A2) \times (A3 \times A4)$. 5.1e: The intersubstitutions of 1a–1d, which all preserve a strict serial organization of the amplifiers, hide the real organization dependence of the Total Amplification Ratio. This can be seen in the rearrangements of four components into series-parallel networks. Assume each box in each circuit has a different amplification ratio. Then to preserve the A.R., the boxes can be interchanged only within organizationally defined equivalence classes defined by crosshatch patterns. Interestingly, these classes often can be aggregated as larger components, as in these cases, where whole clusters with similar patterns can be permuted, as long as they are moved as a cluster (see Wimsatt 1986).

component amplification ratios (figure 5.1a), a composition function which is commutative (figure 5.1b, condition IS), associative (figure 5.1d, condition RA), and qualitatively similar when adding or subtracting parts (figure 5.1c, condition QS). It seems to violate the 4th condition (linearity, condition CI), but we act as if it does not: geometric rates of increase are treated linearly here because it is the exponent which is theoretically significant. Subjective volume grows linearly with the exponent (as our decibels scale reflects), and both of them grow linearly with addition of components to the chain. Staged linear amplifiers demonstrate that aggregativity need not mean "additivity" for all properties—here multiplicative relations seem more appropriate.[6] (Exponential growth is also much more common in biology than linear relations.)

This simple story has some limits however. Amplifiers are themselves integrated functional wholes with differentiated parts—which cannot be permuted with impunity. (That is why we need circuit diagrams to assemble and to understand them—we cannot put them together in just any fashion!) Even the parts are integrated wholes. If you cut randomly through a resistor or capacitor, the pieces will not perform like the original. This is interesting: it shows that *testing these conditions against different ways of decomposing the system is revealing of its organization.*

But also notice that all of the examples so far (in figure 5.1a–d) have the amplifiers arranged in series. This is an implicit organizational constraint on the whole system. (It is readily accepted because our common uses of amplifiers connect them in this way). But we could also connect them differently, as in the three series-parallel networks diagrammed in figure 5.1e, and then the invariances in total amplification ratio, ΣA, disappear. To calculate their amplification ratios, we make three (simplifying) assumptions: (1) branching parallel paths divide currents equally; (2) converging parallels add currents; (3) serial circuits (included aggregated subassemblies) multiply signal strengths (as before). In figure 5.1e, starting with a signal of magnitude 1 unit, and following these rules we get:

For the first circuit, $\Sigma A = [(A1 + A2 + A3)/3] \times A4$.

For the second circuit, $\Sigma A = [(A1 + A2)/2] \times [(A3 + A4)/2]$

For the third circuit, $\Sigma A = [(A1 \times A3)/2] + [(A2 \times A4)/2]$

If all component amplifiers, A1, A2, A3, A4 have equal amplification ratios, $\boldsymbol{a}$, the ΣA's of these circuits are all equal. Because they each have two staged subassemblies, each with a net amplification ratio of $\boldsymbol{a}$, $\Sigma A = \boldsymbol{a} \times \boldsymbol{a} = \boldsymbol{a}^2$. (For these ideal linear amplifiers, parallating at any stage has no effect: since the signal is just multiplied by the same amount along each path, they collectively have the same effect as a single amplifier with the same amplification ratio.) But the decreased amplifying depth reduces ΣA from $\boldsymbol{a}^4$ to $\boldsymbol{a}^2$, so changing to any of these modes from a strictly serial circuit decreases the amplification ratio.

But suppose that the amplification ratios of the component amplifiers are not equal. If A1 = $\boldsymbol{a}$, A2 = $2\boldsymbol{a}$, A3 = $4\boldsymbol{a}$, and A4 = $8\boldsymbol{a}$, then

circuit 1 = $[(7/3)\boldsymbol{a}] \times 8\boldsymbol{a} = (56/3)\boldsymbol{a}^2 = 18.67\boldsymbol{a}^2$.

circuit 2 = $(3/2)\boldsymbol{a} \times 6\boldsymbol{a} = 9\boldsymbol{a}^2$.

circuit 3 = $(5/2)\boldsymbol{a}^2 + (10/2)\boldsymbol{a}^2 = 7.5\boldsymbol{a}^2$.

For comparison, all of them in series (as above) would give $64\boldsymbol{a}^4$. So we see that how the amplifiers are connected together *does* make a difference. Despite first appearances, amplification ratio is not an aggregative property, both because it is not invariant across decompositions internal to the amplifiers, or internal to their parts, and also because it is not invariant across aggregations of the amplifiers which are not serially organized. (This covers manipulations at three levels of organization: rearrangements of parts of the amplifier parts, rearrangements of parts of the amplifiers, and rearrangements of the amplifiers.)

Finally, we have assumed that each sub-amplifier is exactly linear throughout the entire range—from the smallest input to the largest output—required of the entire system. They must multiply input signals of different frequencies and amplitudes by the same amount over this entire range. This is an idealization. Real-world amplifiers are *approximately* linear through given power and frequency ranges of input signals. (Frequency correction curves are published so that linearity can be restored by the user by "boosting" different frequencies by different amounts, but these curves are themselves functions of the amplitude of the input signal.) The amplifiers—not perfectly linear to begin with—become increasingly nonlinear outside these ranges. They are most commonly limited on the low side by insensitivity to inputs below a certain value, and on the high side by not being able to put out enough power to keep the transformation linear. So with real amplifiers the order of the amplifiers *does* matter, even in the serial circuit.

Thirty years ago hi-fi systems had separate pre-amplifiers and amplifiers. Hooked up in the right order, they worked just fine. In the wrong order, the amplifier would be too insensitive to detect the signal from the phono-cartridge, but small amounts of "white noise" it generated internally would be amplified enough to "fry" the downstream preamp. When we call amplifiers linear, we really mean *approximately linear over a given range, within a specified tolerance, ε*. Different uses may require different tolerances, so a system might be regarded as linear for some purposes, but not for others.

## 5.3 Heuristic and Other Uses of Limited Aggregativity

We usually evaluate aggregativity relative to assumed constraints on how components are rearranged or treated—but often forget the constraints. When we do so, we underestimate how much system properties depend upon how the parts are organized. *In an absolute sense, the circuits discussed in the preceding case are not aggregative at all, since the supposed invariance in the composition function depends upon maintaining the serially connected circuit structure, and not decomposing it in a variety of problematic ways, in addition to assuming a host of linearizing assumptions and conditions*. Blindness to assumed constraints is common. Similar qualifications for supposedly aggregative properties arise for additive fitness components, and the fine- versus coarse-grained patterns of environmental change (Wimsatt 1997, 1998) derived and discussed by Levins (1968). The common appearance of unqualified aggregativity is a chimera.

But a few properties are paradigmatic aggregative properties. The great conservation laws of physics—for *mass, energy* (in our ordinary domains where they are effectively separated),[7] *momentum*, and *net charge* indicate that these properties actually do fill the bill. They are aggregative under any and all decompositions, and apply to the most complex of living systems. If you bring a steer to a butcher and find that the meat you get back weighs several hundred pounds less, you blame the butcher, not vanished emergent interactions! But if you want to see the list for the rest of the aggregative properties, that's all there are. This list is very revealing: perhaps we should expect that *anything which was invariant across so many transformations would have a major conservation law associated with it*, and I cannot think of any others.[8]

Some properties which you might have expected to be aggregative are not. From the classical list of primary qualities, volume is not aggregative, if we consider solvent-solute interactions in chemistry. Dissolve salt in water and the volume of the water + salt will be less than the water before the salt was added. (*Sometimes the whole is **less** than the sum of its parts!*)

Different conditions may be met separately for some decompositions of a given system, but not for others, and with varying degrees of accuracy. This has critical importance in theory construction: *it allows implicit use of these criteria as heuristics in evaluating decompositions of a system for further analysis. We look for invariances*. In experimenting with alternative descriptions and manipulations, we tend to look for ways to make the conditions work—decomposing, cutting, pasting, and adjusting the decompositions until these conditions are satisfied to the greatest degree possible. *Decompositions for which more of the conditions for aggregativity are more closely met seem "natural,"* because they provide simpler and less context-dependent regularities, theory, and mathematical models involving these aspects of their behavior.

Applied mathematician and theoretical biologist Jack Cowan (personal observation) loves to tell the difference between a biophysicist and a theoretical biologist: "Take an organism and homogenize it in a Waring blender. The biophysicist is interested in those properties which are invariant under that transformation." This reflects the criteria for aggregativity—and their applicability for a range of conditions—directly. And Cowan's bittersweet joke has a point: blenders are widely used in molecular labs to prepare samples, which are then "fractionated" by centrifugation. Biophysicists get lots of money to study the structure of macromolecules too small to be "disrupted" by this procedure, and more easily extracted if one disrupts everything larger. Depending on their purposes they may take special care to avoid disrupting them any further, or proceed to lower levels. One can almost classify the level of organization of a science by asking what levels it is permissible to "disrupt" (or otherwise ignore) in one's experimental designs and investigations. And we often generate supposedly aggregative behavior under special conditions or strong constraints or special idealizations on the system and its environment, which are then forgotten. This has deep connections with the reductionistic problem-solving heuristics discussed in Wimsatt 1980 and 1998, and can easily lead to a variety of reductionistic biases.

## 5.4 Aggregativity, Vulgar Reductionism, and Detecting Organizational Properties

These cases of aggregativity or partial aggregativity suggest interactions among the development of theory, methods of decomposition, and experimental design, and what we make of what we have found. Whether a property is aggregative or not might seem primarily an ontological question. But it is more. Together these cases give a different picture of the nature and uses of aggregativity and have further implications for the assessment (and biases) of reductionistic methodologies. These ontological questions can be productively turned to ones of substantial heuristic utility.

1. Very few system properties are aggregative, suggesting that emergence, defined as failure of aggregativity, is extremely common—the rule, rather than the exception. If this is too weak a notion of emergence for some, fine. At least the enormous range of cases it allows us to classify and to understand (cases commonly spoken of as emergent) can now be removed from the playing field to focus on the few demonstrable remaining cases, if that is our focus. But there are other consequences which many would regard as much more interesting.

2. This analysis produces a curious inversion of the conclusion from the opposition of emergence and reduction supposed at the beginning of this essay. Will supposed cases of emergence disappear as we come to know more? This suggests the reverse: we tend to start with simple models of complex systems—models according to which the parts are more homogeneous, have simpler interactions, and in which many differentiated parts and relationships are ignored (Wimsatt 1980)—models which are more aggregative. But then as our models grow in realism, we should both capture more properties and see more of them as organization dependent—or emergent.

3. Aggregativity provides natural criteria for choosing among decompositions: we will tend to see more aggregative decompositions as *natural* decompositions, and their parts

as *natural kinds*, individuated using *natural properties* because they will provide simpler and less context-dependent regularities, theory, and mathematical models for these aspects of their behavior. These are then particularly revealing cuts on nature.

4. If aggregativity connects this closely with natural kinds and natural properties, we have a better understanding of the temptations of vulgar reductionisms, "*Nothing but*"-isms and their fallacies. We regularly see statements such as "genes are the only units of selection," "the mind is nothing but neural activity," or "social behavior is nothing more than the actions of individuals." If total aggregativity is so rare, why are claims like these so common? These are particularly pervasive *functional localization fallacies*—moves from the power of a particular decomposition to claims that its entities, properties, and forces are *all* that matters (Wimsatt 1974, Bechtel and Richardson 1993). "Nothing-but" claims are false and methodologically misleading for suggesting that one shouldn't bother to construct models or theories of the system at levels or with methods other than those keyed to the parts in question, or that these are the only "real" ones (Wimsatt 1994). Such properties *look* aggregative for some decompositions, but reveal themselves as emergent or organization-dependent for others or under other conditions.

If we focus too strongly on the preferred decompositions—which we will tend to do if those parts are the elements from which we build our theoretical frameworks—they may come to seem to be everything, or at least everything important. With only one such powerful decomposition, descriptions will tend to be referred to or translated into its preferred entities. (How often human sociobiologists or new-look evolutionary psychologists talk hypothetically about "*the gene* for X" without substantial evidence that 'X' is genetic, or if genetic, a single-locus trait!) This may persist even when several powerful decompositions cross-cut the system in different ways or at different levels (Wimsatt 1974, 1994). (This is part of the lesson of the units of selection controversy—Wimsatt 1980.) Attending to such practices naturally changes our focus from how to specify relations between the system properties and the parts' properties (ontological questions) to looking at the reasons for, process of, and idealizations used in choosing a decomposition, and broader effects of that choice (a set of methodological and heuristic questions).

5. Aggregativity should be a powerful tool, but one that is honored more often in the breach. I have argued the virtues of false models (Wimsatt 1987) as pattern templates to lay down on the phenomena—the better to see the form of the residuals. We are too much taken with things which fit our models, but often have more to learn from edges that do not fit, especially if the templates have a useful form, and particularly crisp edges. Aggregativity provides tools for this kind of analysis in compositional systems—a demanding set of detectors for kinds of organizational interaction. Invariance claims are particularly sharp-edged tools for detecting, calibrating, and classifying

failures of invariance. Yes, we are something more than quarks, atoms, molecules, genes, cells, neurons, and utility maximizers. And now we have some means to count the ways.

## Notes

This chapter originally appeared in *Philosophy of Science* 64 (1997), S372–S384.

1. Herbert Simon (1996, Ch. 7) adopts this view of emergence. For more examples, see Dewan 1976 and Kauffman 1993.

2. Glennan (1996) offers a complementary account of causation and mechanistic explanation.

3. Coupled oscillations arise easily in diverse systems. Dewan (1976) tells a parallel story for voltage and phase regulation in power networks and surveys a variety of cases. Wiener's power network with the "virtual governor" described there is the inspiration for the "virtual metronome." McClintock (1971) describes similar phenomena with the emergence of menstrual synchrony among individuals who interact more frequently in womens' dormitories.

4. These are relevant and important criteria, but may not be completely independent. I believe they are jointly sufficient.

5. Some philosophers distinguish synthetic identities from realizations or instantiations. This distinction does not matter here, but see Wimsatt 1994, 228–229, n. 30.

6. Addition, multiplication and other operations (e.g., logical disjunction) could be appropriate in other contexts. The parts properties, system property, question being asked, purposes of the investigation, and the relevant applicable theories can all play important roles in such judgments.

7. This qualifier indicates another approximation qualifying apparent aggregativities. As specific energies of material systems increase (e.g., when velocities become significant fractions of the speed of light), mass and energy can no longer be treated as separately conserved quantities. Here again, aggregativities depend upon conditions.

8. I do not discuss spin and other things which have conservation laws but no obvious macroscopic correlates.

## References

Bechtel, W. and R. C. Richardson (1993), *Discovering Complexity: Decomposition and Localization as Scientific Research Strategies*. Princeton: Princeton University Press.

Cartwright, N. (1983), *How the Laws of Physics Lie*. Oxford: Clarendon Press.

Dewan, E. (1976), "Consciousness as an Emergent Causal Agent in the Context of Control System Theory," in G. G. Globus, G. Maxwell, and I. Savodnik (eds.), *Consciousness and the Brain: Scientific and Philosophic Strategies*. New York: Plenum Press, pp. 181–198.

Glennan, S. S. (1996), "Mechanism and the Nature of Causation," *Erkenntnis* 44: 49–71.

Huygens, Christiaan, ([1673] 1986), *The Pendulum Clock, or Geometrical Demonstrations Concerning the Motion of Pendula as applied to Clocks.* (*Horologium oscillatorium*; translated with notes by R. J. Blackwell. Ames: Iowa State University Press.

Kauffman, S. A. (1971), "Articulation of Parts Explanations in Biology and the Rational Search for Them," in R. Buck and R. S. Cohen (eds.), *PSA 1970*. Dordrecht: Reidel, pp. 257–272.

———. (1993), *The Origins of Order: Self-Organization and Selection in Evolution*. New York: Oxford University Press.

Levins, R. (1968), *Evolution in Changing Environments*. Princeton: Princeton University Press.

McClintock, M. (1971), "Menstrual Synchrony in Humans," *Nature* 229: 244–245.

Nagel, E. (1961), *The Structure of Science*. New York: Harcourt Brace and World.

Simon, H. A. ([1962] 1996), "The Architecture of Complexity," Reprinted in his 1996, *The Sciences of the Artificial*, 3rd ed. Cambridge, MA: MIT Press, pp. 183–216.

Strobeck, K. (1975), "Selection in a Fine-Grained Environment," *American Naturalist* 109: 419–425.

Taylor, P. J. and J. Haila (eds.) (1997), *Natural Contradictions: Perspectives on Ecology and Change*. Albany: State University of New York Press, forthcoming.

Wimsatt, W. C. (1974), "Complexity and Organization," in K. F. Schaffner and R. S. Cohen (eds.), *PSA 1972*. Dordrecht: Reidel, pp. 67–86.

———. (1976), "Reductive Explanation—A Functional Account," in A. C. Michalos, C. A. Hooker, G. Pearce, and R. S. Cohen, (eds.), *PSA 1974*. Dordrecht: Reidel, pp. 671–710.

———. (1980), "Reductionist Research Strategies and Their Biases in the Units of Selection Controversy," in T. Nickles (ed.), *Scientific Discovery: Case Studies*. Dordrecht: Reidel, pp. 213–259.

———. (1981), "Robustness, Reliability and Overdetermination," in M. Brewer and B. Collins (eds.), *Scientific Inquiry and the Social Sciences*. San Francisco: Jossey-Bass, pp. 124–163.

———. (1986), "Forms of Aggregativity," in A. Donagan, A. N. Perovich, and M. Wedin (eds.), *Human Nature and Natural Knowledge*. Dordrecht: Reidel, pp. 259–291.

———. (1987), "False Models as Means to Truer Theories," in Matthew Nitecki and Antoni Hoffman (eds.), *Neutral Models in Biology*. New York: Oxford University Press, pp. 23–55.

———. (1994), "The Ontology of Complex Systems: Levels, Perspectives, and Causal Thickets," in Mohan Matthen and R. X. Ware (eds.), *Biology and Society: Reflections on Methodology, Canadian Journal of Philosophy*, supplementary volume 20, pp. 207–274.

———. (1997), "Emergence as Non-Aggregativity and the Biases of Reductionism(s)," in J. Haila and P. Taylor (eds.), *Natural Contradictions: Perspectives on Ecology and Change*. Albany: SUNY Press, forthcoming.

———. (1998), *Piecewise Approximations to Reality: Engineering a Realist Philosophy for Limited Beings*. Cambridge, MA: Harvard University Press, forthcoming.

# 6 How Properties Emerge

**Paul Humphreys**

## 6.1. Introduction

Lurking in the shadows of contemporary philosophy of mind is an argument widely believed to produce serious problems for mental causation. This argument has various versions, but one particularly stark formulation is this:

(1) If an event x is causally sufficient for an event y, then no event x* distinct from x is causally relevant to y (*exclusion*).

(2) For every physical event y, some physical event x is causally sufficient for y (*physical determinism*).

(3) For every physical event x and mental event x*, x is distinct from x* (*dualism*).

(4) So: for every physical event y, no mental event x* is causally relevant to y (*epiphenomenalism*). (Yablo 1992, 247–248)[1]

This *exclusion argument*, as it is usually called, has devastating consequences for any position that considers mental properties to be real, including those nonreductive views that suppose mental properties to supervene upon physical properties. For if mental properties are causally impotent vis-a-vis physical properties, the traditional worry about epiphenomenalism confronts us: What is the point of having them in our ontology if they are idle? Abstract objects escape this worry, for we do not expect them to do causal work, but mental properties are retained in part because we believe them to affect the course of the world. If the exclusion argument is sound, then ratiocination, qualia, and the hopes and fears of mankind are simply smoke on the fire of brain processes.

This is bad enough, but there is a second argument, devised by Jaegwon Kim, that in conjunction with the exclusion argument seems to render nonreductive physicalism not merely uncomfortable but untenable, for it has as a conclusion that nonreductive physicalism is committed to the view that some mental properties must cause physical properties. Kim's argument, which I shall call *the downwards causation argument* was originally levelled against both nonreductive and emergentist approaches:

> But why are emergentism and nonreductive physicalism committed to downward causation, causation from the mental to the physical? Here is a brief argument that shows why. At this point we know that, on emergentism, mental properties must have novel causal powers. Now, these powers must manifest themselves by causing either physical properties or other mental properties. If the former, that already is downward causation. Assume then that mental property M causes another mental property M*. I shall show that this is possible only if M causes some physical property. Notice first that M* is an emergent; this means that M* is instantiated on a given occasion only because a certain physical property P*, its emergence base, is instantiated on that occasion. In view of M*'s emergent dependence on P*, then, what are we to think of its causal dependence on M? I believe that these two claims concerning why M* is present on this occasion must be reconciled, and that the only viable way of accomplishing it is to suppose that M caused M* by causing its emergence base P*. In general, the principle involved here is this: *the only way to cause an emergent property to be instantiated is by causing its emergence base property to be instantiated.* And this means that the "same-level" causation of an emergent property presupposes the downward causation of its emergent base. That briefly is why emergentism is committed to downward causation. I believe that this argument remains plausible when emergence is replaced by physical realization at appropriate places. (Kim 1992, 136)

Conjoin this second argument with the first, and you have more than mere trouble for nonreductive physicalism and emergentism, you have contradictory conclusions. (4) entails that there is no downwards causation from the mental to the physical; the downwards causation argument concludes that nonreductive physicalism and emergentism require such downwards causation.

Something must go if mental properties are to survive, and that something is both arguments. Much is wrong with the exclusion argument, but what it shares with the downward causation argument is a pinched commitment to a dualist ontology, a laudable but usually unargued allegiance to the causal closure of the physical realm, and (nowadays) the idea that supervenience is the right way to represent the relation between the lower and higher levels of the world's ontology. Each of these is popular and each is wrong.

## 6.2. A Wider Perspective

The first thing to note about these arguments is that they are extremely general—they do not seem to rely on anything that is characteristic of mental properties, such as intentionality, lack of spatial location, having semantic content, and so on, and indeed it is often mentioned in passing that the arguments can be generalized to apply to a hierarchically ordered set of properties, each level of which is distinct from every other level.[2] If the exclusion argument does generalize to such hierarchies, and if, for example, chemical and biological events occupy higher levels than do physical events, then no chemical or biological event could ever causally influence a physical event, and if both arguments so generalize, then nonreductive physicalism leads to inconsistencies when applied to the general realm of the natural sciences too.

The situation is in fact more extreme than this, because most of our physical ontology lies above the most fundamental level, and in consequence only the most basic physical properties can be causally efficacious if these arguments are correct. Indeed, unless we have already isolated at least some of the most fundamental physical properties, every single one of our causal claims within contemporary physics is false and consequently there are at present no true physical explanations that are grounded in causes.

All of these are, of course, surprising and unwelcome consequences. We tend to believe that current elementary particle physics does reliably describe the formative influences on our world, that chemical corrosion can cause the physical failure of an aircraft wing, and that the growth of a tree can cause changes in the ground temperature beneath it. Yet our beliefs on this score may well be wrong. Perhaps the correct formulation of physicalism is a strict one: if, as many hold, everything is composed of elementary particles or fields, and the laws that govern those objects ultimately determine all else that happens, then there are fundamental physical events that are causally sufficient for aircraft wing failures, changes in soil temperatures, and all other physical events. This is a position that requires serious consideration, and indeed those versions of physicalism that require all non-physical phenomena to supervene upon physical phenomena can be easily and naturally adapted to this kind of strict position. So we need to examine in detail a generalized version of each argument to see whether this radical physicalist conclusion can be supported and whether the problems for nonreductive physicalism extend all the way down to the penultimate level.

There is a second reason for generalizing the arguments. It is preferable to examine arguments in a form in which the contextual assumptions are as transparent as possible. The relation between molecular chemistry and physics, say, is much more easily assessed than is the relation between physics and psychology because we have clearly articulated theories for the first pair, together with a reasonably clear set of constraints on the degree of reducibility of molecular chemistry to particle physics, but those relations are much murkier in the case of physics and psychology. In addition, by focusing on the lower levels of the hierarchy, we can avoid difficult problems involving the mental that are here irrelevant simply by recasting the arguments in a form that avoids reference to specifically mental properties.[3]

To generate the general arguments, we need a hierarchy of levels. For present purposes, I shall simply accept that this can be done in whatever way the reader finds most congenial. (The strata must correspond to some real differences in ontological levels rather than to a mere set of epistemic distinctions.) For concreteness, we can think in terms of this assumption: (L) There is a hierarchy of levels of properties

$L_0, L_1, \ldots L_n, \ldots$ of which at least one distinct level is associated with the subject matter of each special science, and $L_j$ cannot be reduced to $L_i$ for any $i < j$.

I shall not try here to give any additional criteria for distinguishing levels. Rather, I am simply adopting, for the purposes of the argument, the abstract assumption of both arguments that there is indeed some such hierarchy. At the very least, one would have to consider this an idealization of some kind, and we shall see that the assumption that there is a discrete hierarchy of levels is seriously misleading and probably false. It seems more likely that even if the ordering on the complexity of structures ranging from those of elementary physics to those of astrophysics and neurophysiology is discrete, the interactions between such structures will be so entangled that any separation into levels will be quite arbitrary.

## 6.3. The Generalized Exclusion Argument

As a general principle, premise (1) is false as we originally formulated it, for it asserts that, first, causal antecedents of x and, second, events causally intermediate between x and y are causally irrelevant to *y*, and this is obviously wrong.

So, let us use the preliminary definition:

An event z is *causally connected* to a second event x if and only if x causes z or z causes x. z is *causally disconnected* from x just in case z is not causally connected to x. The proper formulation of (1) is then:

(1′) If an event x is causally sufficient for an event y, then no event x* distinct from x and causally disconnected from x is causally relevant to y. (exclusion)[4]

To make the revised principle true requires a strict criterion of event identity so that in particular, the exact time and way in which an event occurs is crucial to that event having the identity it does. This criterion is needed to exclude cases where x is sufficient for y but x*, which is causally disconnected from x, brings about (a somewhat earlier analogue of) y before the connecting process from x has brought about y. Such a strict criterion, called "fragility" by David Lewis,[5] is controversial, and rightly so, but I shall accept it here simply because its adoption avoids distracting issues and does not affect the essential features of the argument.

Suppose now that we accept principle (L) and ask whether an analogue of premise (2) is plausible for each level L. What I shall call i-determinism (which is what a generalization of (2) asserts) is different from i-closure, the thesis that all events that are causally relevant to a given i-level event are themselves i-level events: i-determinism is compatible both with upwards and with downwards causation, where an i-level event causes another i-level event through a chain involving events at other levels. To see that (2) is not true when generalized, consider (2) with "physical" replaced with "biological." It might now be true that for every future biological event there is some biological antecedent that guarantees it. But to assert biological determinism for every biological event in the history of the universe would be to rule out what is commonly believed, which is that biological phenomena were not always present during the development of the universe.[6] The first biological event, under whatever criterion of "biological" you subscribe to, must have had a non-biological cause. So a generalization of (2) is not plausible for any level above the most fundamental level of all, which we call the 0-level, and so we shall restrict ourselves to a formulation of (2) for that level only, the level of whatever constitutes the most fundamental physical properties, *viz.*:

(2′) For every 0-level event y, some 0-level event x is causally sufficient for y. (0-level determinism)[7]

The third premise requires a criterion of distinctness of events (as does (1′)). This cannot be done in terms of spatiotemporal distinctness alone, because on supervenience accounts the supervening event is spatiotemporary coincident with the subvenient event(s). So the burden of characterizing distinctiveness will have to lie on principle (L) and we shall take as a sufficient condition for two events being distinct that they occupy different levels in the hierarchy. With this understanding we have:

(3′) For every 0-level event x and every i-level event $x_i^*$ $(i > 0)$ x is distinct from $x_i^*$, (pluralism)

Then it follows immediately that:

(4′) For every 0-level event y, no i-level event $x_i^*$ $(i > 0)$ that is causally disconnected from every 0-level event antecedent to y is causally relevant to y.

This modified conclusion shows how nonreductive physicalism can avoid the conclusion of the simple version of the exclusion argument and hence can also avoid the overall contradiction with the conclusion of the second argument, because the conclusion (4′) allows higher level events to causally affect 0-level events if the former are part of causal chains that begin and end at the 0-level.[8] In order for us to use that possibility to argue for emergent properties, we need to address the second argument, which amongst other things, precludes higher level causal chains that do not involve 0-level events.

## 6.4. Generalizations of the Downwards Causation Argument

In looking at generalizations of the downwards causation argument, it is worthwhile to again lay out explicitly the assumptions which underlie it. The first is a supervenience assumption that permeates the contemporary literature on nonreductive physicalism and is retained by Kim for emergent properties.[9]

(5) Every emergent property is supervenient upon some set of physical properties.

One natural generalization of this is:

(5′) Every j-level property (j > 0) is supervenient upon some set of i-level properties, for i < j.

Strict physicalists might want to insist that all higher level properties supervene upon 0-level properties. However, unlike the first argument, where premise (2) was true only for 0-level properties, here we can maintain full generality. In fact, if there are emergent properties, the strict physicalist position will be false, and we shall have to leave room for some (non-emergent) properties to supervene upon j-level emergent properties but not upon 0-level properties alone. Next, the assumption explicitly cited by Kim:

(6) The only way to cause an emergent property to be instantiated is by causing its (set of) emergence base properties to be instantiated.

Its generalization will be (assuming that supervenience is a transitive relation):

(6′) The only way to cause a j-level property to be instantiated is by causing a set of i-level properties (i < j), the subvenient basis, to be instantiated.

Premises (5′) and (6′) are closely related, for the plausibility of (6′) rests on accepting something like (5′), the idea being that emergent properties cannot exist separately from whatever physical properties give rise to them.

Then we have the important condition:

(7) A property is emergent only if it has novel causal powers[10]

We can retain this unchanged for the generalized argument.

## 6.5. An Emergentist Answer to the Second Argument

How can we now escape the conclusion of the downwards causation argument? We can begin by refining the event ontology used in the argument, which appeals to properties as causes. This way of speaking, about property causation, is clearly an abbreviation for an instance of one property causing an instance of another property. We shall have to resort here to a certain amount of notational clutter. This will not be pretty, but it has a certain suggestiveness that may be helpful. From here on I shall talk of i-level properties and entities, $P^i_m$ and $x^i_r$, respectively, for $i \geq 0$. I call a property (entity) an i-level property (entity) if i is the first level at which instances of $P^i_m(x^i_r)$ occur. i-level properties may, of course, also have instances at higher levels as for example, physical properties such as mass and volume, do. We need to keep distinct i-level entities and i-level properties, for it is possible that, in general, i-level entities may possess j-level properties, for $i \neq j$. However, for simplicity in what follows I shall assume that i-level properties are instantiated by i-level entities. So we have that $P^i_m(x^i_r)(t_1)$ causes $P^i_n(x^i_s)(t_2)$ where $t_2 > t_1$ and $x^i_r$, $x^i_s$, the entities possessing the properties, may or may not be the same.[11] Here $P^i_m$ is the mth i-level property; $x^i_r$ is the rth i-level entity, and $P^i_m(x^i_r)(t_k)$ denotes the instantiation of $P^i_m$ by $x^i_r$ at time $t_k$.

Suppose now that the i-level properties constitute a set $I = P^i_1, \ldots, P^i_n \ldots$[12] and that these i-level properties are complete in the sense that $I$ is exhaustive of all the i-level properties. Now introduce a *fusion operation* [.*.], such that if $P^i_m(x^i_r)(t_1)$, $P^i_n(x^i_s)(t_1)$ are i-level property instances, then $[P^i_m(x^i_r)(t_1)*P^i_n(x^i_s)(t_1)]$ is an i + 1-level property instance, the result of fusing $P^i_m(x^i_r)(t_1)$ and $P^i_n(x^i_s)(t_1)$. I want to emphasize here that it is the fusion operation on the property instances that has the real importance for emergence. Usually, the fusion operation acting on objects will merely result in a simple concatenation of the objects, here represented by $[(x^i_r) + (x^i_s)]$, within which the individuals $x^i_r$ and $x^i_s$ retain their identities. However, as we noted earlier, for full generality we would need to allow for the possibility of new i + 1-level objects.[13] Moreover, fusion usually is not instantaneous, and we should represent that fact. So I shall represent the action of fusion by $[P^i_m(x^i_r)(t_1)*P^i_n(x^i_s)(t_1)] = [P^i_m*P^i_n][(x^i_r) + (x^i_s)](t'_1)$.[14] For simplicity, I shall assume here that * itself is an i-level operation (i.e., that it is an operation of the same level as the property instances which it fuses.)

*By a fusion operation, I mean a real physical operation, and not a mathematical or logical operation on predicative representations of properties.* That is, * is neither a logical operation such as conjunction or disjunction nor a mathematical operation such as set formation.[15] * need not be a causal interaction, for it can represent interactions of quite different kinds.

The key feature of $[P^i_m*P^i_n][(x^i_r) + (x^i_s)](t'_1)$ is that it is a unified whole in the sense that its causal effects cannot be correctly represented in terms of the separate causal effects of $P^i_m(x^i_r)(t_1)$ and of $P^i_n(x^i_s)(t_1)$. Moreover, within the fusion $[P^i_m*P^i_n][(x^i_r) + (x^i_s)](t'_1)$ the original property instances $P^i_m(x^i_r)(t_1)$, $P^i_n(x^i_s)(t_1)$ no longer exist as separate entities and they do not have all of their i-level causal powers available for use at the (i + 1)st level.[16] Some of them, so to speak, have been "used up" in forming the fused property instance. Hence, these i-level property instances no longer have an independent existence within the fusion. In the course of fusing they become the i + 1-level property instance, rather than realizing the i + 1-level property in the way that supervenience theorists allow the subvenient property instances to continue to exist at the same time as the supervenient property instance. For example, the cusped panelling and brattishing that makes the fan vaulting of the King's College chapel architecturally transcendent exists simultaneously with the supervenient aesthetic glories of that ceiling. In contrast, when emergence occurs, the lower level property instances go out of existence in producing the higher level emergent instances. This is why supervenience approaches have great difficulty in properly representing emergent effects. To see this, consider the following formulation of strong supervenience:[17]

A strongly supervenes upon B just in case, necessarily, for each x and each property F in A, if x has F, then there is a property G in B such that x has G and necessarily if any y has G, it has F. (Kim 1984, 165)

Now let A be the fusion $[P^i_m{*}P^i_n][(x^i_r)+(x^i_s)](t'_1)$. Upon what can this supervene? Because we are here considering only the abstract possibility of emergent features, consider a very simple world in which $P^i_m(x^i_r)(t_1)$ and $P^i_n(x^i_r)(t_1)$ are the only i-level property instances occurring at $t_1$, and in which there are no i-level property instances at $t'_1$. Then, trivially, there is nothing at $t'_1$ at the i-level upon which $[P^i_m{*}P^i_n][(x^i_r)+(x^i_s)](t'_1)$ can supervene. Faced with this, the supervenience advocate could try a different strategy, one that relies on the fact that the definition of strong supervenience does not require the supervenient and subvenient instances to be simultaneous. So one could use the earlier instances $P^i_m(x^i_r)(t_1)$ and $P^i_n(x^i_s)(t_1)$ themselves as the base upon which the later $[P^i_m{*}P^i_n][(x^i_r)+(x^i_s)](t'_1)$ supervenes. Yet once one has allowed this temporal gap, the supervenience relation is in danger of collapsing into an ordinary causal relation. In order for the base instances to (nomologically) necessitate the fusion instance, the absence of all intervening defeaters will have to be included in the subvenient base, and this will give us a base that looks very much like a Millian unconditional cause.[18] Whatever the supervenience relation might be, the way it is used in nonreductive physicalism is surely not as a causal relation, because that would immediately convert nonreductive physicalism into old-fashioned epiphenomenalism.

A second reason why supervenience seems to be an inappropriate representation of certain cases of fusion is given by the physical examples in section 6.6, and I thus refer the reader to that section. However, because my purpose here is not to attack supervenience, but rather to provide a solution to the problem of upwards and downwards causation, I now return to that issue.

It has to be said that the unity imposed by fusion might be an illusion produced in all apparent cases by an epistemic deficit, and that when properly represented all chemical properties, for example, might be representable in terms of the separable (causal) properties of their chemical or physical constituents. But whether this can be done or not is, of course, the issue around which emergentism revolves and I shall address it explicitly in a moment in terms of some examples. What I maintain here is this: that one comprehensible version of emergentism asserts that at least some $i+1$-level property instances exist, that they are formed by fusion operations from i-level property instances, and that the $i+1$-level property instances are not supervenient upon the i-level property instances. With this representation of the emergent $i+1$-level property, let us add a claim that is characteristic of many versions of non-reductive physicalism, especially those motivated by multiple realizability considerations; the token identity of $i+1$-level property instances and fusions of i-level property instances. That is, although we cannot identify the property $P_1^{i+1}$ with $[P_m^i{*}P_n^i]$ when $P_1^{i+1}$ is multiply realizable, we *can* identify some instances of $P_1^{i+1}$ with some instances of $[P_m^i{*}P_n^i]$. It is important to remember for the purposes of the present argument that we are concerned with causal and other interactions, and not with the problem of how properties themselves are related across levels. The latter is the focus of greatest concern within nonreductive physicalism, but our problems can be solved without addressing the interlevel relationships of the properties themselves. In fact, given that the higher level properties are emergent, there is no reason to identify them with, or to reduce them to, combinations of lower level properties. Coupled with our previous reminder that it is property instances that are involved in causal relations, and not properties directly, we now have a solution to the downward causation problem for the case of emergent properties.[19]

Suppose that $P_1^{i+1}(x_1^{i+1})(t_1')$ causes $P_k^{i+1}(x_k^{i+1})(t_2')$, where both of these instances are at the $i+1$-level. What we have is that the i-level property instances $P_m^i(x_r^i)(t_1)$ and $P_n^i(x_s^i)(t_1)$ fuse to produce the $i+1$-level property instance $[P_m^i{*}P_n^i][(x_r^i)+(x_s^i)](t_1')$, which is identical with $P_1^{i+1}(x_1^{i+1})(t_1')$. This $i+1$-level property instance then causes the second $i+1$-level property instance $P_k^{i+1}(x_k^{i+1})(t_2')$. This second $i+1$-level property instance, *if it is also emergent*, will be identical with, although not result from, a fusion of i-level property instances $[P_r^i{*}P_s^i][(x_u^i)+(x_v^i)](t_2')$. But there is no direct causal link from the individual property instances $P_m^i(x_r^i)(t_1)$ and $P_n^i(x_s^i)(t_1)$ to the individual decomposed property instances $P_r^i(x_u^i)(t_3)$ and $P_s^i(x_v^i)(t_3)$.

Diagrammatically, we then have:

| | | |
|---|---|---|
| $P_1^{i+1}(x_1^{i+1})(t_1')$ | —causes⟶ | $P_k^{i+1}(x_k^{i+1})(t_2')$ |
| (is identical with) | | (is identical with) |
| $[P_m^i{*}P_n^i][(x_r^i)+(x_s^i)](t_1')$ | | $[P_r^i{*}P_s^i][(x_u^i)+(x_v^i)](t_2')$ |
| ↗ ↔ ↖ | | ↙ ↔ ↘ |
| (fuses) | | (decomposes) |
| $P_m^i(x_r^i)(t_1)$ $P_n^i(x_s^i)(t_1)$ | | $P_r^i(x_u^i)(t_3)$ $P_s^i(x_v^i)(t_3)$ |

I note here that decomposition does not have to occur. The system might stay at the $(i+1)$st level while it produces further $(i+1)$-level effects. Nor is it necessary that $P_k^{i+1}$ be an emergent property. When it is not, the identity on the right side of the diagram will not hold, and the lower two tiers on the right will be missing. Perhaps some $i+1$ property instances are primitive in this way, but this is doubtful given what we know of the evolution of our universe.

One further source of concern exists and it relates directly to assumption (6′) of the downward causation argument. Is it possible for i + 1-level instances to directly produce other i + 1-level instances without synthesizing them from lower level instances? These higher level instances are usually emergent, and so it might be thought that they must themselves be formed by fusion from lower level instances and not by direct action at the higher level. This concern fails to give sufficient credit to the ontological autonomy of emergent property instances. Recall that i-level instances no longer exist within i + 1-level instances—the higher level instances act as property instance atoms even though they may, under the right circumstances, be decomposed into lower level instances. It is perfectly possible for an i + 1-level instance to be directly transformed into a different i + 1-level instance (often with the aid of other property instances) or to directly transform another, already existing, i + 1-level property instance (again usually with the aid of other property instances.) Simply because the i-level instances no longer exist, they can play no role in this causal transformation.

We can now see what is wrong with premises (5′) and (6′). (5′) misrepresents the way in which emergent property instances are produced, for as we have seen, the relationship between the higher and lower levels is not one of supervenience. A last reply is available to supervenience advocates insisting on the need for relations between properties. If the emergent instance is produced by a causal interaction, they can insist that causes require laws and that the generality inherent in laws requires properties as well as instances. This is not something that a thoroughgoing ontological approach needs to accept. Singular causes can be taken as fundamental,[20] and whether or not causal laws (or their statements) can then be formulated in terms of relations between properties (predicates) depends upon the complexity of the part of the world involved. Sometimes they can be, often they cannot.

We have already seen the problem with (6′): it is false to say that the i-level property instances co-occur with the (i + 1)st level property instance. The former no longer exist when they fuse to form the latter. That is why the notation used here is not entirely adequate, but this is almost inescapable given that formal syntax is ordinarily compositional. $[P^i_r{*}P^i_s][(x^i_u) + (x^i_v)](t'_2)$ is not "composed of" $P^i_r(x^i_u)(t_3)$ and $P^i_s(x^i_v)(t_3)$: a physical process of decomposition (better "defusion") is required to create the last two.

So, we have given a construal of what i + 1-level emergent properties might be, shown how they can have causal properties that are new in the sense that they are not possessed by their i-level origins, and provided the appropriate sense in which the emergent properties are instantiated only because lower level properties are instantiated. But there is now no sense in which, as the exclusion argument claims, i-level property instances are overdetermined by a combination of previous i-level instances and i + 1-level instances. Nor are the i + 1-level instances always required to be brought into being by some simultaneous set of i-level property instances, as the downward causation argument asserts. I believe that there are two reasons why a solution of this kind is not immediately obvious when reading these arguments. First, by representing the emergence base within the downwards causation argument as a single, unanalyzed property P*, the fact that emergent property instances are formed by a fusion process on lower level instances cannot be accurately represented. Secondly, by representing the situation in terms of properties rather than property instances, supervenience seems to be forced upon us, when in fact a treatment purely in terms of instances is open to us.

We do, of course, lose the causal closure of i-level property instances. This is not something that should disturb us overmuch. If a picture like the one I have described is roughly correct, then it turns out that an undifferentiated commitment to "physicalism" is too crude and that both the "mental" and the "physical" are made up on multilayered sets of strata, each level of which is emergent from and (probably) only arbitrarily separable from the layers beneath it.

To sum up what we have established: the claim that an i + 1-level emergent property is instantiated only because its i-level emergence base is instantiated is wrong—the "emergence base" is not the reason the emergent property is instantiated—it is the move from the i-level to the i + 1-level by fusion that gives us emergence. Second, the problem of overdetermination that is lurking behind downwards causation is now less problematical. We may maintain that all i-level events are determined by i-level antecedents but often this will be by way of j-level intermediaries.

So, to conclude: we have seen that two intriguing and widely canvassed arguments against emergent properties do not succeed in establishing their conclusions. It is more important, though, to emphasize that a robustly ontic attitude towards emergent properties, rather than the more common logical approaches, can give us a sense of what emergent features might be like. Most important of all, I think, is to stop thinking of these issues exclusively in terms of mental properties, and to look for examples in more basic sciences.

In addition, construing the arguments in terms of multiple layers of property instances reminds us of two things: that there are many, many levels of properties between the most fundamental physical level and the psychological, and that it is left unacceptably vague in much of the philosophical literature just what is meant by "physical." Being reminded of the large variety of property levels below the psychological, some of which are arguably emergent, should at least make us aware of the need to be more explicit on that score, and that *some* of the mysteries surrounding the physical/mental cleavage are perhaps the result of an inappropriate dichotomy.

## 6.6. From Metaphysics to Physics and Back

Thus far, the discussion has been completely abstract. My intention has been simply to show how one sort of emergent feature can avoid various difficulties inherent in supervenience treatments. What we have thus shown is the possibility of property instances being emergent, free from the difficulties stemming from the exclusion and downwards causation arguments. Even if there were no actual examples of fusion, the account of emergence given here would be useful because it provides a coherent account of a particular kind of emergence that is devoid of the mysteries associated with earlier attempts to explicate the concept. There is, of course, the further question of whether our world contains examples of emergent property instances. The answer to this is a reasonably confident "yes." I shall here only sketch the form that such examples take, referring interested readers to more detailed sources for a richer description.

It frequently has been noted that one of the distinctive features of quantum states is the inclusion of non-separable states for compound systems, the feature that Schrödinger called "quantum entanglements."[21] That is, the composite system can be in a pure state when the component systems are not, and the state of one component cannot be completely specified without reference to the state of the other component. Furthermore, the state of the compound system determines the states of the constituents, but not vice versa.[22] This last fact is exactly the reverse of what supervenience requires, which is that the states of the constituents of the system determine the state of the compound, but when the supervening properties are multiple realizable, the converse does not hold. I believe that the interactions which give rise to these entangled states lend themselves to the fusion treatment described in the earlier part of this paper, because the essentially relational interactions between the "constituents" (which no longer can be separately individuated within the entangled pair) have exactly the features required for fusion. One might be hesitant to use quantum entanglements as an argument by themselves because of the notorious difficulties involved in providing a realist interpretation for the theory. But what seems to me to be a powerful argument in favor of the existence of these emergent features is that these quantum entanglements are the source of macroscopic phenomena that are directly observable. In particular, the phase transitions that give rise to superconductivity and superfluidity in helium are a direct result of nonseparable states.[23]

It is a question of considerable interest whether, and to what extent, fusion occurs in other areas. It would be a mistake to speculate on such matters, because the existence of such interactions is a contingent matter, to be settled by scientific investigation. There is indeed, within that part of metaphysics that can be (partially) naturalized, an important but neglected principle: *certain metaphysical questions cannot be answered (yet) because we do not know enough.* On the basis of this principle, those readers who want an answer as to whether, for example, mental phenomena are emergent in the above sense will, I am afraid, have to be patient.

There is one other way in which fusion can occur, and it is neither a matter of speculation nor something directly amenable to empirical inquiry. To see it, one simply needs to be reminded of something that has been lost in the avalanche of logical reconstructions of causation and other concepts in this century. It is that singular causal interactions between property instances, construed realistically, provide "horizontal" examples of the kind of novelty that has here and elsewhere been discussed in "vertical" terms. By "construed realistically," I mean "taken to be *sui generis* features of the world, the properties of which are fundamentally misrepresented by reductive analyses or (humean) supervenience treatments of causation."[24] This is not the place to persuade readers of the benefits of the realist singular view,[25] so I shall simply note that the issues of "horizontal" and "vertical" novelty are connected, for the explaining away of the former, especially by supervenience, tends to gain its plausibility from a sparse ontology of spacetime points possessing a restricted set of primitive physical properties. If you believe, in contrast, that solid state physics (for example) is more than just advanced elementary particle physics, you will begin to ask how phenomena from the two fields interact. You should then be prepared to find that emergence may be complicated, but that it is neither mysterious nor uncommon.[26]

## Acknowledgments

Previous versions of this paper were read at Virginia Polytechnic Institute, Duke University. The University of Virginia, the University of Pittsburgh, the British Society for the Philosophy of Science, and an IUHPS meeting in Warsaw. Comments and suggestions from those audiences and the two anonymous referees from *Philosophy of Science* were very helpful in improving the paper. I am also grateful for conversations and correspondence with Robert Almeder, James Bogen, Richard Burian, John Forge, David Henderson, Martin Jones, Jaegwon Kim, James Klagge, Ken Olson, Fritz Rohrlich, and Abner Shimony. Research for this paper was conducted partly under NSF grant SBR-9311982 and the support is gratefully acknowledged.

## Notes

This chapter originally appeared in *Philosophy of Science* 64 (1997), 1–17.

1. See Yablo 1992, 247, fn. 5 for a partial list of versions of this argument that have appeared in the philosophical literature. I note here that he does not endorse the simple version of the exclusion argument because it is unsound.

2. See, e.g., Yablo 1992, 247, fn 5.

3. This allows, of course, that the realm of the mental does generate additional peculiar problems. The exclusion and downwards causation arguments are entirely independent of those peculiarities, however, except perhaps for a prejudice against the mental. Were it not for that prejudice, the exclusion argument could be run in reverse to exclude physical causes, a feature emphasized to me by Peter Dlugos.

4. I am here setting aside overdetermining events as genuine examples of causation. Although some discussions of the exclusion argument see acceptance of overdetermination as a way out of the difficulty, this is not a convincing move, and I shall not follow that route. Not the least reason for this is that cases of simultaneous overdetermination are exceedingly rare.

5. See Lewis 1986, Postscript E. Lewis rejects extreme versions of the fragility approach.

6. This temporal development gives rise to evolutionary emergence. I shall not pursue that topic here but the reader can easily develop such an account from materials in this paper.

7. Premise (2′) is, on current evidence, false because fundamental physics appears to be indeterministic in certain respects. I have preserved the original form of the argument to keep things simple, but to allay worries about the truth status of (2′), one can reformulate the argument thusly: Replace "causally sufficient" in (1′) and (2′) by "causally complete," "event" by "set of events," and x by $\{x_i\}$, where "causally complete" means either that all events necessary in the circumstances for y are included in the set or that all events that are probabilistically relevant to y are in the set. See, e.g., Lewis 1986 for one account of the former, Humphreys 1989 for one account of the latter. (3), (4) remain as they are. (2′) will be false if the universe had a first uncaused event, but that fact is irrelevant here.

8. More complex versions of the argument obviously allow similar possibilities for causal chains beginning and ending at higher levels than 0.

9. Kim actually examines both the realizability and the supervenience approaches. I restrict myself to the latter.

10. In fact, as Martin Jones pointed out to me (pers. comm.), the novelty of the causal powers seems to play no role in Kim's central argument. Even if mental properties produced familiar physical consequences that could also be brought about by physical properties, the argument would still hold. The use of novelty is primarily in characterizing the difference between emergent and nonemergent properties, for it is an essential feature of emergent properties that they be new.

11. Kim asserts (1992, 123) that the emergentists of the 1920s held that no new entities emerged at new levels of the hierarchy, only new properties. Allowing that to be historically accurate, it is as well to allow, at least notationally, that we might have new entities emerging as well as properties.

12. The cardinality of this set is unrestricted—the integer subscripts are used for convenience only.

13. Supervenience advocates have also recognized the need for this. See, e.g., Kim 1988.

14. There is a notationally harmless but metaphysically important ambiguity here between fusion operations on property instances and on properties. The latter is metaphysically derivative from the former in that when $[P^i_m(x^i_r)(t_1)*P^i_n(x^i_s)(t_1)]$ exists, then there is by virtue of this an instance of a novel property, signified by $P_m*P_n$, at level $i+1$. This disambiguation sits most happily with the position that fusion brings into being new properties, a position that seems to fit well with the idea of emergence. Those who subscribe to the view that there are eternal emergent properties that are uninstantiated prior to some time can think of $P_m*P_n$ as a mere notational device indicating a move to a previously uninstantiated property at a higher level.

15. In contrast, it is standard in the literature on supervenience to construe the subvenient basis in terms of sets of properties, or in terms of a disjunctive normal form of properties, where it is assumed that it makes sense to perform logical operations on properties. These devices are inappropriate for characterizing emergent properties and are a legacy of the continuing but, in my view, fruitless attempt to reconstruct causation and associated concepts logically rather than ontologically.

16. As mentioned earlier, the objects themselves will often retain their separate identities.

17. This argument carries over, with simple modifications, to the definition of weak supervenience.

18. See Humphreys 1989, section 25, for a discussion of this condition.

19. I want to emphasize here that what follows is to be construed only as a representation of the correct relationship between emergent property instances and property instances on lower levels when causal sequences are involved. It is not to be construed as an argument that in all cases where we have different levels of property instances, emergentism holds. Supervenience does have some restricted uses.

20. See Humphreys 1989, section 25.

21. The discussion of nonseparability goes back at least to Schrödinger 1935. d'Espagnat 1965 is another early source and more recent discussions can be found in Teller 1986, Shimony 1987, French 1989, Healey 1991, and d'Espagnat 1995, among many others.

22. See, e.g., Beltrametti and Cassinelli 1981, 65–72.

23. See Shimony 1993, 221.

24. I focus on causal interactions here only because of their familiarity. Other kinds of interactions can, one assumes, produce genuine novelty.

25. For a partial account, see Humphreys 1989, section 25.

26. Since this paper was first drafted in 1991 I have realized that the term "fusion" has a standard use in the mereological literature that is almost opposite to its use here. I believe that my use is better justified etymologically, but mereology was there first. The reader is hereby advised never to confuse the two uses.

## References

Beltrametti, E. and G. Gassinelli (1981), *The Logic of Quantum Mechanics*. Reading, MA, Addison-Wesley.

d'Espagnat, B. (1965), *Conceptions de la physique contemporaine*. Paris: Hermann.

———. (1995), *Veiled Reality*. Reading, MA, Addison-Wesley.

French, S. (1989), "Individuation, Supervenience, and Bell's Theorem," *Philosophical Studies* 55: 1–22.

Healey, R. (1991), "Holism and Nonseparability," *Journal of Philosophy* 88: 393–421.

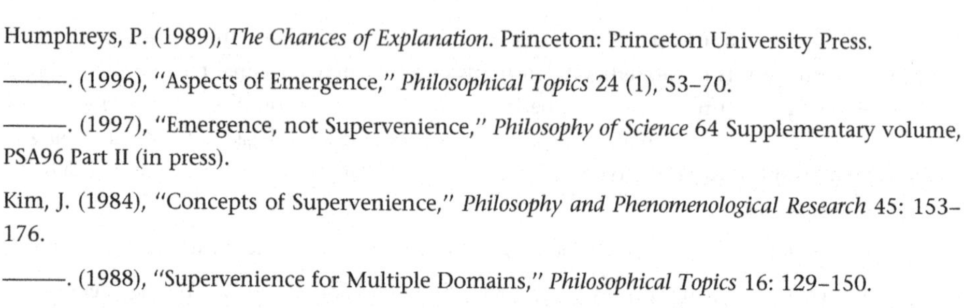

Humphreys, P. (1989), *The Chances of Explanation.* Princeton: Princeton University Press.

———. (1996), "Aspects of Emergence," *Philosophical Topics* 24 (1), 53–70.

———. (1997), "Emergence, not Supervenience," *Philosophy of Science* 64 Supplementary volume, PSA96 Part II (in press).

Kim, J. (1984), "Concepts of Supervenience," *Philosophy and Phenomenological Research* 45: 153–176.

———. (1988), "Supervenience for Multiple Domains," *Philosophical Topics* 16: 129–150.

———. (1992), "'Downward Causation' in Emergentism and Nonreductive Physicalism," in A. Beckermann, H. Flohr, and Jaegwon Kim (eds.), *Emergence or Reduction: Essays on the Prospects of Nonreductive Physicalism.* New York: Walter de Gruyter, pp. 119–138.

———. (1993), "The Nonreductivists's Troubles with Mental Causation," in J. Heil and A. Mele (eds.), *Mental Causation.* Oxford: Oxford University Press, pp. 189–210.

Lewis, D. (1986), "Causation" and "Postscripts to Causation," in D. Lewis, *Philosophical Papers, Volume II.* Oxford: Oxford University Press, pp. 159–213.

Schrödinger, E. (1935), "Discussion of Probability Relations between Separated Systems," *Proc. of the Cambridge Phil. Soc.* XXXI: 555–563.

Shimony, A. (1987), "The Methodology of Synthesis: Parts and Wholes in Low-Energy Physics," in R. Kargon and P. Achinstein (eds.). *Kelvin's Baltimore Lectures and Modern Theoretical Physics.* Cambridge, MA.: MIT Press, pp. 399–423.

———. (1993), "Some Proposals Concerning Parts and Wholes," in A. Shimony, *Search for a Naturalistic World View, Volume II.* Cambridge: Cambridge University Press, pp. 218–227.

Teller, P. (1986), "Relational Holism and Quantum Mechanics," *British Journal for the Philosophy of Science* 37: 71–81.

Yablo, S. (1992), "Mental Causation," *The Philosophical Review* 101: 245–280.

# 7 Making Sense of Emergence

Jaegwon Kim

## 7.1

It has been about a century and half since the ideas that we now associate with emergentism began taking shape.[1] At the core of these ideas was the thought that as systems acquire increasingly higher degrees of organizational complexity they begin to exhibit novel properties that in some sense transcend the properties of their constituent parts, and behave in ways that cannot be predicted on the basis of the laws governing simpler systems. It is now standard to trace the birth of emergentism back to John Stuart Mill and his distinction between "heteropathic" and "homopathic" laws,[2] although few of us would be surprised to learn that the same or similar ideas had been entertained by our earlier philosophical forebears.[3] Academic philosophers—like Samuel Alexander and C. D. Broad in Britain, A. O. Lovejoy and Roy Wood Sellars in the United States—played an important role in developing the concept of emergence and the attendant doctrines of emergentism, but it is interesting to note that the fundamental idea seems to have had a special appeal to scientists and those outside professional philosophy. These include the British biologist C. Lloyd Morgan, a leading theoretician of the emergentist movement early in this century, and, more recently, the noted neurophysiologist Roger W. Sperry.

In spite of its obvious and direct relevance to some of the central issues in the philosophy and methodology of science, however, emergentism failed to become a visible part of the Problematik of the mainstream philosophy of science. The main reason for this, I believe, is that philosophy of science during much of the middle half of this century, from the 1930s to the '60s—at least, in the analytic tradition—was shaped by the positivist and hyper-empiricist view of science that dominated the Anglo-American philosophy at the time. Influential philosophers of science during this period—for example, Carl Hempel and Ernest Nagel[4]—claimed that the classic idea of emergence was confused and incoherent, often likening it to neo-vitalism, and what they saw as the only salvageable part of the emergence concept—the part that they could state in their own postivist/formalist idiom—usually turned out to be largely trivial, something that could be of little interest for serious philosophical purposes.

But the idea of emergence refused to die, continuing to attract a small but steady stream of advocates from both the philosophical and the scientific ranks, and it now appears to be making a strong comeback. This turn of events is not surprising, given the nearly total collapse of positivistic reductionism and the ideal of unified science which was well underway by the early '70s. The lowly fortunes of reductionism have continued to this day, providing a fertile soil for the reemergence of emergentism. Classic emergentists like Morgan and Alexander thought of themselves as occupying a moderate intermediate position between the extremes of "mechanistic" reductionism on one hand and explicit dualisms like Cartesianism and neo-vitalism on the other. For them everything that exists is constituted by matter, or basic material particles, there being no "insertion" of alien entities or forces from the outside. It is only that complex systems aggregated out of these material particles begin to exhibit genuinely novel properties that are irreducible to, and neither predictable nor explainable in terms of, the properties of their constituents. It is evident that emergentism is a form of what is now standardly called "nonreductive materialism," a doctrine that aspires to position itself as a compromise between physicalist reductionism and all-out dualisms. It is no wonder then that we now see an increasing, and unapologetic, use of expressions like "emergent property," "emergent phenomenon," and "emergent law," substantially in the sense intended by the classic emergentists, not only in philosophical writings but in primary scientific literature as well.[5]

Does this mean that emergentism has returned—as an ontological doctrine about how the phenomena of this world are organized into autonomous emergent levels and as a metascientific thesis about the relationship between basic physics and the special sciences? I think the answer is a definite yes. The fading away of reductionism and the enthronement of nonreductive materialism as the new orthodoxy simply *amount to* the resurgence of emergentism—not all of its sometimes quaint and quirky ideas but its core ontological and methodological doctrines. The return of emergentism is seldom noticed, and much less openly celebrated; it is clear, however, that the fortunes of reductionism correlate inversely with those of emergentism (*modulo* the rejection of substantival dualism). It is no undue exaggeration to say that we have been under the reign of emergentism since the early 1970s.

I have argued elsewhere[6] against nonreductive materialism, urging that this halfway house is an inherently unstable position, and that it threatens to collapse into either reductionism or more serious forms of dualism. But in this chapter I am not primarily concerned with the truth or tenability of emergentism or nonreductive materialism; rather, my main concern is with making sense of the idea of emergence—the idea that certain properties of complex systems are emergent while others are not. Even if we succeed with the conceptual task of giving a coherent sense to emergence, it is another question whether any particular group of properties is emergent—for example, whether intentional or qualitative mental properties are emergent relative to neural/biological properties, or whether biological properties are emergent relative to physicochemical properties—or indeed whether there are any emergent properties at all.

In trying to make emergence intelligible, it is useful to divide the ideas usually associated with the concept into two groups. One group of ideas are manifest in the statement that emergent properties are "novel" and "unpredictable" from knowledge of their lower-level bases, and that they are not "explainable" or "mechanistically reducible" in terms of their underlying properties. The second group of ideas I have in mind comprises the specific emergentist doctrines concerning emergent properties, and, in particular, claims about the causal powers of the emergents. Prominent among them is the claim that the emergents bring into the world new causal powers of their own, and, in particular, that they have powers to influence and control the direction of the lower-level processes from which they emerge. This is a fundamental tenet of emergentism, not only in the classic emergentism of Samuel Alexander, Lloyd Morgan, and others but also in its various modern versions. Emergentists often contrast their position with epiphenomenalism, dismissing the latter with open scorn. On their view, emergents have causal/explanatory powers in their own right, introducing novel, and hitherto unknown, causal structures into the world.

In this chapter I will adopt the following strategy: I am going to take the first group of ideas as constitutive of the idea of an emergent property, and try to give a unified account of emergence on the basis of a model of reduction that, although its basic ideas are far from new, is significantly different from the classic Nagelian model of reduction that has formed the background of debates in this area. I will then consider the doctrines that I take to constitute emergentism, focusing on the claims about the causal powers of emergent properties, especially the idea of "downward causation."

## 7.2

The concepts of explanation, prediction, and reduction figure prominently at several critical junctures in the development of the doctrine of emergence. Most importantly, the concept of explanation is invoked in the claim that emergent phenomena or properties, unlike those that are merely "resultant," are not *explainable*, or *reductively explainable*, on the basis of their "basal conditions," the lower-level conditions out of which they emerge. This is frequently coupled with the claim that emergent phenomena are *not predictable* even from the most complete and exhaustive knowledge of their emergence base. I believe that emergentists took the two claims to be equivalent, or at least as forming a single package.

Let us assume that every material object has a unique complete microstructural description: that is, any physical system can be exhaustively described in terms of (i) the basic particles that constitute it (this assumes classic atomism, which the early emergentists accepted); (ii) all the intrinsic properties of these particles; and (iii) the relations that configure these particles into a structure (with "substantial unity," as some emergentists would say). Such a description will give us the total "relatedness" of basal constituents; it also gives us what we may call the *total microstructural (or micro-based) property* of the system—that is, a macro-property (macro since it belongs to the system as a whole) constituted by the system's basic micro-constituents, their intrinsic properties, and the relations that structure them into a system with unity and stability as a substance.[7]

I would expect most emergentists to accept mereological supervenience, in the following form:

[Mereological supervenience] Systems with an identical total microstructural property have all other properties in common.[8] Equivalently, all properties of a physical system supervene on, or are determined by, its total microstructural property.

It is a central claim of classic emergentism that among these properties supervenient on a system's total microstructural property, some have the special character of being "emergent," while the rest are only "resultant." What is the basis for this distinction? Lloyd Morgan says this:

> The concept of emergence was dealt with (to go no further back) by J. S. Mill... The word "emergent," as contrasted with "resultant," was suggested by G. H. Lewes... Both adduce examples from chemistry and from physiology; both deal with properties; both distinguish those properties (a) which are additive and subtractive only, and predictable, from those (b) which are new and unpredictable.[9]

There is no need to interpret the talk of "additivity" and "subtractability" literally; I believe these terms were used to indicate that resultant properties are simply and straightforwardly calculated and predicted from the base properties. But obviously ease and simplicity of calculation as such is of no relevance here; predictability is not lost or diminished if calculationally complex mathematical/logical procedures must be used.[10] I believe that predictability is the key idea here, and that an appropriate notion of predictability must be explained in terms that are independent of addition or subtraction, or the simplicity of mathematical operations.

In any case, resultant properties are to be those that are predictable from a system's total microstructural property, but emergent properties are those that are not so predictable. Morgan's (b) above introduces the idea of "newness," or "novelty," an idea often invoked by the emergentists. Is he using "new" and "unpredictable" here as expressing more or less the same idea, or is he implying, or at least hinting, that emergent properties are unpredictable *because* they are new and novel properties? I believe that "new" as used by the emergentists has two dimensions: an emergent property is new because it is unpredictable, and this is its epistemological sense; and, second, it has a metaphysical sense, namely that an emergent property brings with it new causal powers, powers that did not exist before its emergence. We will discuss the causal issue in the latter part of this chapter.

In speaking of predictability, it is important to distinguish between *inductive predictability* and *theoretical predictability*, a distinction that I believe the emergentists were clearly aware of. Even emergent properties are inductively predictable: Having observed that an emergent property, *E*, emerged whenever any system instantiated a microstructural property *M*, we may predict that this particular system will instantiate *E* at *t*, given our knowledge or belief that it will instantiate, *M*, at *t*.[11] More generally, on the basis of such empirical data we may have a well-confirmed "emergence law" to the effect that whenever a system instantiates basal condition *M* it instantiates an emergent, *E*. What is being denied by emergentists is the theoretical predictability of *E* on the basis of *M*: we may know all that can be known about *M*—in particular, laws that govern the entities, properties and relations constitutive of *M*—but this knowledge does not suffice to yield a prediction of *E*. This unpredictability may be the result of our not even having the *concept* of *E*, this concept lying entirely outside the concepts in which our theory of *M* is couched. In cases where *E* is a phenomenal property of experiences (a "quale"), we may have no idea what *E* is like before we experience it.[12] But this isn't the only barrier to predictability. It may be that we know what *E* is like—we have already experienced *E*—but we may be powerless to predict whether or not *E*—or whether *E* rather than another emergent *E**—will emerge when a complex is formed with a novel microstructure *M** that is similar to *M* is some significant respects. In such a case the emergence law "Whenever a system instantiates *M*, it instantiates *E*" would have to be taken as a primitive, stating a brute correlation between *M* and *E*.

It is clear that we can inductively predict—in fact, we do this all the time—the occurrences of conscious states in the sense just explained, but, if the emergentists were right about anything, they were probably right about the phenomenal properties of conscious experience: these properties appear not to be theoretically predictable on the basis of a complete knowledge of the neurophysiology of the brain. This is reflected in the following apparent difference between phenomenal properties and other mental properties (including cognitive/intentional properties): We can imagine designing and constructing novel physical systems that will instantiate certain cognitive capacities and functions (e.g., perception, information processing, inference and reasoning, and using information to guide behavior)—arguably, we have already designed and fabricated such devices in robots and other computer-driven mechanisms. But it is difficult to imagine our designing novel devices and structures that will have phenomenal experiences; I don't think we have any idea where to begin. The only way we can hope to manufacture a mechanism with phenomenal consciousness is to produce an appropriate physical duplicate of a system that is known to be conscious. Notice that this involves inductive prediction, whereas theoretical prediction is what is needed to design new physical devices with consciousness. The emergentists were wrong in thinking that sundry chemical and biological properties were emergent;[13] but this was an understandable mistake given the state of the sciences before the advent of solid-state physics and molecular biology. The interest of the ideas underlying the emergentist's distinction between the two kinds of properties need not be diminished by the choice of wrong examples.

## 7.3

As was noted at the start of our discussion, another idea that is closely related to the claimed unpredictability of emergents is the doctrine that the emergence of emergent properties cannot be *explained* on the basis of the underlying processes, and that emergent properties are not *reducible* to the basal conditions from which they emerge. These two claims can be combined into one: Emergent properties are not *reductively explainable* in terms of the underlying processes. Some may wish to distinguish the issue of reduction from that of reductive explanation;[14] we will address this issue later. I will now turn to the task of describing a model of reduction that connects and makes sense of these three ideas, namely that emergent properties are *not predictable* from their basal conditions, that they are *not explainable* in terms of them, and that they are *not reducible* to them.

Let me begin with an example—an idealized, admittedly somewhat simplistic example. To reduce the gene to the DNA molecule, we must first prime the target property, by giving it a *functional* interpretation—that is, by construing it in terms of the causal work it is to perform. Briefly, the property of being a gene is the property of having some property (or being a mechanism) that performs a certain causal function, namely that of transmitting phenotypic characteristics from parents to offsprings. As it turns out, it is the DNA molecule that fills this causal specification ("causal role"), and we have a theory that explains just how the DNA molecule is able to perform this causal work. When all of this is in, we are entitled to the claim that the gene has been reduced to the DNA molecule.

We can now formulate a general model to accommodate reductions of this form. Let **B** be the domain of properties (also phenomena, facts, etc., if you wish) serving as the reduction base—for us, these contain the basal conditions for our emergent properties. The reduction of property *E* to **B** involves three steps:[15]

Step 1 *E* must be *functionalized*—that is, *E* must be construed, or reconstrued, as a property defined by its causal/nomic relations to other properties, specifically properties in the reduction base **B**.

We can think of a functional definition of *E* over domain **B** as typically taking the following (simplified) form:

Having $E =_{\text{def}}$ Having some property $P$ in **B** such that (i) $C_1, \ldots, C_n$[16] cause $P$ to be instantiated, and (ii) $P$ causes $F_1, \ldots, F_m$ to be instantiated.

(We allow either (i) or (ii) to be empty.) The main point to notice is that the functionalization of $E$ makes $E$ nonintrinsic and relational—relational with respect to other properties in **B**. $E$'s being instantiated is for a certain property $P$ to be instantiated, with this instantiation bearing causal/nomic relations to the instantiations of a specified set of properties in the base domain. We may call any property $P$ in **B** that satisfies the causal specification (i) and (ii) a "realizer" or "implementer" of $E$. Clearly, multiple realizers for $E$ are allowed on this account; so multiply realizable properties fall within the scope of the present model of reduction. A functionalization of property $E$ in the present sense is to be taken as establishing a conceptual/definitional connection for $E$ and the selected causal role. An important part of this procedure is to decide how much of what we know (or believe) about $E$'s nomic/causal involvement should be taken as *defining*, or *constitutive of*, $E$ and how much will be left out. We should keep in mind that such conceptual decisions can be and often are based on empirical knowledge, knowledge of the causal/nomic relations in which $E$ is embedded, and can be constrained by theoretical desiderata of various sorts, and that in practice the boundary between what's conceptual and what isn't is certain to be a vague and shifting one.

Step 2 Find realizers of $E$ in **B**. If the reduction, or reductive explanation, of a particular instance of $E$ in a given system is wanted, find the particular realizing property $P$ in virtue of which $E$ is instantiated on this occasion in this system; similarly, for classes of systems belonging to the same species or structure types.

This of course is a scientifically significant part of the reductive procedure; it took many years of scientific research to identify the DNA as a realizer of the gene.

Step 3 Find a theory (at the level of **B**) that explains how realizers of $E$ perform the causal task that is constitutive of $E$ (i.e., the causal role specified in Step 1). Such a theory may also explain other significant causal/nomic relations in which $E$ plays a role.

We presumably have a story at the microbiological level about how DNA molecules manage to code and transmit genetic information. When temperature, for gases, is reduced to mean translational kinetic energy of molecules (another over-simplified stock example[17]), we have a theory that explains the myriad causal/nomic relations in which temperature plays a role. Steps 2 and 3 can be expected to be part of the same scientific research: ascertaining realizers of $E$ will almost certainly involve theories about causal/nomic interrelations among lower-level properties in the base domain.

Notice how this functional conception of reduction differs from the classic Nagel model of intertheoretical reduction[18]—in particular, there is no talk of "bridge laws" or "derivation" of laws. The question whether appropriate bridge laws are available that connect the domain to be reduced with the base domain—more specifically, whether or not there are bridge laws providing for each property to be reduced a nomically coextensive property in the base domain—has been at center stage in debates over reduction and reductionism. However, from the emergentist point of view, the bridge laws, far from being the enablers of reduction (as they are in Nagel reductions), are themselves among the targets of reduction. For it is these bridge laws, laws that state that whenever certain specified basal conditions are present a certain novel property is manifested, that the emergentists were anxious to have explained. Why is it that pain, not itch or tickle, occurs when a certain neural condition (e.g., C-fiber stimulation) holds? Why doesn't pain accompany conditions of a different neural type? Why does *any* phenomenal consciousness occur when these neural conditions are present? These are the kinds of explanatory/reductive questions with which the emergentists were preoccupied. And I think they were right. The "mystery" of consciousness is not dispelled by any reductive procedure that, as in Nagel reduction, takes these bridge laws as brute unexplained primitives.

The philosophical emptiness of Nagel reduction, at least in contexts like mind-body reduction, if it isn't already evident, can be plainly seen from the following fact: a Nagel reduction of the mental to the physical is consistent with, and sometimes even entailed by, many dualist mind-body theories, such as the double-aspect theory, the theory of preestablished harmony, occasionalism, and epiphenomenalism. It is not even excluded by the dualism of mental and physical substances (although Descartes' own interactionist version probably excludes it). This amply shows that the antireductionist argument based on the unavailability of mind-body bridge laws—most importantly, the multiple realization argument of Putnam and Fodor—is irrelevant to the real issue of mind-body reduction or the possibility of giving a reductive explanation of mentality. Much of the debate over the past two decades about reductionism has been carried on in terms of an inappropriate model of reduction, and now appears largely beside the point for issues of real philosophical significance.

## 7.4

Let us now try to see how the functional model of reduction can meet the explanatory/predictive/ontological demands that reductions of genuine philosophical interest must meet. Let $E$ be the property targeted for reduction, where $E$ has been functionalized as the property of having some property $P$ meeting causal specification $C$.

### 7.4.1 The Explanatory Question

Why does this system exhibit $E$ at $t$? Because having $E$ is, by definition, having a property with causal role $C$, and the system, at $t$, has property $Q$, which fills causal role $C$

(and hence realizes *E*). Moreover, we have a theory that explains exactly how *Q* manages to fill *C*.

Why do systems exhibit *E* whenever they instantiate *Q*? Because *E* is a functional property defined by causal role *C*, and *Q* is a realizer of *E* for these systems. And there is a theory that explains how *Q* realizes *E* in these systems.

Suppose that pain could be given a functional definition—something like this: being in pain is being in some state (or instantiating some property) caused by tissue damage and causing winces and groans. Why are you experiencing pain? Because being in pain *is* being in a state caused by tissue damage and causing winces and groans, and you are in neural state *N*, which is one of those states (in you, or in systems like you) that are caused by tissue damage and that cause winces and groans. Why do people experience pain when they are in neural state *N*? Because *N* is implicated in these causal/nomic relations, and being in pain is being in some state with just these causal/nomic relations. It is clear that in this way all our explanatory demands can be met. There is nothing further to be explained about why pain occurs, or why pain occurs when neural condition *N* is present.

But is this a *reductive* explanation? This question is connected with the question whether, and in what sense, the proposed model is a model of reduction, a question that will be considered below.

It is of course another question whether pain can be functionalized. We will briefly return to this issue later, but our concern here is to give a clear sense to what it is to "reduce" pain.

### 7.4.2 The Predictive Question

Will this system exhibit *E* at time *t*? Can we predict this from knowledge of what goes on in the base domain? Yes, because, given the functional definition of *E*, we can in principle identify the realizers of *E* for the system solely on the basis of knowledge of the causal/nomic relations obtaining in the base domain. Given this knowledge of *E*'s realizers for this system, we can predict whether or not the system will, at *t*, instantiate property *E* from our knowledge, or warranted belief, that it will, or will not, instantiate a realizer of *E* at *t*.

Clearly, what enables the ascent from the reduction base to higher properties is the conceptual connections generated by the functionalization of the higher properties. This is in sharp contrast to Nagelian reduction with bridge laws taken as auxiliary premises. These laws are standardly conceived as empirical and contingent, and must be viewed as net additions to our theory about the reduction base, which means that *the base theory so augmented is no longer a theory exclusively about the originally given base domain*. This is why bridge laws only enable inductive predictions, whereas functionalization makes theoretical predictions possible.

These reflections seem to give us an answer to a question we raised earlier—why we seem to lack the ability to design novel physical devices that will exhibit phenomenal consciousness: it is because brute bridge laws may be all we can get to connect phenomenal properties with physical properties, whereas what is required is an ability to make theoretical predictions of qualia solely on the basis of knowledge of the base domain, namely physics, chemistry, biology, and the like. The functionalization of phenomenal experience would give us such an ability.

### 7.4.3 The Ontological Question

In what sense is the functional model a model of *reduction*? What does it reduce, and how does it do it? Central to the concept of reduction evidently is the idea that what has been reduced need not be countenanced as an *independent* existent beyond the entities in the reduction base—that if *X* has been reduced to *Y*, *X* is not something "over and above" *Y*. From an ontological point of view, reduction must mean *reduction*—it must result in a simpler, leaner ontology. Reduction is not necessarily elimination: reduction of *X* to *Y* need not do away with *X*, for *X* may be conserved as *Y* (or as part of *Y*). Thus, we can speak of "conservative" reduction (some call this "retentive" reduction), reduction that conserves the reduced entities, as distinguished from "eliminative" reduction, which rids our ontology of reduced entities. Either way we end up with a leaner ontology. Evidently, conservative reduction requires identities, for to conserve *X as Y* means that *X is Y*, whereas eliminative reduction has no need for reductive identities.

Our question, then, is in what ways the model of reduction being recommended here serves the cause of ontological simplification. Two cases may be distinguished: the first concerns instances of property *E*; the second concerns property *E* itself.

First, consider property instances: system *s* has *E*, in virtue of *s*'s instantiating one of its realizers, say *Q*. Now, *s*'s having *E* on this occasion just is its having some property meeting causal specification *C*, and in this particular instance, *s* has *Q*, where *Q* meets specification *C*. Thus, *s*'s having *E* on this occasion is identical with its having *Q* on this occasion. There is no fact of the matter about *s*'s having *E* on this occasion over and above *s*'s having *Q*. Each instance of *E*, therefore, is an instance of one of *E*'s realizers, and all instances of *E* can be partitioned into $Q_1$-instances, $Q_2$-instances, . . . , where the *Q*'s are *E*'s realizers. Hence, the *E*-instances reduce to the $Q_i$-instances.

Suppose someone were to object as follow: There is no good reason to identify this instance of *E* with the instance of *Q* in virtue of which *E* is realized on this occasion. Rather, *s*'s having *E* should be identified with *s*'s having some property or other meeting causal specification *C*, and this latter instance is not identical with *s*'s having *Q*. For having some property or other meeting *C* is not the same property as having *Q*; that is, property $E \neq$ property $Q$. How should we counter this line of argument? I think it will be helpful to consider the causal picture, and ask: What are the *causal powers* of *this instance of E*, namely *s*'s having *E* on this occasion? If *s* has *E* in virtue of *E*'s realizer *Q*, it is difficult to see how we could avoid saying this: the causal powers of this instance of *E* are exactly the causal powers of this instance of *Q*. This is what I have elsewhere called the "causal inheritance principle":

If a functional property *E* is instantiated on a given occasion in virtue of one of its realizers, *Q*, being instantiated, then the causal powers of this instance of *E* are identical with the causal powers of this instance of *Q*.

If this principle is accepted, the *E*-instance and the *Q*-instance have identical causal properties, and this exerts powerful pressure to identify them. What good would it do to count them as different? If they were different, the difference could not even be detected.

This means that on the present picture *E*-instances are conservatively reduced to *Q*-instances, instances of *E*'s realizers. Let us now turn to the reduction of *E*, the property itself. Here we need to come to terms with *E*'s having multiple realizers, $Q_1, Q_2, \ldots$ There are three possible approaches here.

First, one may choose to defend *E* as a legitimate higher-level property irreducible to its realizers, the *Q*'s. This is the position taken by many functionalists: psychological properties are functional properties defined in terms of input/output correlations, with internal physical/biological properties as realizers, and yet they are irreducible to their realizers, constituting an autonomous domain for the special science of psychology (cognitive science, or whatever).

Second, one may choose to identify *E* with the disjunction of its realizers, $Q_1 \vee Q_2 \vee \cdots$[19] Notice, though, that this identity is not necessary—it does not hold in every possible world—since whether or not a property realizes *E* depends on the laws that prevail at a given world. The reason is that *E* is defined in terms of a causal/nomic condition, and whether something satisfies such conditions depends on the laws that are in force at a given world. This means that in another world with different laws, *E* may have a wholly distinct set of realizers, and in still others *E* may have no realizers at all. So the identity, $E = Q_1 \vee Q_2 \vee \cdots$ is metaphysically contingent, although nomologically necessary, and "*E*" becomes nonrigid, although it remains nomologically rigid or "semirigid" (as we may say). For example, in a world with laws quite different from those prevailing in this world, molecules of another kind, not the DNA molecules, may perform the causal task of coding and transmitting genetic information.[20]

Third, we may give up *E* as a genuine property and only recognize the expression "*E*" or the concept *E*. As it turns out, many different properties are picked out by the concept *E*, depending on the circumstances—the kind of structures involved and the nomological nature of the world under consideration. One could argue that by forming "second-order" functional *expressions* by existentially quantifying over "first-order" properties, we cannot be generating new properties (possibly with new causal powers), but only new ways of indifferently picking out, or grouping, first-order properties, in terms of causal specifications that are of interest to us.[21] As noted, the concept is only nomologically rigid: it picks out the same properties only across worlds that are similar in causal/nomological respects.

Here I will not argue my points in detail. It is clear, however, that the second and third approach effectively reduce the target property *E*: the second is a conservative reduction, retaining *E* as a disjunction of properties in the base domain. In contrast, the third is eliminative: it recommends the elimination of *E* as a property, retaining only the concept *E* (which may play a practically indispensable role in our discourse, both ordinary and scientific). The first approach, as I said, is one that is widely accepted: many philosophers, in spite of (or, in their view, on account of) multiple realization, want to argue that *E* is an irreducible property that nonetheless can be a property playing an important role in a special, "higher-level," science. I believe, however, that this position cannot be sustained. For if the "multiplicity" or "diversity" of realizers means anything, it must mean that these realizers are causally and nomologically diverse. Unless two realizers of E show significant causal/nomological diversity, there is no clear reason why we should count them as two, not one. It follows then that multiply realizable properties are ipso facto causally and nomologically heterogeneous. This is especially obvious when one reflects on the causal inheritance principle. All this points to the inescapable conclusion that *E*, because of its causal/nomic heterogeneity, is unfit to figure in laws, and is thereby disqualified as a useful scientific property. On this approach, then, one could protect *E* but not as a property with a role in scientific laws and explanations. You could insist on the genuine propertyhood of *E* as much as you like, but the victory would be empty.[22] The conclusion, therefore, has to be this: as a significant scientific property, *E* has been reduced—eliminatively.

What I hope I have shown is this: Functionalization of a property is both necessary and sufficient for reduction (sufficient as a first conceptual step, the rest being scientific research). This accords well with the classic doctrines of emergentism: as I argued, it nicely explains why reducible properties are predictable and explainable, and correlatively it explains why irreducible properties are neither predictable nor explainable on the basis of the underlying processes. I believe this makes good sense of the central ideas that make up the concept of emergence.

However, emergentism may yet be an empty doctrine. For there may not be any emergent properties, all properties being physical properties or else functionalizable and therefore reducible to physical properties. Physical properties include not only basic physical magnitudes and the properties of microparticles but microstructual properties of larger complexes of basic particles. So are there emergent properties?

Many scientists have argued that certain "self-organizing" phenomena of organic, living systems are emergent. But it is not clear that these are emergent in our sense of nonfunctionalizability.[23] And, as I said earlier, the classic emergentists were mostly wrong in putting forward examples of chemical and biological properties as emergent. It seems to me that if anything is going to be emergent, the phenomenal properties of consciousness, or "qualia," are the most promising candidates. Here I don't want to rehearse the standard arguments pro and con, but merely affirm, for what it's worth, my own bias toward the pro side: qualia are intrinsic properties if anything is, and to functionalize them is to eliminate them as intrinsic properties.[24]

## 7.5

The doctrine of emergence has lately been associated quite closely with the idea of "downward causation." It is not only that emergent properties are to have their own distinctive causal powers but also that they be able to exercise their causal powers "downward"—that is, with respect to processes at lower-levels, levels from which they emerge. The claim that emergents have causal powers is entirely natural and plausible if you believe that there are such properties. For what purpose would it serve to insist on the existence of emergent properties if they were mere epiphenomena with no causal or explanatory relevance?

The very idea of downward causation involves vertical directionality—an "upward" direction and a "downward" direction. This in turn suggests an ordered hierarchy of domains that gives meaning to talk of something being located at a "higher" or "lower" or "the same" position in relation to another item on this hierarchy. As is familiar to everyone, positions on such a hierarchy are usually called "levels," or sometimes "orders." In fact, talk of "levels"—as in "level of description," "level of explanation," "level of organization," "level of complexity," "level of analysis," and the like—has thoroughly penetrated not only writings about science, including of course philosophy of science, but also the primary scientific literature of many fields.

The emergentists of the early 20th century were among the first to articulate what may be called "the layered model" of the world, although a general view of this kind is independent of emergentism and has been espoused by those who are opposed to emergentism.[25] In fact, a model of this kind provides an essential framework needed to formulate the emergentist/reductionist debate. In any case, the layered model takes the natural world as stratified into levels, from lower to higher, from the basic to the constructed and evolved, from the simple to the more complex. All objects and phenomena have each a unique place in this ordered hierarchy. Most early emergentists, such as Samuel Alexander and C. Lloyd Morgan, viewed this hierarchy to have evolved historically: In the beginning there were only basic physical particles, or just a spacetime framework (as Alexander maintained), and these have evolved into increasingly more complex structures—atoms, molecules, unicellular organisms, multicellular organisms, organisms with consciousness and mentality, and so on. Contemporary interest in emergence and the hierarchical model is focused not on this kind of quasi-scientific and quasi-metaphysical history of the world, but rather on what it says about the synchronic structure of the world—how things and phenomena at different levels hang together in a temporal cross section of the world, or over small time intervals. We want to know whether, and how, the emergentist ideas can help us in understanding the interlevel relationships between items at the adjacent levels on this hierarchy, and ultimately how everything is related to the items at the bottom physical level (if there is such a level).

The layered model gives rise to many interesting questions: for example, how are these levels to be defined and individuated? Is there really a single unique hierarchy of levels that encompasses all of reality or does this need to be contextualized or relativized in certain ways? Does a single ladder-like structure suffice, or is a branching tree-like structure more appropriate? Exactly what ordering relations generate the hierarchical structures? But these questions go well beyond the scope of this paper. Here we will work with a fairly standard, intuitive notion of levels that is shared by most of us.[26] This will not significantly compromise the discussion to follow.

Although, as one would expect, there has been no universal agreement among the emergentists, the central doctrines of emergentism are well known. For our present purposes, we will take them to include the following claims:

1. Emergence of complex higher-level entities Systems with a higher-level of complexity emerge from the coming together of lower-level entities in new structural configurations (the new "relatedness" of these entities).

This claim is by no means unique to emergentism; it is completely at home with universal physical reductionism (what the early emergentists called "mechanism"), the view that all things and phenomena are physical, and are explainable and predictable ultimately in terms of fundamental physical laws. A characteristically emergentist doctrine makes its appearance in the idea that some of the properties of these complex systems, though physically grounded, are nonphysical, and belong outside the physical domain. The following three propositions unpack this idea.

2. Emergence of higher-level properties All properties of higher-level entities arise out of the properties and relations that characterize their constituent parts. Some properties of these higher, complex systems are "emergent," and the rest merely "resultant."

Instead of the expression "arise out of," such expressions as "supervene on" and "are consequential upon" could have been used. In any case, the idea is that when appropriate lower-level conditions are realized in a higher-level system (that is, the parts that constitute the system come to be configured in a certain relational structure), the system will necessarily exhibit certain higher-level properties, and, moreover, that no higher-level property will appear unless an appropriate set of lower-level conditions is realized. Thus, "arise" and "supervene" are neutral with respect to the emergent/resultant distinction: both emergent and resultant properties of a whole supervene on, or arise out of, its microstructural, or micro-based, properties.

The distinction between properties that are emergent and those that are merely resultant is a central component of emergentism. As we have already seen, it is standard to characterize this distinction in terms of predictability and explainability.

3. The unpredictability of emergent properties Emergent properties are not predictable from exhaustive information concerning their "basal conditions." In contrast, resultant properties are predictable from lower-level information.

4. The unexplainability/irreducibility of emergent properties Emergent properties, unlike those that are merely resultant, are neither explainable nor reducible in terms of their basal conditions.

Earlier in this paper we saw how it is possible to give unity to these claims on the basis of an appropriate model of reduction. More specifically, by identifying emergent properties with irreducible properties, on the functional model of reduction, it is possible to explain why emergent properties are neither explainable nor predictable on the basis of the conditions from which they emerge, whereas nonemergent (or resultant) properties are so explainable and predictable.

Our present concern, however, lies with the question what emergent properties, after having emerged, can *do*—that is, how they are able to make their special contributions to the ongoing processes of the world. It is obviously very important to the emergentists that emergent properties can be active participants in causal processes involving the systems they characterize. None perhaps understood this better than Samuel Alexander, who made the following pointed comment on epiphenomenalism, the doctrine that mental properties are wholly lacking in causal powers:

> [Epiphenomenalism] supposes something to exist in nature which has nothing to do, no purpose to serve, a species of *noblesse* which depends on the work of its inferiors, but is kept for show and might as well, and undoubtedly would in time be abolished.[27]

We may, therefore, set forth the following as the fifth doctrine of emergentism:

5. The causal efficacy of the emergents Emergent properties have causal powers of their own—novel causal powers irreducible to the causal powers of their basal constituents.

In what ways, then, can emergent properties manifest their causal powers?

This of course is where the idea of "downward causation" enters the scene. But when we view the situation with the layered model in mind, we see that the following three types of inter- or intra-level causation must be recognized: (i) *same-level causation*, (ii) *downward causation*, and (iii) *upward causation*. Same-level causation, as the expression suggests, involves causal relations between two properties at the same level—including cases in which an instantiation of one emergent property causes another emergent property to be instantiated. Downward causation occurs when a higher-level property, which may be an emergent property, causes the instantiation of a lower-level property; similarly, upward causation involves the causation of a higher-level property by a lower-level property. I believe that, for the emergentist,[28] there is good reason to believe that downward causation is fundamental and of crucial importance in understanding causation. For it can be shown that both upward and same-level causation (except same-level causation at the ultimate bottom level, if there is such a level and if there are causal relations at this level) presupposes the possibility of downward causation.

Here is an argument that shows why this is so.[29] Suppose that a property $M$, at a certain level $L$, causes another property $M^+$, at level $L+1$. Assume that $M^+$ emerges, or results, from a property $M^*$ at level $L$ ($M^*$ therefore is on the same level as $M$). Now we immediately see a tension in this situation when we ask: "What is responsible for this occurrence of $M^+$? What explains $M^+$'s instantiation on this occasion?" For in this picture there initially are two competing answers: First, $M^+$ is there because, *ex hypothesi*, $M$ caused it; second, $M^+$ is there because its emergence base $M^*$ has been realized. Given its emergence base $M^*$, $M^+$ must of necessity be instantiated, no matter what conditions preceded it; $M^*$ alone suffices to guarantee $M^+$'s occurrence on this occasion, and without $M^*$, or an appropriate alternative base, $M^+$ could not have occurred. This apparently puts $M$'s claim to have caused $M^+$ in jeopardy. I believe that the only coherent description of the situation that respects $M$'s causal claim is this: $M$ causes $M^+$ *by causing its base condition* $M^*$. But $M$'s causation of $M^*$ is an instance of same-level causation. This shows that upward causation entails same-level causation; that is, upward causation is possible only if same-level causation is possible.

As an example, consider this: physical/mechanical work on a piece of marble ($M$) causes the marble to become a beautiful sculpture ($M^+$). But the beauty of the sculpture emerges from the physical properties ($M^*$ consisting in shape, color, texture, size, etc.) of the marble piece. Notice how natural, and seemingly unavoidable, it is to say that the physical work on the marble caused the beauty of the marble piece *by causing it to have the right physical properties*. This of course is an instance of same-level causation. Another example: a bee sting causes a sharp pain. But pain emerges from a certain neural condition $N$ (say, C-fiber excitation). I believe that we want to say, and must say, that the bee sting caused the pain by causing $N$ (the firing of C-fibers). This again is same-level causation.

An exactly similar argument will show that same-level causation presupposes downward causation. Briefly, this can be shown as follows: Suppose $M$ causes $M^*$, where $M$ and $M^*$ are both at level $L$. But $M^*$ itself arises out of a set of properties $M^-$ at level

$L-1$. When we ponder the question how $M^*$ gets to be instantiated on this occasion, again we come to the conclusion that $M$ caused $M^*$ to be instantiated on this occasion *by causing* $M^-$, its base condition, to be instantiated. But $M$'s causation of $M^-$ is downward causation. This completes the argument.

A general principle is implicit in the foregoing considerations, and it is this:

To cause any property (except those at the very bottom level) to be instantiated, you must cause the basal conditions from which it arises (either as an emergent or as a resultant).

We may call this "the principle of downward causation".

## 7.6

Even the early emergentists were explicit on the importance they attached to downward causation, although of course it is unlikely that they were influenced by anything like the argument of the preceding section. The following statement by C. Lloyd Morgan is typical:

> Now what emerges at any given level affords an instance of what I speak of as a new kind of relatedness of which there are no instances at lower levels... But when some new kind of relatedness is supervenient (say at the level of life), *the way in which the physical events which are involved run their course is different in virtue of its presence—different from what it would have been if life had been absent.*[30]

Compare this with what Roger Sperry says over 50 years later:

> ...the conscious subjective properties in our present view are interpreted to have causal potency in regulating the course of brain events; that is, the mental forces or properties exert a regulative control influence in brain physiology.[31]

Both Morgan and Sperry are saying that life and consciousness, emergent properties out of physicochemical and neural properties respectively, have a causal influence on the flow of events at the lower levels, levels from which they emerge. That of course is downward causation.

The appearance of an emergent property signals, for the emergentists, a genuine change, a significant evolutionary step, in the history of the world, and this requires emergent properties to be genuine properties with causal powers. They are supposed to represent novel additions to the ontology of the world, and this could be so only if they bring with them *genuinely new* causal powers; that is, they must be capable of making novel causal contributions that go beyond the causal powers of the lower-level basal conditions from which they emerge.

But how do emergent properties exercise their novel causal powers? How is that possible? According to the argument presented in the preceding section, they can do so only by causally influencing events and phenomena at lower-levels—that is, through downward causation. That was what we called the principle of downward causation. But is downward causation possible? The idea of downward causation has struck some thinkers as incoherent, and it is difficult to deny that there is an air of paradox about it: After all, higher-level properties arise out of lower-level conditions, and without the presence of the latter in suitable configurations, the former could not even be there. So how could these higher-level properties causally influence and alter the conditions from which they arise? Is it coherent to suppose that the presence of *X* is entirely responsible for the occurrence of *Y* (so *Y*'s very existence is totally dependent on *X*) and yet *Y* somehow manages to exercise a causal influence on *X*? I believe a train of thought like this is behind the suspicions surrounding the idea of downward causation. But if downward causation is incoherent, that alone will do serious damage to emergentism. For the principle of downward causation directly implies that if emergent properties have no downward causal powers, they can have no causal powers at all, and this means that emergent phenomena would just turn out to be epiphenomena, a prospect that would have severely distressed Alexander, Morgan, and Sperry.

But we need to analyze whether the kind of intuitive argument in the preceding paragraph against downward causation has any real force. For cases in which higher-level entities and their properties prima facie causally influence lower-level entities and their properties seem legion. The celadon vase on my desk has a mass of 1 kilogram. If it is dropped out the window of my second floor office, it will crash on the paved sidewalk, causing myriads of molecules of all sorts to violently fly away in every which direction. Even before it hits the ground, it will cut a rapid downward swath, causing all sorts of disturbance among the local air molecules. And these effects are surely micro and lower-level in relation to the fall of an object with a mass of 1 kilogram. Note that we cannot think of this case as one in which the "real" causal process occurs at the micro-level, between the micro-constituents of the vase and the air molecules, for the simple reason that no micro-constituents of the vase, in fact no proper part, of my celadon vase has a mass of 1 kilogram. There is no question that the vase, in virtue of having this mass, has a set of causal powers that none of its micro-constituents have; the causal powers that this property represents cannot be reduced to the causal powers of micro-constituents of its bearers. Of course, emergentists would not consider mass an emergent property; they would say that the mass of an object is a resultant property, a property that is merely "additive or subtractive." But this simple example suffices to show that there need not be anything strange or incoherent in the idea of downward causation as such—the idea that complex systems, in virtue of their macrolevel properties, can cause changes at lower microlevels.

However, the idea of downward causation advocated by some emergentists is stronger and more complex than what is suggested by our example. Here again is Sperry:

> The subjective mental phenomena are conceived to influence and govern the flow of nerve impulse traffic by virtue of their encompassing emergent properties. Individual nerve impulses and other excitatory components of a cerebral activity pattern are simply carried along or shunted this way and that by the prevailing overall dynamics of the whole active process (in principle—just as drops of water are carried along by a local eddy in a stream or the way the molecules and atoms of a wheel are carried along when it rolls down hill, regardless of whether the individual molecules and atoms happen to like it or not).[32]

Sperry has used these and other similar analogies elsewhere; in particular, the rolling wheel seems to have been one of his favorites. What is distinctive about this form of downward causation appears to be this: Some activity or event involving a whole $W$ is a cause of, or has a causal influence on, the events involving its *own* microconstituents. We may call this *reflexive downward causation,* to distinguish it from the more mundane *nonreflexive* kind, involved in the example of the falling vase above, in which an event involving a whole causes events involving lower-level entities that are not among its constituents.

But downward causation must be viewed in the context of the doctrine that emergent properties arise out of their basal conditions (claim 2 in section V). For Sperry himself recognizes this in his claim that there is also *upward determination* in this situation. The paragraph quoted above from Sperry continues as follows:

> Obviously, it also works the other way around, that is, the conscious properties of cerebral patterns are directly dependent on the action of the component neural elements. Thus, a mutual interdependence is recognized between the sustaining physico-chemical processes and the enveloping conscious qualities. The neuro-physiology, in other words, controls the mental effects, and the mental properties in turn control the neurophysiology.[33]

After all, an eddy is there because the individual water molecules constituting it are swirling around in a circular motion in a certain way; in fact, an eddy *is nothing but* these water molecules moving in this particular pattern. Take away the water molecules, and you have taken away the eddy: there cannot be a disembodied eddy still swirling around without any water molecules! Thus, reflexive downward causation is combined with upward determination. When each and every molecule in a puddle of water begins to move in an appropriate way—and only then—will there be an eddy of water. But in spite of this, Sperry says, it remains true that the eddy is moving the molecules around "whether they like it or not."

Thus, reflexive downward causation is combined with upward determination. Schematically, the situation looks like this: a whole, $W$, has a certain (emergent) property $M$; $W$ is constituted by parts, $a_1, \ldots, a_n$, and there are properties $P_1, \ldots, P_n$ respectively of $a_1, \ldots, a_n$ and a certain relation $R$ holding for the $a_i$s. The following two claims make explicit what Sperry seems to have in mind (I do not want to rule out other possible interpretations of Sperry):

(i) [Downward causation] $W$'s having property $M$ causes some $a_j$ to have $P_j$; but

(ii) [Upward determination] each $a_i$'s having $P_i$ and $R$ holding for the $a_i$s together determine $W$ to have $M$—that is, $W$'s having $M$ depends wholly on (or is wholly constitued by) the $a_i$s having the $P_i$ respectively and being related by $R$.

The question is whether or not it is possible, or coherent, to hold both (i) and (ii).

## 7.7

As I said, downward causation as such presents us with no special problems; however, what Sperry wants (also there is a hint of this in the quotation from Lloyd Morgan above) is the reflexive variety of downward causation. But how is it possible for the whole to causally affect its constituent parts on which its very existence and nature depend? If causation or determination is transitive, doesn't this ultimately imply a kind of self-causation, or self-determination—an apparent absurdity? It seems to me that there is reason to worry about the coherence of the whole idea.

Let us see if it is possible to make reflective downward causation intelligible. To sharpen the issues we should distinguish two cases:

Case 1 At a certain time $t$, a whole, $W$, has emergent property $M$, where $M$ emerges from the following configuration of conditions: $W$ has a complete decomposition into parts $a_1, \ldots, a_n$; each $a_i$ has property $P_i$; and relation $R$ holds for the sequence $a_i, \ldots, a_n$. For some $a_j$, $W$'s having $M$ at $t$ causes $a_j$ to have $P_j$ at $t$.

Note that the time $t$ is fixed throughout, and both the downward causation and upward emergence (or determination) hold for states or conditions occurring at the very same time. We may, therefore, call this "synchronic reflexive downward causation."[34] A whole has a certain emergent property, $M$, at a given time, $t$, and the fact that this property emerges at $t$ is dependent on its having a certain micro-configuration at $t$, and this includes a given constituent of it, $a_j$, having $P_j$ at $t$. That is, unless $a_j$ had $P_j$ at $t$, $W$ could not have had its emergent property $M$ at $t$. Given this, it makes one feel uncomfortable to be told *also* that $a_j$ is caused to have $P_j$ at that very time, $t$, by the whole's having $M$ at $t$.

But what exactly is the source of this metaphysical discomfort? Why does this picture seem in some way circular and incoherent? Moreover, what is it about causal circularity that makes it unacceptable? One possible explanation, something I find plausible myself, is that we tacitly subscribe to a metaphysical principle like the following:

For an object, $x$, to exercise, at time $t$, the causal/determinative powers it has in virtue of having property $P$, $x$ must *already* possess $P$ at $t$. When $x$ is caused to acquire $P$ at $t$, it does not already possess $P$ at $t$ and is not capable of exercising the causal/determinative powers inherent in $P$.

If a name is wanted, we may call this "the causal-power actuality principle." The reader will have noticed that this principle has been stated in terms of an object "acquiring" property $P$ at a time. In Case 1 above, we said that the whole, $W$, causes one of its proper parts, $a_j$, to "have" $P$. If there is real downward causation, from $W$'s having $M$ to $a_j$'s having $P$, this "having" must be understood as "acquiring." For if $a_j$ already has $P_j$ at $t$, what role can W's having $M$ at $t$ play in causing it to have $P_j$ at $t$? Obviously, none.

In any case, it is now easy to see the incoherence involved in Case 1: the assumption that $W$'s having $M$ at $t$ causes $a_j$ to have $P_j$ at $t$ implies, together with the causal-power actuality principle, that $a_j$ does not already have $P_j$ at $t$. This means, again via the causal-power actuality principle, that $a_j$ cannot, at $t$, exercise the causal/determinative power it has in virtue of having $P_j$, which in turn implies that the assumed emergence base of $W$'s having $M$ at $t$ has vanished and $W$ cannot have $M$ at $t$. Case 1, therefore, collapses.

If you are willing to reject the causal-power actuality principle and live with causal circularity (perhaps even celebrate it in the name of "mutual causal interdependence"), then Case 1 could serve as a model of downward causation for you. Speaking for myself, I think there is a good deal of plausibility in the principle that says that for properties to exercise their causal/determinative powers they must actually be possessed by objects at the time; it cannot be that the objects are in the process of acquiring them at that time. So let's try another model.

Case 2 As before, $W$ has emergent property $M$ at $t$, and $a_j$ has $P_j$ at $t$. We now consider the causal effect of $W$'s having $M$ at $t$ on $a_j$ at a *later time* $t + \Delta t$. Suppose, then, that $W$'s having $M$ at $t$ causes $a_j$ to have $Q$ at $t + \Delta t$.

This, therefore, is a case of *diachronic* reflexive downward causation. It is still reflexive in that a whole causes one of its micro-constituents to change in a certain way. Notice, however, that the mysteriousness of causal reflexivity seems to have vanished. The reason is obvious: the time delay between the putative cause and effect removes the potential circularity, and the causal-power actuality principle does not apply. $W$'s having $M$ at $t$ causes $a_j$ to have $Q$ at $t + \Delta t$. But $a_j$'s having $Q$ at $t + \Delta t$ is not part of the basal conditions out of which $M$ emerges in $W$ at $t$; so there can be no problem of circular reciprocal causation/determination. This becomes particularly clear if we consider the four-dimensional (or "time slice") view of persisting things. On this view, $W$'s having $M$ at $t$ turns out to be $W$ at $t$ having $M$—that is, the time slice of $W$ at $t$ having $M$. Let us use "$[x, t]$" to denote the time slice of $x$ at $t$ (if $t$ is an instant, $[x, t]$ is a temporal cross section). Diachronic downward causation, then, comes to this: $[W, t]$ having $M$ causes $[a_j, t + \Delta t]$ to have $Q$, where, of course, $t < t + \Delta t$. The point to notice is that $[a_j, t + \Delta t]$ is *not* a constituent of $[W, t]$, and this gets rid of the hint of reflexivity present in Case 2.

Examples falling under Case 2 are everywhere. I fall from the ladder and break my arm. I walk to the kitchen for a drink of water and ten seconds later, all my limbs and organs have been displaced from my study to the kitchen. Sperry's bird flies into the blue yonder, and all of the bird's cells and molecules, too, have gone yonder. It doesn't seem to me that these cases present us with any special mysteries rooted in self-reflexivity, or that they show emergent causation to be something special and unique. For consider Sperry's bird: for simplicity, think of the bird's five constituent parts, its head, torso, two wings, and the tail. For the bird to move from point $p_1$ to point $p_2$ *is* for its five parts (together, undetached) to move from $p_1$ to $p_2$. The whole bird is at $p_1$ at $t_1$ and moving in a certain direction, and this causes, let us suppose, its tail to be at $p_2$ at $t_2$. There is nothing mysterious or incoherent about this. The case—the bird's being at $p_1$ at $t_1$ and moving in a certain way—includes its tail's being at $p_1$ at $t_1$ and moving in a certain way. But that's all right: we expect an object's state at a given time to be an important causal factor for its state a short time later. And it is clear that Sperry's other examples, such as the water eddy and the rolling wheel, can be similarly accommodated.

We must conclude then that of the two types of reflexive downward causation, the diachronic variety poses no special problems but perhaps for that reason rather unremarkable as a type of causation, but that the synchronic kind *is* problematic and it is doubtful that it can be given a coherent sense. This may be due to its violation of what I called the causal-power actuality principle, but apart from any recondite metaphysical principle that might be involved, one cannot escape the uneasy feeling that there is something circular and incoherent about this variety of downward causation.

## 7.8

Emergentists like C. Lloyd Morgan will likely point out that the Sperry-style cases do not really involve downward causation by emergent properties, since the motion of the bird as a whole is the same kind of event as the motion of its constituent parts. The properties implicated in causal relations in these cases are one and the same, namely motion, and this shows that these cases simply are not cases of emergent causation, whether downward or upward. (The same will be said about the example of the falling celadon vase.) It would seem, then, that contrary to what Sperry seems to suggest, emergent downward causation should not simply be identified with causation from properties of the whole to properties of its own parts, that is, reflexive downward causation.

One reason that downward causation is thought interesting and important is that mental-to-physical causation is commonly supposed to be a special case of it, the mental occupying a higher emergent level relative to the physical level. So let us turn to mind-body causation. Here again we may consider two varieties, synchronic reflexive downward causation and its diachronic counterpart. Can my experience of pain at a given time causally influence its basal neural process (C-fiber excitation, say) at the very same time? Here we encounter exactly the same difficulties that we saw in Sperry's examples of the water eddy and the like (taken as cases of synchronic downward causation), and I do not believe that classical emergentists, like Alexander, Morgan, and C. D. Broad, would necessarily have insisted on it. Nor do I see why Sperry himself, as an emergentist, should need it; it isn't at all clear that Sperry's overall position on the mind-body relation requires a commitment to this dubious variety of emergent causation.

This leaves diachronic downward causation as the only player on the scene—up to this point, at any rate. One might say that this is all that the emergentists need—the diachronic causal influence of emergent phenomena on lower-level phenomena. But the problem is that even this apparently unproblematic variety of downward causation is beset with difficulties. On my view, the difficulties boil down to a single argument to be sketched below. The critical question that motivates the argument is this: If an emergent, *M*, emerges from basal condition *P*, why can't *P* displace *M* as a cause of any putative effect of *M*? Why can't *P* do all the work in explaining why any alleged effect of *M* occurred?[35] As you may recall, I earlier argued that any upward causation or same-level causation of effect $M^*$ by cause *M* presupposes *M*'s causation of $M^*$'s lower-level base, $P^*$ (it is supposed that $M^*$ is a higher-level property with a lower-level base; $M^*$ may or may not be an emergent property). But if this is a case of downward emergent causation, *M* is a higher-level property, and as such it must have an emergent base, *P*. Now we are faced with *P*'s threat to preempt *M*'s status as a cause of $P^*$ (and hence of $M^*$). For if causation is understood as nomological (law-based) sufficiency, *P*, as *M*'s emergence base, is nomologically sufficient for it, and *M*, as $P^*$'s cause, is nomologically sufficient for $P^*$. Hence, *P* is nomologically sufficient for $P^*$ and hence qualifies as its cause. The same conclusion follows if causation is understood in terms of counterfactuals—roughly, as a condition without which the effect would not have occurred. Moreover, it is not possible to view the situation as involving a *causal chain* from *P* to $P^*$ with *M* as an intermediate causal link. The reason is that the emergence relation from *P* to *M* cannot properly be viewed as causal.[36] This appears to make the emergent property *M* otiose and dispensable as a cause of $P^*$; it seems that we can explain the occurrence of $P^*$ simply in terms of *P*, without invoking *M* at all. If *M* is to be retained as a cause of $P^*$, or of $M^*$, a positive argument has to be provided, and we have yet to see one. In my opinion, this simple argument has not so far been overcome by an effective counter-argument.

If higher-level property $M$ can be reduced to its lower-level base, $M$'s causal status can be restored. As may be recalled from our earlier discussion, however, if $M$ is emergent, this is precisely what cannot be done: emergent properties, by definition, are not reducible to their lower-level bases. The conclusion, therefore, isn't encouraging to emergentists: If emergent properties exist, they are causally, and hence explanatorily, inert and therefore largely useless for the purpose of causal/explanatory theories.

If these considerations are correct, higher-level properties can serve as causes in downward causal relations only if they are reducible to lower-level properties.[37] The paradox is that if they are so reducible, they are not really "higher-level" any longer. If they are reducible to properties at level $L$, they, too, must belong to $L$. Does this make the idea of downward causation useless? Not necessarily. For example, we may try to salvage downward causation by giving it a *conceptual* interpretation. That is, we interpret the hierarchical levels as levels of concepts and descriptions, or levels within our representational apparatus, rather than levels of properties and phenomena in the world. We can then speak of downward causation when a cause is described in terms of higher-level concepts, or in a higher-level language, higher in relation to the concepts in which its effect is represented. On this approach, then, the same cause may be representable in lower-level concepts and languages as well, and a single causal relation would be describable in different languages. The conceptual approach may not save real downward causation, and it brings with it a host of new questions; however, it may be a good enough way of saving *downward causal explanation*, and perhaps that is all we need or should care about.[38]

## Notes

This chapter originally appeared in *Philosophical Studies* 95 (1999), 3–36.

1. For helpful historical surveys of emergentism see Brian McLaughlin, "The Rise and Fall of British Emergentism," and Achim Stephan, "Emergence—A Systematic View on Its Historical Facets," both in *Emergence or Reduction*? ed. A. Beckermann, H. Flohr, and J. Kim (Berlin: De Gruyter, 1993).

2. In *A System of Logic* (1843), Bk. III, ch. vi.

3. It appears that Galen (AD 129–*c.* 200) had a clear statement of the distinction between emergent and nonemergent properties of wholes; see *On the Elements according to Hippocrates*, 1.3, 70.15–74.23. I owe this reference to Victor Caston.

4. See Hempel's "Studies in the Logic of Explanation" (with Paul Oppenheim), reprinted in his *Aspects of Scientific Explanation* (New York: The Free Press, 1965), and Nagel, *The Structure of Science* (New York: Harcourt Brace and World, 1961), ch. 11. It is interesting to note that another early positivist philosopher of science, Karl Popper, became in the final stages of his career a strong defender of emergentism; see John C. Eccles and Karl R. Popper, *The Self and Its Brain* (Berlin and New York: Springer International, 1977).

5. E.g., John Searle, *The Rediscovery of the Mind* (Cambridge: MIT Press, 1992); Francisco Varella, Evan Thompson, and Elearnor Rosch, *The Embodied Mind* (Cambridge: MIT Press, 1993).

6. In particular, "The Myth of Nonreductive Materialism," reprinted in *Supervenience and Mind* (Cambridge: Cambridge University Press, 1993).

7. Some will complain that this picture is inextricably wedded to the now defunct prequantum classical particle physics; that may be, but it is the picture the British emergentists worked with. Moreover, it is an open question, I believe, whether anything of substance would change if the issues were set in a quantum-mechanical framework.

8. Obviously extrinsic/relational/historical properties (e.g., being 50 miles to the south of Boston) must be excluded, and the statement is to be understood to apply only to the intrinsic properties of systems. There is also a tacit assumption that the intrinsic properties of a system determine its causal powers.

9. C. Llyod Morgan, *Emergent Evolution* (London: Williams and Norgate, 1923), pp. 2–3.

10. What I believe to be more appropriate here is William C. Wimsatt's notion of aggregativity defined in terms of certain invariance conditions; see his "Emergence as Non-Aggregativity and the Biases of Reductionisms" (forthcoming in *Natural Contradictions: Perspectives on Ecology and Change*, ed. P. J. Taylor and Jrjo Haila). However, I bypass these considerations here in favor of a simpler and more straightforwrad notion of predictability. See also Paul Humphreys' interesting paper, "How Properties Emerge" (forthcoming in *Philosophy of Science*). I cannot discuss here Humphrey's interesting proposals, but I believe everything of any significance I say here is consistent with his views.

11. Cf. Morgan: "Lewes says that the nature of emergent characters can only be learnt by experience of their occurrence; hence they are unpredictable before the event," *Emergent Evolution*, p. 5.

12. See, e.g., what Michael Tye calls "perspectival subjectivity," in his *Ten Problems of Consciousness* (Cambridge: MIT Press, 1995). And of course the situation here reminds one of Frank Jackson's much discussed case of the blind superneurophysiologist Mary.

13. As noted by McLaughlin in "The Rise and Fall of British Emergentism."

14. For example, David Chalmers, *The Conscious Mind* (New York: Oxford University Press, 1996), p. 43.

15. The fundamental ideas for this view of reduction are present in David Armstrong's *A Materialist Theory of Mind* (New York: Humanities Press, 1964), and David Lewis's "An Argument for the identity Theory," *Journal of Philosophy* 67 (1970): 203–211. However, neither Armstrong nor Lewis, to my knowledge, explicitly associate these ideas directly with models of reduction. The idea of functional analysis of mental terms or properties is of course the heart of the functionalist approach to mentality; it is interesting, therefore, to note that most functionalists have regarded their approach as essentially antireductionist. For similar views on reduction see Robert Van Gulick, "Nonreductive Materialism and the Nature of Intertheoretic Constraint," in *Emergence or Reduction?* ed. A. Beckermann, H. Flohr, and J. Kim; Joseph Levine, "On Leaving Out What It Is Like," in *Consciousness*, ed. Martin Davies and Glyn W. Humphreys (Oxford: Blackwell, 1993).

See also Chalmers' discussion of "reductive explanation," in *The Conscious Mind*, ch. 2. I discuss these issues in greater detail in *Mind in a Physical World* (forthcoming).

16. For brevity we will often speak of a property causing another property—what is meant of course is that an instantiation of a property causes another property to be instantiated.

17. See Lawrence Sklar, *Physics and Chance* (Cambridge: Cambridge University Press, 1993).

18. See Ernest Nagel, *The Structure of Science* (New York: Harcourt Brace and World, 1961).

19. In "Reduction with Autonomy" (forthcoming in *Philosophical Perspectives*, 1997) Lousie Antony and Joseph Levine advance interesting arguments against the disjunctive approach.

20. This point is valid whether or not *E* has single or multiple realizers in the actual world. A property may have a single realizer here but multiple realizers in other worlds, and vice versa.

21. For more details on this approach see my "The Mind-Body Problem: Taking Stock After 40 Years," forthcoming in *Philosophical Perspectives*, 1997, and *Mind in a Physical World* (forthcoming).

22. For more details see my "Multiple Realization and the Metaphysics of Reduction," reprinted in *Supervenience and Mind*.

23. This point is argued by David Chalmers; see his *The Conscious Mind*, p. 129.

24. More details and an overview of the philosophical terrain involved, see Chalmers, *The Conscious Mind*, ch. 3. Two early papers arguing this point are Joseph Levine, "Materialism and Qualia: the Explanatory Gap," *Pacific Philosophical Quarterly* 64 (1983): 354–361, and Frank Jackson, "Epiphenomenal Qualia," *Philosophical Quarterly* 32 (1982): 127–136.

25. See, e.g., Paul Oppenheim and Hilary Putnam, "Unity of Science as a Working Hypothesis," in *Minnesota Studies in Philosophy of Science*, vol. 2, ed. Hervert Feigl, Michael Scriven, and Grover Maxwell (Minneapolis: University of Minnesota Press, 1958). As the title of the paper suggests, Oppenheim and Putnam advocate a strong physical reductionism, a doctrine that is diametrically opposed to emergentism.

26. For an informative discussion of the issues in this area see William C. Wimsatt, "Reductionism, Levels of Organization, and the Mind-Body Problem," in *Consciousness and the Brain*, ed. Gordon G. Globus, Grover Maxwell, and Irwin Savodnik (New York and London: Plenum Press, 1976), and "The Ontology of Complex Systems: Levels of Organization, Perspectives, and Causal Thickets," *Canadian Journal of Philosophy*, Supplementary Volume 20 (1994): 207–274.

27. *Space, Time, and Deity*, vol. 2 (London: Macmillan, 1927), p. 8.

28. As I have argued elsewhere, this holds for certain positions other than emergentism, e.g., the view that higher properties supervene on lower properties, and the view that higher properties are realized by lower properties.

29. I first presented this argument in "'Downward Causation' in Emergentism and Nonreductive Physicalism," in *Emergence or Reduction*?

30. *Emergent Evolution* (London: Williams & Norgate, 1927), pp. 15–16. Emphasis added.

31. "Mental Phenomena as Causal Determinants in Brain Function," in *Consciousness and the Brain*, ed. Globus, Maxwell, and Savodnik, p. 165.

32. "A Modified Concept of Consciousness," *Psychological Review* 76 (1969): 532–536.

33. Ibid.

34. This case, therefore, involves the controversial idea of simultaneous causation (where a cause and its effect occur at the same time). However, this is a general metaphysical issue, and in the present context it will be unproductive to focus on this aspect of the situation.

35. I raised this question earlier in "'Downward Causation' in Emergentism and Nonreductive Physicalism," in *Emergence or Reduction*? The argument can be generalized to the supervenience and realization views of the mind-body relation. For more details see my "The Nonreductivist's Troubles with Mental Causation" (reprinted in *Supervenience and Mind*) and *Mind in a Physical World*. See also Timothy O'Connor, "Emergent Properties," *American Philosophical Quarterly* 31 (1994): 91–104, for an attempt to counter the argument.

36. C. Lloyd Morgan explicitly denies that emergence is a form of causation, in *Emergent Evolution*, p. 28.

37. Here I must enter some caveats. As the reader may recall, I earlier said that there is no special problem of downward causation, citing such examples as my celadon crashing on the pavement of the sidewalk. Cases like this are not the cases of downward causation that most emergentists have in mind, for like Sperry's example of the flying bird they don't seem to involve genuine "higher-level" properties. In general, complex systems obviously can bring new causal powers into the world, powers that cannot be identified with causal powers of more basic, simpler systems. Among them are the causal powers of microstructural, or micro-based, properties of a complex system. Note that these properties are not themselves emergent properties; rather, they form the basal conditions from which further properties emerge (for example, that consciousness is not itself a microstructural property of an organism, although it may emerge from one). If all this sounds too complicated, you could regard the argument in the text to be restricted to consciousness and other standard examples of emergent properties. For further discussion, see my *Mind in a Physical World*.

38. This paper is largely based on the following two papers of mine: "Explanation, Prediction, and Reduction in Emergentism," forthcoming in *Intellectica*, and "Making Sense of Downward Causation," forthcoming in a volume of essays on emergence and downward causation, ed. Peter Boegh Andersen et al.

# 8 Downward Causation and Autonomy in Weak Emergence

Mark A. Bedau

## 8.1 The Problem of Emergence

Emergence is a perennial philosophical problem. Apparent emergent phenomena are quite common, especially in the subjects treated by biology and psychology, but emergent phenomena also seem metaphysically objectionable. Some of these objections can be traced to the autonomy and downward causation that are distinctive of emergent phenomena. Emergence is receiving renewed attention today, in part because the notion repeatedly arises in certain contemporary approaches to understanding complex biological and psychological systems; I have in mind such approaches as neural networks, dynamical systems theory, and agent-based models—what for simplicity I'll call *complexity science*. For anyone interested in understanding emergence, two things about complexity science are striking. First, it aims to explain exactly those natural phenomena that seem to involve emergence; the range of phenomena covered by complexity science are about as broad as the examples of apparent emergence in nature. Second, the models in complexity science are typically described as emergent, so much so that one could fairly call the whole enterprise the science of emergence (e.g., Holland 1998, Kauffman 1995). A good strategy, then, for understanding emergence is to turn to complexity science for guidance. A few years ago I introduced the notion of weak emergence to capture the sort of emergence involved in this scientific work (Bedau 1997). This chapter expands on that project.

There are a variety of notions of emergence, and they are contested. We can provide some order to this controversy by distinguishing two hallmarks of how macro-level emergent phenomena are related to their micro-level bases:

(1) Emergent phenomena are *dependent* on underlying processes.

(2) Emergent phenomena are *autonomous* from underlying processes.

These two hallmarks are vague. There are many ways in which phenomena might be dependent on underlying processes, and there are also many ways in which phenomena might be autonomous from underlying processes. Any way of simultaneously meeting both hallmarks is a candidate notion of emergence. The hallmarks structure and unify these various notions and provide a framework for comparing them.

Taken together, the two hallmarks explain the controversy over emergence, for viewing macro phenomena as both dependent on and autonomous from their micro bases seems metaphysically problematic: inconsistent or illegitimate or unacceptably mysterious. It is like viewing something as both transparent and opaque. The problem of emergence is to explain or explain away this apparent metaphysical unacceptability.

We should not assume that there is just one solution to the problem of emergence. Some philosophers search for the one true account of emergence and for the one correct solution to the problem of emergence, but that is not my goal. For one thing, while the two hallmarks set boundary conditions on notions of emergence, different notions may fit this bill in different ways. So different concepts of emergence might provide different useful perspectives on the problem of emergence. Capturing a distinctive feature of the phenomena explained by complexity science is the utility of my preferred notion of emergence. Furthermore, I doubt that there is a single, specific, useful, pre-theoretical concept of emergence, so traditional conceptual analysis is of questionable value in this context. Defining a metaphysically acceptable and scientifically useful notion of emergence might involve inventing new concepts that revise our view of the world. My project is open to what Peter Strawson termed "revisionary" rather than "descriptive" metaphysics (Strawson 1963).

The problem has two main kinds of solutions. One concludes that emergence has no legitimate place in our understanding of the real world. This strategy construes apparent emergent phenomena as misleading appearances to be explained away. The other strategy treats apparent emergent phenomena as genuine. Success with the second strategy requires explicating a precise notion of emergence, showing that it applies to apparent emergent phenomena, and then explaining away the appearance of problematic metaphysics. I defend a version of this second strategy.

The proper application of the term "emergence" is controversial. Does it apply properly to properties, objects, behavior, phenomena, laws, whole systems, something else? My answer is pluralistic; I think we can apply the term in all these ways and more. Being alive, for example, might be an emergent property, an organism might be an emergent entity, and the mental life of an organism might be an emergent phenomenon. These different subjects of emergence can be related in a straightforward way—for example, an entity with an emergent property is an emergent entity and an emergent phenomenon involves an emergent entity possessing an emergent property—and they all can be traced back to the notion of an emergent property. So I will first explain the notion of an emergent property, and then extend the notion of emergence to other contexts. This will allow me to talk of emergent properties, entities, phenomena, etc., as the context suggests.

Before explaining my preferred notion of emergence, I will sketch a broader canvas containing different kinds of emergence. Then I will explain my notion of weak emergence and illustrate it with cellular automata—a typical kind of system studied in complexity science.[1] Finally, I will examine downward causation and autonomy in the context of weak emergence—two connected problems that tend to pull in opposite directions. In the end we will see that weak emergence avoids the problems of downward causation, and that a certain kind of robust weak emergence has an interesting metaphysical autonomy. I conclude that this robust weak emergence is philosophically acceptable and scientifically illuminating; it is all the emergence to which we are now entitled.

## 8.2 Three Kinds of Emergence

It is useful to distinguish three kinds of emergence: nominal, weak, and strong.[2] These are not narrow definitions but broad conceptions each of which contains many different instances. My classification is not exhaustive. It ignores some views about emergence, such as the view that attributes emergence on the subjective basis of observer surprise (Ronald et al. 1999). The classification's utility is that it captures three main objectivist approaches to emergence. Emphasizing the underlying similarities within each view and the differences between the contrasting views highlights the strengths and weaknesses of the view that I will defend.

The classification of kinds of emergence assumes a distinction between a micro level and a macro level, and the issue is to specify what it is for the macro to emerge from the micro. We might be interested in how an individual cell in an organism emerges out of various biomolecules and their chemical interactions, or we might be interested in how an organism emerges out of various cells and their biological interactions. As this example shows, a macro level in one context might be a micro level in another; the macro/micro distinction is context dependent and shifts with our interests. In addition, a nested hierarchy of successively greater macro levels gives rise to multiple levels of emergence. Any final theory of emergence must clarify what such levels are and how they are related.

Macro entities and micro entities each have various kinds of properties. Some of the kinds of properties that characterize a macro entity can also apply to its micro constituents; others cannot. For example, consider micelles. These are clusters of amphiphillic polymers arranged in such a way that the polymers' hydro-phillic ends are on the outside and their hydro-phobic tails are on the inside. Those polymers are themselves composed out of hydro-phyllic and -phobic monomers. In this context, the micelles are macro objects, while the individual monomeric molecules are micro objects.[3] The micelles and the monomers both have certain kinds of physical properties in common (having a location, mass, etc.). By contrast, some of the properties of micelles (such as their permeability) are the kind of properties that monomers simply cannot possess. Here is another example: The constituent molecules in a cup of water, considered individually, cannot have properties like fluidity or transparency, though these properties do apply to the whole cup of water.

This contrast illustrates a core component of all three kinds of emergence: the notion of a kind of property that can be possessed by macro objects but cannot be possessed by micro objects. The simplest and barest notion of an emergent property, which I term mere *nominal emergence*, is simply this notion of a macro property that is the kind of property that cannot be a micro property. Nominal emergence has been emphasized by Harré (1985) and Baas (1994), among others. It should be noted that the notion of nominal emergence does not *explain* which properties apply to wholes and not to their parts. Rather, it *assumes* we can already identify those properties, and it simply terms them nominally emergent. Full understanding of nominal emergence would require a general theory of when macro entities have a new kind of property that their constituents cannot have.

Nominal emergence easily explains the two hallmarks of emergence. Macro-level emergent phenomena are dependent on micro-level phenomena in the straightforward sense that wholes are dependent on their constituents; and emergent phenomena are autonomous from underlying phenomena in the straightforward sense that emergent properties do not apply to the underlying entities. When dependence and autonomy are understood in these ways, there is no problem in seeing how emergent phenomena could simultaneously be both dependent on and autonomous from their underlying bases.

The notion of nominal emergence is very broad. It applies to a large number of intuitive examples of emergent phenomena and corresponds to the compelling picture of reality consisting of a hierarchy of levels. Its breadth is its greatest weakness, though, for it applies to all macro-level properties that are not possessed by micro-level entities. Macro-properties are traditionally classified into two kinds: genuine emergent properties and mere "resultant" properties, where resultant properties are those that can be predicted and explained from the properties of the components. For example, a circle consists of a collection of points, and the individual points have no shape. So being a circle is a property of a "whole" but not its constituent "parts"—that is, it is a nominal emergent property. However, if you know that all the points in a geometrical figure are equidistant from a given point, then you can derive that the figure is a circle. So being a circle is a resultant property. To distinguish emergent from resultant properties one must turn to more restricted kinds of emergence. The two more restricted kinds of emergence simply add further conditions to nominal emergence.[4]

The most stringent conception of emergence, which I call *strong emergence*, adds the requirement that emergent properties are supervenient properties with irreducible causal powers.[5] These macro-causal powers have effects at both macro and micro levels, and macro-to-micro effects are termed "downward" causation. We saw above that micro determination of the macro is one of the hallmarks of emergence, and supervenience is a popular contemporary interpretation of this determination. Supervenience explains the sense in which emergent properties depend on their underlying bases, and irreducible macro-causal power explains the sense in which they are autonomous from their underlying bases. These irreducible causal powers give emergent properties the dramatic form of ontological novelty that many people associate with the most puzzling kinds of emergent phenomena, such as qualia and consciousness. In fact, most of the contemporary interest in strong emergence (e.g., O'Connor 1994, Kim 1992, 1997, 1999, Chalmers 1996) arises out of concerns to account for those aspects of mental life like the qualitative aspects of consciousness that most resist reductionistic analysis.

The supervenient causal powers that characterize strong emergence are the source of its most pressing problems. One problem is the so-called "exclusion" argument emphasized by Kim (1992, 1997, 1999). This is the worry that emergent macro-causal powers would compete with micro-causal powers for causal influence over micro events, and that the more fundamental micro-causal powers would always win this competition. I will examine downward emergent causation at length later in this chapter. The exclusion argument aside, the very notion of strong emergent causal powers is problematic to some people. By definition, such causal powers cannot be explained in terms of the aggregation of the micro-level potentialities; they are primitive or "brute" natural powers that arise inexplicably with the existence of certain macro-level entities. This contravenes *causal fundamentalism*—the idea that macro causal powers supervene on and are determined by micro causal powers, that is, the doctrine that "the macro is the way it is in virtue of how things are at the micro" (Jackson and Pettit 1992, p. 5). Many naturalistically inclined philosophers (e.g., Jackson and Pettit) find causal fundamentalism compelling, so they would accordingly be skeptical about any form of emergence that contravenes causal fundamentalism. Still, causal fundamentalism is not a necessary truth, and strong emergence should be embraced if it has compelling enough supporting evidence. But this is where the final problem with strong emergence arises. All the evidence today suggests that strong emergence is scientifically irrelevant. Virtually all attempts to provide scientific evidence for strong emergence focus on one isolated moribund example: Sperry's explanation of consciousness from over thirty years ago (e.g., Sperry 1969). There is no evidence that strong emergence plays any role in contemporary science. The scientific irrelevance of strong emergence is easy to understand, given that strong emergent causal powers must be brute natural phenomena. Even if there were such causal powers, they could at best play a primitive role in science. Strong emergence starts where scientific explanation ends.

Poised between nominal and strong emergence is an intermediate notion, which I call *weak emergence*.[6] It involves more than mere nominal emergence but less than strong emergence. Something could fail to exhibit weak emergence in two different ways: either by being merely resultant or by being strongly emergent. Weak emergence refers to the aggregate global behavior of certain systems. The system's global behavior derives just from the operation of micro-level processes, but the micro-level interactions are interwoven in such a complicated network that the global behavior has no simple explanation. The central idea behind weak emergence is that emergent causal powers can be derived from micro-level information but only in a certain complex way. As Herbert Simon puts it, "given the properties of the parts and the laws of their interaction, it is not a trivial matter to infer the properties of the whole" (1996, p. 184). In contrast with strong emergence, weak emergent causal powers can be explained from the causal powers of micro-level components; so weak and strong emergence are mutually exclusive. In contrast with mere nominal emergence, those explanations must be of a certain complicated sort; if the explanation too simple, the properties will be merely resultant rather than weakly emergent. Weak emergence is a proper subset of nominal emergence, and there are different specifications of the special conditions involved (e.g., Wimsatt 1986, 1997, Newman 1996, Bedau 1997, Rueger 2000).

The strengths and weaknesses of weak emergence are both due to the fact that weak emergent phenomena can be derived from full knowledge of the micro facts. Weak emergence attributes the apparent underivability of emergent phenomena to the complex consequences of myriad non-linear and context-dependent micro-level interactions. These are exactly the kind of micro-level interactions at work in natural systems that exhibit apparent emergent phenomena, so weak emergence has a natural explanation for these apparent emergent phenomena. Weak emergence also has a simple explanation for the two hallmarks of emergence. Weakly emergent macro phenomena clearly depend on their underlying micro phenomena. So weak emergent phenomena are *ontologically* dependent on and reducible to micro phenomena; their existence consists in nothing more than the coordinated existence of certain micro phenomena. Furthermore, weakly emergent causal powers can be explained by means of the composition of context-dependent micro causal powers. So weakly emergent phenomena are also *causally* dependent on and reducible to their underlying phenomena; weak emergence presumes causal fundamentalism. (More on this below.) At the same time, weakly emergent macro phenomena are autonomous in the sense that they can be derived only in a certain non-trivial way. In other words, they have *explanatory* autonomy and irreducibility, due to the complex way in which the iteration and aggregation of context-dependent micro interactions generate the macro phenomena. (Section 8.6 develops the ramifications of distinguishing two forms of this explanatory autonomy.) There is nothing metaphysically illegitimate about combining this explanatory autonomy (irreducibility) with ontological and causal dependence (reducibility), so weak emergence dissolves the problem of emergence.

Some apparent emergent macro phenomena like consciousness still resist micro explanation, even in principle. This might reflect just our ignorance, but another possibility is that these phenomena are strongly emergent. The scope of weak emergence is limited to what has a micro-level derivation (of a certain complex sort). So those who hope that emergence will account for irreducible phenomena will find weak emergence unsatisfying.

My project in this chapter is to develop and defend a version of weak emergence that is ubiquitous in complexity science. My main aim is to explain how it avoids the problems of downward causation and how it can involve metaphysical autonomy. My arguments may generalize (with some modifications) to other versions of weak emergence, but I will not explore those generalizations here because I think my preferred notion of weak emergence has the greatest general utility in understanding emergence in nature.

## 8.3 Weak Emergence as Underivability Except by Simulation

For ease of exposition, I will first explain weak emergence in a certain simple context and then extend it more broadly. Assume that some system has micro and macro entities. Assume also that all the macro entities consist of nothing more than appropriate kinds of micro entities appropriately configured and arranged. (The micro entities might be constituted by entities at a yet lower level, but we can ignore that here.) All of the ultimate constituents of any macro entity are simply micro entities; macro entities are ontologically dependent on and reducible to micro entities. The system's micro and macro entities have various kinds of properties. Some of the macro properties might be nominally emergent, i.e., not the kind of property found at the micro level. Nevertheless, we assume that all the macro properties are structural properties, in the sense that they are constituted by micro entities possessing appropriate micro-level properties. That is, a macro entity has a macro property only in so far as its constituent micro entities have an appropriate structure (are appropriately related to each other) and have the appropriate micro properties. The state of a micro entity consists of its location and its possession of intrinsic properties, and its state changes if these change. A macro entity also has a state, and this consists simply in the aggregation of the states of all its component micro entities and their spatial relations. The fundamental micro-level causal dynamics of the system—its "physics"—is captured in a set of explicit rules for how the state of a micro entity changes as a function of its current state and the current states of its local neighboring entities. Macro entities and their states are wholly constituted by the states and locations of their constituent micro entities, so the causal dynamics involving macro objects is wholly determined by the underlying micro dynamics. Thus, causal fundamentalism reigns in such a system; macro causal powers are wholly constituted and determined by micro causal powers. The micro dynamics is context sensitive since a micro entity's state depends on the states of its micro-level neighbors. The context sensitivity of the system's underlying causal dynamics entails that understanding how a micro entity behaves in isolation or in certain simple contexts does not enable one to understand how that entity will behave in all contexts, especially those that are more complicated. *Locally reducible* systems are those that meet all the conditions spelled out in this paragraph.

The notion of weak emergence concerns the way in which a system's micro facts determine its macro facts. A system's micro facts at a given time consist of its micro dynamics and the states and locations of all its micro elements at that time. If the system is open, then its micro facts include the flux of micro entities that enter or leave the system at that time. Its micro facts also include the micro-level accidents at that time, if the system's micro dynamics is nondeterministic. Since causal fundamentalism applies to locally reducible systems, the micro facts in such systems determine the system's subsequent evolution at all levels. Given all the system's micro facts, an explicit simulation could step through the changes of state and location of each micro element in the system, mirroring the system's micro-level causal dynamics. Since macro entities and states are constituted by the locations and states of their constituent micro entities, this explicit simulation would reflect the evolution over time of the system's macro facts. Such an explicit simulation amounts to a special kind of *derivation* of the system's macro properties from its micro facts. It is an especially "long-winded" derivation because it mirrors each individual step in the system's micro-level causal dynamics. A locally reducible system's macro properties are always derivable from the micro facts by a simulation. However, in some situations it is possible to construct a quite different "short-cut" derivation of a system's macro properties, perhaps using a simple mathematical formula for the evolution of certain macro properties arbitrarily far into the future. Such short-cut derivations are the bread and butter of conventional scientific explanations. They reveal the future behavior of a system without explicitly simulating it.

It is now easy to define weak emergence. Assume that *P* is a nominally emergent property possessed by some locally reducible system *S*. Then *P* is weakly emergent if and only if *P* is derivable from all of *S*'s micro facts but only by simulation. Weak emergence also applies to systems that are not locally reducible, when they contain locally reducible subsystems that exhibit weak emergence.[7] Notice that the notion of weak emergence is relative to a choice of macro and micro levels. A macro property could be weakly emergent with respect to one micro level but not with respect to another (although in my experience this is just an abstract possibility). It is usually obvious which levels are appropriate to choose in each context, so I will usually leave this implicit.

My goal here is not a complete account of weak emergence but just an analysis of some paradigmatic cases. It is natural to extend in various ways the core notion of an emergent property exhibited by a system given complete micro facts. Note that the core definition allows a given property to be weakly emergent in one context with one set of micro facts, but not weakly emergent in another context with different micro facts. Abstracting away from any particular context, one could define the notion of an emergent property in a system as a kind of property that is emergent in that system in some context. It is natural to think of certain macro objects or entities as emergent, and the natural way to define these is as objects with some weak emergent property.[8] A weak emergent phenomenon can be defined as a phenomenon that involves emergent properties or objects, and a weak emergent system can be defined as one that exhibits some weak emergent phenomenon, object, or property. A weak emergent law could be defined as a law about weak emergent systems, phenomena, objects, or properties. The notion of weak emergence can be extended into further contexts along similar lines.

I have been speaking of underivability except by simulation as if there were a sharp dividing line separating weak emergent properties from merely resultant properties, but this is an oversimplification (Assad and Packard 1992). One can define various sharp distinctions involving underivability except by simulation, but focusing on one to the exclusion of the others is somewhat arbitrary. The underlying truth is that properties come in various degrees of derivability without simulation, so there is a spectrum of more or less weak emergence. A core concept of weak emergence concerns properties that in principle are underivable except by finite feasible simulation. A slightly weaker notion of emergence concerns properties that in principle are derivable without simulation, but in practice must be simulated. A slightly stronger notion of emergence concerns properties that are underivable except by simulation, but the requisite simulation is unfeasible or infinite. A variety of even weaker and stronger notions also exist. Nevertheless, the paradigm concept along this scale is weak emergence as defined above.

It is important to recognize that my notion of weak emergence concerns how something *can* be derived, not whether it *has* been derived. It concerns which derivations exist (in the Platonic sense), not which have been discovered. Perhaps nobody has ever worked through a short-cut derivation of some macro property. Nevertheless, if there is such a derivation, then the macro property is not weakly emergent. If a genius like Newton discovers a new short-cut derivation for macro properties in a certain class of system, this changes what properties we *think* are weakly emergent but not which properties *are* weakly emergent. Notice also that weak emergence does not concern some human psychological or logical frailty. It is not that human minds lack the power to work through simulations without the aid of a computer. Nor is it that available computing power is too limited (e.g., detailed simulations of the world's weather are beyond the capacity of current hardware). Rather, it involves the formal limitations of any possible derivation performed by any possible device or entity. To dramatize this point, consider a Laplacian supercalculator that could flawlessly perform calculations many orders of magnitude faster than any human. Such a supercalculator would be free from any anthropocentric or hardware-centered limitation in reasoning speed or accuracy. Nevertheless, it could not derive weakly emergent properties except by simulation. The Laplacian supercalculator's derivations of weak emergence might look instantaneous to us, but their logical form would be just like the logical forms of our derivations. Each derivation iterates step by step through the aggregation of local interactions among the micro elements.

The phrase "derivation by simulation" might seem to suggest that weak emergence applies only to what we normally think of as simulations, but this is a mistake. Weak emergence also applies directly to natural systems, whether or not anyone constructs a model or simulation of them. A derivation by simulation involves the temporal iteration of the spatial aggregation of local causal interactions among micro elements. That is, it involves the local causal processes by which micro interactions give rise to macro phenomena. The notion clearly applies to natural systems as well as computer models. So-called "agent-based" or "individual-based" or "bottom-up" simulations in complexity science have exactly this form.[9] They explicitly represent micro interactions, with the aim of seeing what implicit macro phenomena are produced when the micro interactions are aggregated over space and iterated over time. My phrase "derivation by simulation" is a technical expression that refers to temporal iteration of the spatial aggregation of such local micro interactions. We could perhaps use the phrase "derivation by iteration and aggregation," but that would be cumbersome. Since "simulation" is coming to mean exactly this kind of process (Rasmussen and Barrett 1995), I adopt the more economical phrase "derivation by simulation." Derivation by simulation is the process by which causal influence typically propagates in nature. Macro processes in nature are caused by the iteration and aggregation of micro causal interactions. The iteration and aggregation of local causal interactions that generate natural phenomena can be viewed as a computation (Wolfram 1994), just like the causal processes inside a computer. These intrinsic natural computations are a special case of derivation by simulation. Natural systems compute their future behavior by aggregating the relevant local causal interactions and iterating these effects in real time. They "simulate" themselves, in a trivial sense. Thus, derivation by simulation and weak emergence apply to natural systems just as they apply to computer models.

The behavior of weakly emergent systems cannot be determined by any computation that is essentially simpler than the intrinsic natural computational process by which the system's behavior is generated. Wolfram (1994) terms these systems "computationally irreducible." The point can also be expressed using Chaitin's (1966, 1975) notion of algorithmic complexity and randomness: roughly, the macro is random with respect to the micro, in the sense that there is no derivation of the macro from the micro that is shorter than an explicit simulation. Computational irreducibility—that is, weak emergence—is characteristic of complex systems and it explains why computer simulations are a necessary tool in their study.

## 8.4 Weak Emergence and Reduction in Cellular Automata

Some examples can make the ideas of weak emergence and derivation by simulation more concrete. The examples also illustrate the sort of systems studied in complexity science.[10] One advantage of such systems is that we have exact and total knowledge of the fundamental laws govern the behavior of the micro elements. The examples are all cellular automata, consisting of a two-dimensional lattice of cells, like an infinitely large checker board. Each cell can be in either of two states, which we'll refer to as being alive and being dead. (You can think of them equivalently as being in state 0 and 1, or black and white.)

Time moves forward in discrete steps. The state of each cell at a given time is a simple function of its own state and the states of its eight neighboring cells at the previous moment in time; this rule is called the system's "update function." Assume that one of these systems is started with some initial configuration of living and dead cells. (These initial states could be chosen by somebody or determined randomly.) The next state of each cell is completely determined by its previous state and the previous state of its neighbors, according to the update function. Notice that causal fundamentalism holds in cellular automata. The only primitive causal interactions in the system are the interactions between neighboring cells, as specified by the system's update function. If there are any higher-level causal interactions in the system, they all can be explained ultimately by the interactions among the system's elementary particles: the individual cells.

The only difference between the cellular automata that we will consider is their update functions. The first updates the state of each cell as follows:

All Life A cell is alive at a given time whether or not it or any of its neighbors were alive or dead at the previous moment.

My name for this update function should be obvious, and so should its behavior. No matter what configuration of living and dead cells the system has initially, at the next moment and for every subsequent moment every cell in the system is alive. Given this update function, it is a trivial matter to derive the behavior of any individual cell or clump of cells in the system at any point in the future. All regions at all times in the future consist simply of living cells.

Part of what makes the All Life rule so trivial is that a cell's state does not make a difference to its subsequent state. Living and dead cells alike all become alive. The second update rule is slightly more complicated, as follows:

Spreading Life A dead cell becomes alive if and only if at least one of its neighbors were alive at the previous moment; once a cell becomes alive it remains alive.

The behavior of this system is also quite trivial to derive, and its name reflects this behavior. Life spreads at the speed of light (one cell per moment of time) in all directions from any living cell. Once a dead cell is touched by a living cell, it becomes alive and then remains alive forever after. Life spreads from a single living cell in a steadily growing square. If the initial configuration contains a random sprinkling of living cells, a square of life spreads from each at the speed of light. Eventually these spreading squares overlap to form a connected shape growing at the speed of light.

The third system we will consider is the most famous of all cellular automaton: the so-called "Game of Life" devised in the 1960s by John Conway (Berlekamp et al. 1982; see also Gardner 1983 and Poundstone 1985). It has the following update rule:

Game of Life A living cell remains alive if and only if either two or three of its neighbors were alive at the previous moment; a dead cell becomes alive if and only if exactly three of its neighbors was alive at the previous moment.

The Game of Life's update rule is more complicated than the rules for All Life and Spreading Life, but it is still quite simple. It is easy to calculate the subsequent behavior of many initial configurations. For example, an initial configuration consisting of a single living cell will turn to all dead cells after one tick of the clock, and it will remain that way forever. Or consider a $2 \times 2$ block of living cells. Each of the cells in this initial configuration has three living neighbors, so it remains alive. Each of the dead cells that border the block has at most two living neighbors, so it remains dead. Thus the $2 \times 2$ block alone remains unchanging forever—an example of what is called a "still life" in the Game of Life. Another interesting configuration is a vertical strip of living cells three cells long and one cell wide. The top and bottom cells in this strip die at the first clock tick, since each has only one living neighbor. The middle cell remains alive, since it has two living neighbors. But this is not all. The two dead cells adjacent to the middle cell have three living neighbors—the three cells in the strip—so they each become alive. Thus, after one clock tick, there is a horizontal strip of living cells, three cells long and one cell wide. By parity of reasoning, one more clock tick turns this configuration back into the original vertical strip. Thus, with each clock tick this configuration changes back and forth between vertical and horizontal $3 \times 1$ strips—an example of what is called a "blinker" in the Game of Life.

Still lives and blinkers do not begin to exhaust the possibilities. One particular configuration of five living cells changes back into the same pattern in four clock ticks, except that the pattern is shifted one cell along the diagonal. Thus, over time, this pattern glides across the lattice of cells at one quarter the speed of light, moving forever in a straight line along the diagonal—an example of a "glider." Other gliding patterns leave various configurations of living cells in their wake—these are called "puffers." Other configurations periodically spawn a new glider—these are called "glider guns." Still other configurations will annihilate any glider that hits them—these are called "eaters." Gliders moving at ninety degrees to each other sometimes collide, with various kinds of outcome, including mutual annihilation or production of a new glider.

Streams of gliders can be interpreted as signals bearing digital information, and clusters of glider guns, eaters, and other configurations can function in concert just like AND, OR, NOT, and other logic switching gates. These gates can be connnected into circuits that process information and perform calculations. In fact, Conway proved that these gates can even be cunningly arranged so that they constitute a universal Turing machine (Berlekamp et al. 1982). Hence, the Game of Life can be configured in such a way that it can be interpreted as computing literally any possible algorithm operating on any possible input. As Poundstone vividly puts it, the Game of Life can "model every precisely definable aspect of the real world" (Poundstone 1985, p. 25).

For our present purposes, the most important respect in which the Game of Life differs from All Life and Spreading Life is that many properties in the Game of Life are weakly emergent. For example, consider the macro property of indefinite growth (i.e., increase number of living cells). Some initial configurations exhibit indefinite growth, and others do not. Any configuration consisting only of still lifes and blinkers will not exhibit indefinite growth. By contrast, a configuration consisting of a glider gun will exhibit indefinite growth, since it will periodically increase the number of living cells by five as it spawns new gliders. Other configurations are more difficult to assess. The so-called R pentomino—a certain five-cell pattern that resembles the shape of the letter R—exhibits wildly unstable behavior. Poundstone (1985, p. 33) describes its behavior this way: "One configuration leads to another and another and another, each different from all of its predecessors. On a high-speed computer display, the R pentomino roils furiously. It expands, scattering debris over the Life place and ejecting gliders." Now, does the R pentomino exhibit indefinite growth? If the R pentomino continually ejects gliders that remain undisturbed as they travel into the infinite distance, for example, then it would grow forever. But does it? The only way to answer this quetion is let the Game of Life "play" itself out with the R pentomino as initial condition. That is, one has no option but to observe the R pentomino's behavior. As it happends, after 1103 time steps the R pentomino settles down to a stable state consisting of still lifes and blinkers that just fits into a 51-by-109 cell region. Thus, the halt to the growth of the R pentomino is a weakly emergent macro state in the Game of Life.

By contrast, the behavior of any initial configuration in both All Life and Spreading Life are trivial to derive. There is no need to observe the behavior of All Life and Spreading Life to determine whether the R pentomino in those cellular automata exhibits indefinite growth, for example. The same holds for any other macro-property in All Life and Spreading Life. They exhibit no weakly emergent behavior.

It is noteworthy how much of the interesting behavior of the Game of Life depend on the precise details of its cellular birth-death rule. To get a feel for this, consider the time evolution of the Game of Life given a randomly generated initial condition (see figure 8.1). As time progresses we see the pattern slowly growing, with still lifes and blinkers and gliders appearing interspersed between a number of piles of rapidly

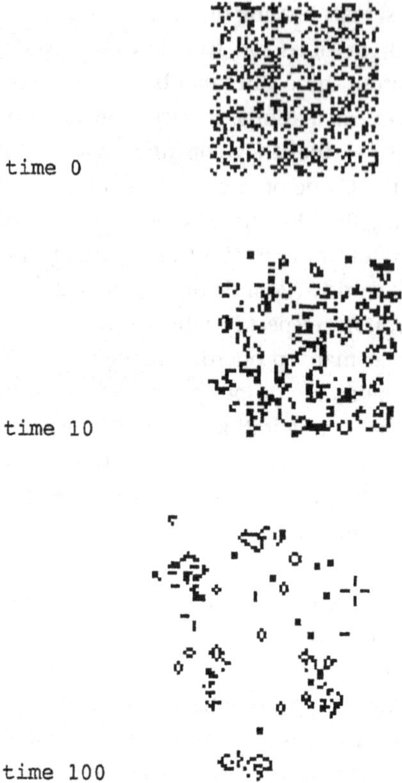


**Figure 8.1**
Time evolution of the Game of Life starting from a 50 × 50 random initial condition in which 30 percent of the cells are alive. By time 10 the pattern has started to thin. By time 100 a number of still lifes and blinkers are evident and a glider leaving from the upper left, but the pattern also contains many randomly structured piles of "muck." By time 300 the pattern has grown slightly, having spawned another glider (right side), preserved some still lifes and blinkers while creating and destroying others, and continuing to roil in two large unstructured piles of "muck." This pattern continues to grow slowly until after over 700 time steps it becomes stable, consisting only of still lifes, blinkers, and a finite number of gliders (out of the picture) that continue to move forever farther into the distance.

time 300

time 700

**Figure 8.1**
(continued)

and irregularly changing and unpredictably spreading and contracting "muck." In this particular case, these piles of muck eventually dissipate and after many hundreds of time steps the pattern stabilizes with over sixty still lifes and blinkers spread out over a region about three times the size of the initial random pattern and a few gliders wiggling off to infinity.

But if we make a minor change in the birth-death rule, the resulting system's behavior changes completely. For example, consider what happens to exactly the same random initial condition if survival is a little harder. (I will adopt the convention of naming an update function with the number of neighbors required to give birth to a new living cell followed by the number of neighbors required to keep a living cell alive.)

3-3 Life A dead cell becomes alive if and only if exactly three of its neighbors were alive at the previous moment; a living cell remains alive if and only if exactly three of its neighbors were alive at the previous moment.

Figure 8.2 shows the behavior of 3-3 Life given the same random initial condition displayed in figure 8.1. In stark contrast to the behavior of the Game of Life in figure 8.1, 3-3 Life quickly reduces this pattern to a small stable configuration of still lifes and blinkers. The interesting thing about 3-3 Life is that all other initial conditions exhibit the same kind of behavior; they all quickly reduce to a small stable pattern consisting of at most some still lifes and blinkers. This collapse to a few isolated periodic subpatterns is a universal generalization about 3-3 Life's global behavior. This is a general macro-level law about 3-3 Life, somewhat analogous to the second law of thermodynamics for our world.

Now, consider a different minimal change of the Game of Life's update function, one that makes birth a little easier but survival a little harder.

2-2 Life A dead cell becomes alive if and only if exactly two of its neighbors were alive at the previous moment; a living cell remains alive if and only if exactly two of its neighbors were alive at the previous moment.

This cellular automaton exhibits a completely different kind of behavior from both the Game of Life and 3-3 Life. A typical example of its behavior is shown in figure 8.3, in which 3-3 Life was started with the same random initial condition used in figures 8.1 and 8.2. In this case, though, a random "slime" of living and dead cells steadily grows and eventually spreads over the entire world. This slime is a continually changing chaotic mixture of living and dead cells. Similar indefinitely growing random slimes proliferate from virtually all other initial configuration in 2-2 Life.[11] This spreading chaos is a general macro-level law about 2-2 Life.

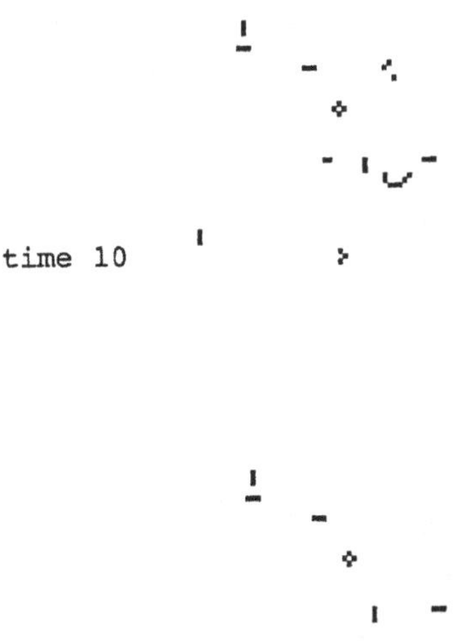


time 14

**Figure 8.2**
Typical time evolution of 3-3 Life, a near cousin of the Game of Life in which a cell survives just in case exactly three of its nearest neighbors were just alive. 3-3 Life was started from exactly the same random initial condition used in figure 8.1. By time 14 the world has collapsed to a small stable configuration with six blinkers and one still life. All other initial conditions in 3-3 Life quickly collapse to equally simple and small stable patterns.

A little experimentation is all it takes to confirm the typical behavior of 3-3 Life and 2-2 Life: collapse to isolated periodicity and spreading chaos. From a statistical point of view, their global behavior is very easy to predict. Changing the state of a cell here or there in an initial condition makes no difference to the quality of their global behavior; indeed, neither does *drastically* changing the initial configuration. (Figure 8.4 shows random "slime" growing from different initial conditions in 2-2 Life.) By contrast, the Game of Life has no typical global behavior. Some configurations quickly collapse into stable periodic patterns. Other very similar configurations continue to change indefinitely. Changing the state of one cell can completely change the system's global behavior. Neither 3-3 Life nor 2-2 Life has the exquisite sensitivity and balance of order and disorder that allows the Game of Life to exhibit complex macro-level patterns such as switching gates, logic circuits, or universal computers.

Nevertheless, all three cellular automata exhibit weak emergence. This is easily recognizable from the fact that their *exact* global behavior (whether statistically predictable or not) can be derived only by simulation—iterating through time the aggregate local effect of the update function across all cells. Sufficient experience with 3-3 Life and 2-2 Life provides empirical evidence for macro-level laws about the *kind* of behavior they

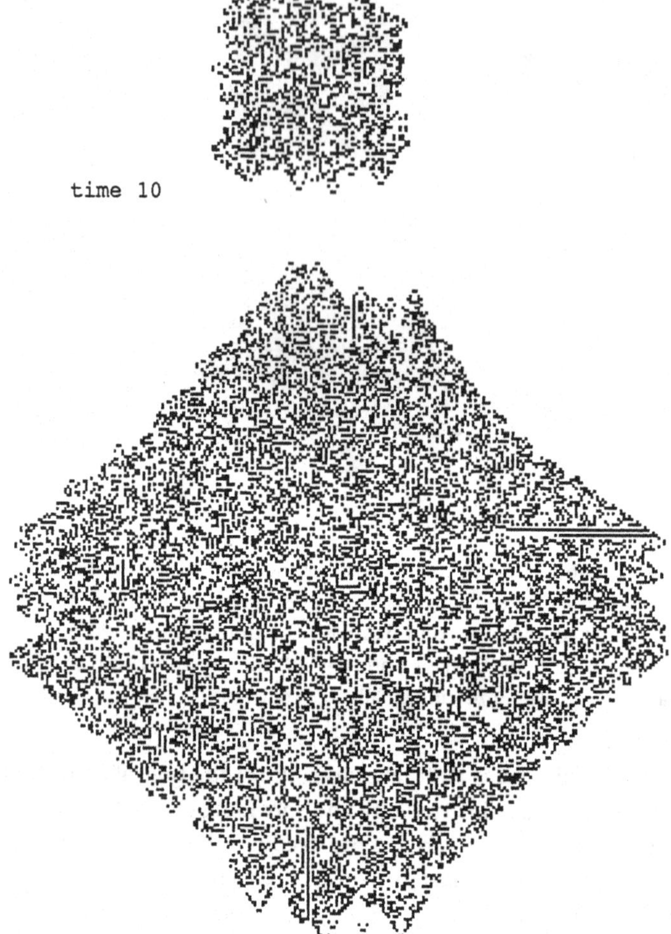


**Figure 8.3**
Typical time evolution of 2-2 Life, another near cousin of the Game of Life, which differs only in that a dead cell becomes alive just in case exactly two (not three) of its nearest neighbors were just alive. In this case, 2-2 Life was started with exactly the same random initial condition in figures 8.1 and 8.2. By time 10 a random "slime" pattern of cells can be seen growing. By time 100 the random slime has increased in size by more than a factor of four. This random slime pattern will continue to grow indefinitely. Similar random slime patterns in 2-2 Life grow from virtually all other initial conditions (see figure 8.4).

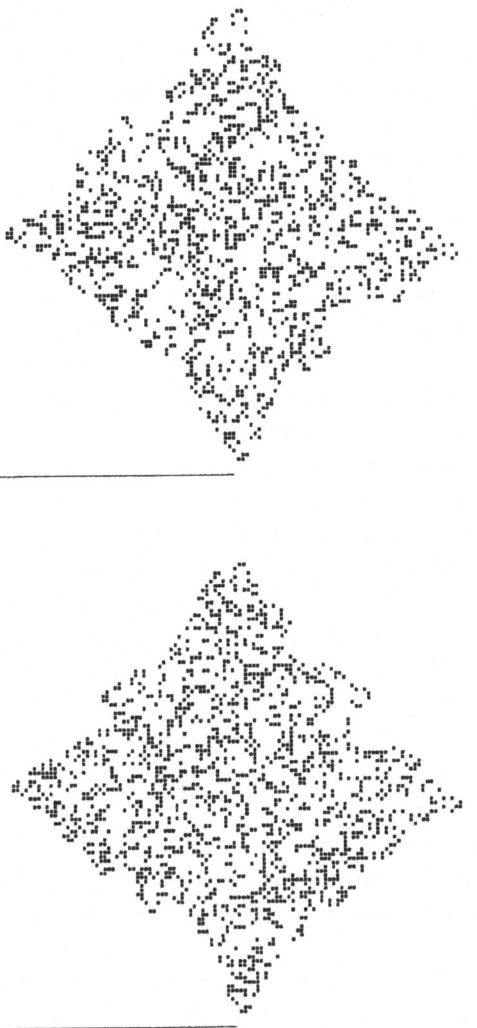

**Figure 8.4**
The random "slime" pattern in 2-2 Life after 50 time steps from two initial configurations consisting of 30 living cells confined with a 10 × 10 region. The only difference between the two initial configurations was the position of one living cell. Note that these two slime patterns and the slime pattern in figure 8.3 are all qualitatively similar.

each exhibit. But the only way to tell exactly *which* instance of that behavior will be produced from a given initial configuration is to watch how the system unfolds in time—i.e., to "simulate" it. Given a random initial configuration, you can be sure that 3-3 Life will quickly reduce to a collection of isolated still lifes and blinkers, and that 2-2 Life will produce a steadily growing chaotically changing mixture of living cells. But the only way to determine exactly which collection of still lifes and blinkers, or exactly which chaotically changing sequence of living cells, is to step through the behavior of the whole system. This is the signature of systems with weak emergent properties.

Earlier we saw that it is often difficult to tell whether a given initial condition in the Game of Life leads to indefinite growth. 3-3 Life and 2-2 Life are different in this respect. Given 3-3 Life's law of collapse to isolated periodicity, we know that 3-3 Life never shows indefinite growth. Likewise, given 2-2 Life's law of spreading chaos, we know that 2-2 Life (virtually) always shows indefinite growth. However, the presence or absence of indefinite growth is still a weak emergent property in 3-3 Life and 2-2 Life. Our knowledge that 3-3 Life never exhibits indefinite growth depends on having learned its law of collapse to periodicity, and analogously for 2-2 Life's law of spreading chaos. But these macro-laws are *emergent laws*—that is, laws about the system's emergent properties—and they are discovered empirically. Our knowledge of these laws comes from our prior empirical observations of how the systems behave under different initial conditions. This is analogous to how we know that a rock can break a window. Weak emergence concerns the derivation of macro-properties, and these derivations involve exact and absolutely certain inferences from the system's micro facts. Empirically grounded generalizations about the system's behavior play no part in such derivations. Thus, in the sense that is relevant to weak emergence, it is not possible to derive the presence or absence of indefinite growth in 3-3 Life or 2-2 Life.

It is important to note that all five of our cellular automata are ontologically and causally on a par, though not all exhibit weak emergence. Each cellular automaton is nothing but a lattice of cells, and the behavior of its cells is wholly determined by a local update function. Any large-scale macro patterns exhibited by the cellular automata are derived from iterating the behavior of each cell over time according to the system's update function and aggregating the cells over the lattice. In other words, the macro behavior of each system is constituted by iterating and aggregating local causal interactions.

Emergence is sometimes contrasted with reduction, but this oversimplifies matters, especially for weak emergence. The three kinds of reduction we distinguished earlier (reduction of ontology, causation, and explanation) need not go hand in hand. Ontological reductionism and causal reductionism hold for all cellular automata—indeed, for all weak emergence. Local causal influence propagates in space and time in the same way in all cellular automata. The distinctive feature of those cellular automata that exhibit weak emergence is not the lack of ontological or causal reductionism, nor the lack of context-sensitive derivation of macro properties. It is simply having micro-level context-sensitive interactions that are complex enough that their aggregate effect has no short-cut derivation. Macro properties in All Life and Spreading Life always have a short-cut derivation. But this is not so for 2-2 Life, 3-3 Life, or the Game of Life.

Embracing ontological and causal reduction permits weak emergence to avoid one of the traditional complaints against emergence. J. J. C. Smart (1963), for example, objected that emergence debarred viewing the natural world as a very complicated mechanism. However, weak emergence postulates just complicated mechanisms with context-sensitive micro-level interactions. Rather than rejecting reduction, it requires (ontological and causal) reduction, for these are what make derivation by simulation possible.

## 8.5 Downward Causation of Weak Emergence

Ordinary macro causation consists of causal relations among ordiinary macro objects. Examples are when a rock thrown at a window cracks it, or an ocean wave hits a sand castle and demolishes it.[12] But macro-level causes can also have micro-level effects. This is termed "downward causation." Downward causation is a straightforward consequence of ordinary macro causation. To see this, choose some micro piece of the macro effect and note that the macro cause is also responsible for the consequent changes in the micro piece of the macro effect. For example, consider the first molecular bond that broke when the window cracked. The rock caused that molecular bond to break. Or consider the violent dislocation of a particular grain of sand at the top of the castle. The wave caused its dislocation.

Emergence is interesting in part because of emergent causal powers. Emergent phenomena without causal powers would be mere epiphenomena. Weak emergent properties, objects, phenomena, etc. often have causal powers. For example, the property of being a glider gun is a weak emergent property of a certain macro-level collection of cells in the Game of Life, and it has the causal power of generating a regular stream of gliders—a macro-level pattern of cells propagating in space. For example, the glider gun shown in figure 8.5 shoots another glider every forty-six time steps. This weak emergent macro-level causation brings downward causation in its train. To pick just one example, as successive gliders are shot from the gun they cause a certain pattern of behavior in a particular individual cell in their path—call it cell 17 (see figure 8.5). When a glider first touches cell 17, the cell becomes alive. While the glider passes, cell 17 remains alive for three more generations. Then it becomes dead and remains so for forty-two more time steps, until the next glider touches it. Clearly, this repeating pattern in cell 17's behavior is caused by the macro-level glider gun.

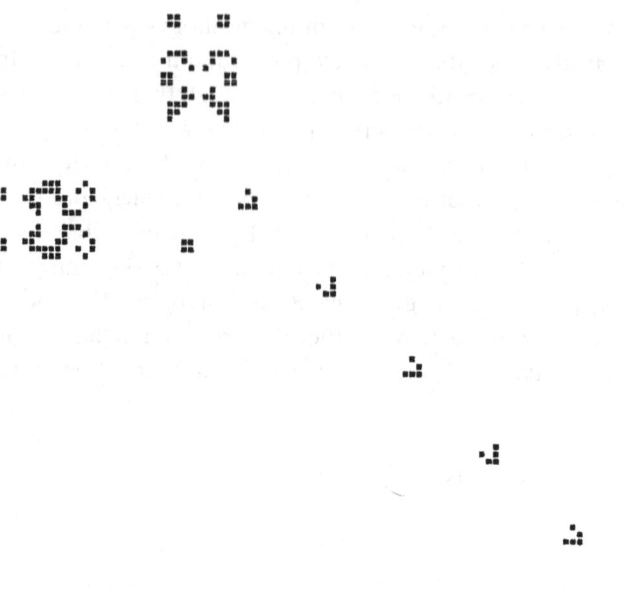

**Figure 8.5**
A glider gun that so far has shot six gliders moving away along the southeast diagonal, illustrating weak downward causation. This gun spawns a new glider every 46 time steps. The left-most cell in the glider closest to the gun—call it cell 17—has just become alive. Cell 17 will remain alive for 4 time steps as the glider passes, and then it will become dead for 42 time steps. This pattern repeats when the next glider passes. This micro-level pattern in the behavior of cell 17 is an example of downward causation, brought about by the glider gun.

Campbell (1974) called attention to emergent downward causation, because he wanted to combat excessive reductionism and bolster the perceived reality of higher-level emergent biological organization. Downward causation is also emphasized recently by advocates of strong emergence (e.g., Kim 1992 and 1999, O'Connor 1994), because the characteristic feature of strong emergence is irreducible downward causal power.

Downward causation is now one of the main sources of controversy about emergence. There are at least three apparent problems. The first is that the very idea of emergent downward causation seems incoherent in some way. Kim (1999, p. 25) introduces the worry in this way:

The idea of downward causation has struck some thinkers as incoherent, and it is difficult to deny that there is an air of paradox about it: After all, higher-level properties arise out of lower-level conditions, and without the presence of the latter in suitable configurations, the former could not even be there. So how could these higher-level properties causally influence and alter the conditions from which they arise? Is it coherent to suppose that the presence of *X* is entirely responsible for the occurrence of *Y* (so *Y*'s very existence is totally dependent on *X*) and yet *Y* somehow manages to exercise causal influence on *X*?

The upshot is that there seems to be something viciously circular about downward causation.

The second worry is that, even if emergent downward causation is coherent, it makes a difference only if it violates micro causal laws (Kim 1997). This worry arises because of a background presumption that micro events are caused by prior micro events according to fundamental micro laws. If emergent downward causation brought about some micro event E, there would be two unattractive possibilities. One is that E is also brought about by some micro cause, in which case the emergent macro cause of E is irrelevant. The other possibility is that the macro and micro causes conflict because micro causation would have brought about an incompatible micro effect, E′, so the downward causation would violate the fundamental micro laws.[13]

Even if emergent downward causation is coherent and consistent with fundamental micro laws, a third worry still arises. This worry also grows out of the fact that micro-level events have sufficient micro-level causes. Any macro-level cause that has a micro-level effect (i.e., any downward causation) will compete for explanatory relevance with the micro-level explanation. But the micro-level explanation is more fundamental. So the micro-level explanation of the micro-level effects will preempt the macro-level explanation. This "exclusion" argument has been emphasized by Kim (1992, 1999), and it has provoked extensive contemporary discussion (e.g., Chalmers 1996).

I want to show that these worries present no problems for weak downward causation. There is a simple two-step argument that shows this. The first step is to note that ordinary downward causation is unproblematic. An ocean wave demolishes a sand castle, causing the violent dislocation of a grain of sand. A vortex in the draining bathtub causes a suspended dust speck to spin in a tight spiral. A traffic jam causes my car's motion to slow and become erratic. I take it as uncontroversial that such ordinary cases of downward causation are philosophically unproblematic. They violate no physical laws, they are not preempted by micro causes, and they are not viciously circular or incoherent. The second step is to note that weak downward causation is simply a species of ordinary downward causation. Many ordinary macro objects with downward causal effects are weakly emergent. Waves, vortices, and traffic jams are all plausible candidates for weak emergence. Their macro causal powers are constituted by the causal powers of their micro constitutents, and these are typically so complicated that the only way to derive their effects is by iterating their aggregate context-dependent effects—i.e., by simulation. In any event, weak emergent properties and objects have the kind of relation to their micro-level bases that ordinary macro-scale physical properties and objects have to their bases. Weak emergent causal powers are constituted by the causal powers of the micro constituents. The weak emergent macro cause is nothing but the iteration of the aggregate micro causes. Ontological and causal reduction holds. Since weak downward causation is just a subset of ordinary macro causation, the one is no more problematic than the other.

This defense of weak downward causation is confirmed when we examine each of the three worries. First, since a weak macro cause is identical with the aggregation and iteration of micro causes, weak macro causation cannot violate micro causal laws. In fact, since weak macro causation is constituted by the appropriate context-sensitive micro causation, weak macro causation *depends* on the micro causal laws. They are the mechanism through which weak macro causation is realized. Second, since a weak macro cause is nothing more than the aggregation of micro causes, macro and micro causes are not two things that can compete with each other for causal influence. One constitutes the other. So, the micro causes cannot exclude weak macro causes. Third, once we see that weak downward causation does not violate fundamental micro explanations and is not preempted by them, the apparent incoherence or vicious circularity of emergent downward causation reduces to the worry that downward causal effects must precede their causes. But weak downward causation is diachronic. Higher-level properties can causally influence the conditions by which they are sustained, but this process unfolds over time. The higher-level properties arise out of lower-level conditions, and without those lower-level conditions the higher-level properties would not be present. But a weak macro cause cannot alter the conditions from which it arose. At most it can alter the conditions for its subsequent survival, and this is neither viciously circular nor incoherent.

These abstract considerations are concretely exemplified by our earlier of weak downward causation in the Game of Life: the glider gun that causes a repeating pattern in cell 17 (figure 8.5). First, the downward causation is diachronic; the micro effects are subsequent to their macro causes. So there is no vicious circularity. Second, this downward causation is brought about simply by aggregating the state changes in each cell, given the appropriate initial condition, and then iterating these aggregated local changes over time. The glider gun (a macro object consisting of a special aggregation of micro elements) creates the context for qualitatively distinctive (macro- and) micro-level effects, but this violates no micro laws. Indeed, it exploits those micro laws. Third, macro glider gun explanation does not compete with micro update-rule explanation. The macro explanation is constituted by iterating the aggregated micro explanation. So explanatory exclusion is no threat.

## 8.6 The Autonomy of Weak Emergence

The preceding discussion of downward causation emphasized that weak emergent phenomena are nothing more than the aggregation of the micro phenomena that constitute them. This prompts a final worry about whether the explanations of weak emergent phenomena are sufficiently autonomous. Consider some weak emergent macro property *P*. This property is brought about by the aggregation of a collection of micro causal histories—the causal histories of all the micro properties that constitute *P*. So, isn't the underlying explanation of *P* just the aggregation of the micro explanations of all the relevant micro elements? If the underlying explanation of the macro phenomena is merely the aggregation of micro explanations, then all the real explanatory power resides at the micro level and the macro phenomena are merely an effect of what happens at the micro level.[14] In this case, weak emergent phenomena have no real macro-level explanatory autonomy.

Some of the plausibility for this line of argument comes from the ontological and causal reducibility of weak emergent phenomena. Since their existence and causal powers are nothing more than the existence and causal powers of the micro elements that instantiate them, wouldn't their real underlying explanation also be at the micro level? Weak emergent macro phenomena have various macro explanations, and these explanations may be convenient and useful for us. In particular, the overwhelming complexity of their aggregate micro explanation typically overwhelms us, preventing us from grasping how they are generated.[15] Hence we resort to computer simulations, observing the resulting macro properties and experimentally manipulating micro causes to see their macro effects. The computer can aggregate micro causal histories fast enough for us to see their weak emergent macro effect. But isn't the explanation of the macro effect exhausted by the micro causal processes?

The nub of this worry is that, if weak emergence has any macro explanatory autonomy, the autonomy is just our inability to follow through the details of the complicated micro causal pathways. It amounts to nothing more than an epistemic obstacle to following the ontological and causal reduction. We study the weak emergent effects of these micro causal processes by observing the macro effects directly (in nature or in computer simulations). But the macro phenomena are mere effects of micro causal processes. This explanatory autonomy is merely epistemological rather than ontological. It reflects just our need for macro explanations of certain phenomena; it does not reflect any distinctive objective structure in reality. In particular, it does not reflect any autonomous and irreducible macro-level ontology.[16] Or, at least, that is the worry.

The correct response to this worry takes different branches for different kinds of weak emergence. In some cases the worry is sound. All weak emergence has a certain epistemic autonomy, for the context-sensitive micro causal interactions can be explained only by iterating the aggregated effect of all the micro interactions.[17] Thus, as a practical matter, we must study them through simulation. Some weak emergence is nothing more than this. Such weak emergent phenomena are mere effects of micro contingencies and their explanatory autonomy is merely epistemological.

One example of such merely epistemological weak emergence is a configuration in the Game of Life that accidentally (so to speak) emits an evenly spaced stream of six gliders moving along the same trajectory. What is crucial is that this configuration contains no glider gun. It's an irregular collection of still lifes, blinkers, and miscellaneous piles of "muck" that happens to emit six gliders. It might be somewhat like the configuration in figure 8.1 at time 100, which has just emitted a glider from the northwest corner, except that it happens to emit five more evenly spaced gliders in the same direction. The configuration is always changing in an irregular fashion, and there is no overarching explanation for why the six gliders stream out. The explanation for the gliders is just the aggregation of the causal histories of the individual cells that participate in the process. The macro-level glider stream is a mere effect of those micro contingencies.

Contrast the accidental glider stream with the configuration of cells shown in figure 8.5. This configuration of cells also emits an evenly spaced stream of six gliders heading in the same direction. Furthermore, the aggregation of the causal histories of the individual cells that participate in the process explains the glider stream. However, there is more to the explanation of this second stream of gliders, because the configuration of cells is a *glider gun* and glider guns always emit evenly spaced gliders in a given direction. The glider gun provides an overarching, macro-level explanation for the second glider stream. Furthermore, this same macro explanation holds for any number of other guns that shoot other gliders. There are many kinds of gliders and many kinds of glider guns. (Figure 8.6 shows two more glider guns.) The aggregate micro explanation of the second glider stream omits this information. Furthermore, this information supports counterfactuals about the stream. The same glider stream would have been produced if the first six gliders had been destroyed somehow (e.g., by colliding with six other gliders). Indeed, the same glider stream would have been produced if the configuration had been changed into any number of ways, as long as the result was a gun that shot the same kind of gliders. Any such macro gun would have produced the same macro effect. Thus, the full explanation of the six gliders in figure 8.5 consists of more than the aggregation of the causal histories of the relevant micro cells. There is a macro explanation that is not reducible to that aggregation of micro histories. If those micro histories had been different, the macro explanation could still have been true. The macro explanation is autonomous from the aggregate micro explanation.

Consider another example: the chaotically changing "slime" that spreads at the speed of light from an initial configuration in 2-2 Life (figures 8.3 and 8.4). These examples illustrate a general macro law that I mentioned earlier: 2-2 Life always generates such random slime, provided the initial configuration is dense enough for any life to grow. Each instance of spreading slime can be explained by aggregating the causal histories of the micro cells that participate in the pattern. But this aggregate micro ex-

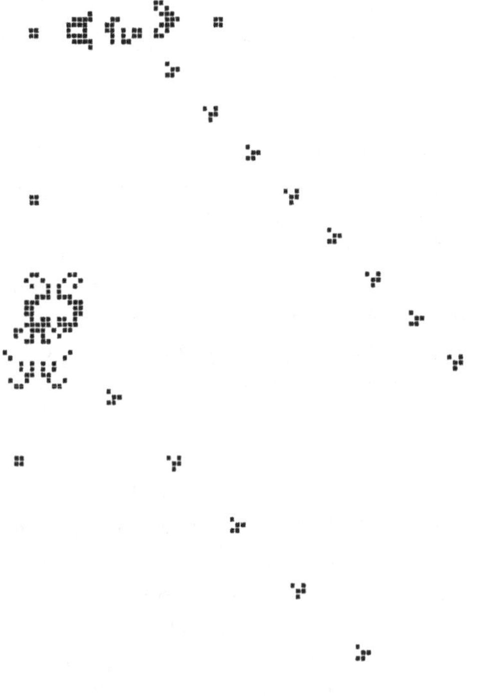

**Figure 8.6**
Two more guns shooting gliders on the southeast diagonal. Note that the configuration of cells constituting these two guns and the gun in figure 8.5 all differ.

planation leaves out an important fact: the random slime macro law. Alter the initial condition (and thus the micro histories) in virtually any way you want, and the same kind of macro behavior would still be generated. The fact that the same kind of behavior would have been produced if the micro details had been different is clearly relevant to the explanation of spreading slime observed in any particular instance. The macro law explanation is autonomous from the aggregation of micro histories in each particular instance.

Notice that weak emergent phenomena in the real world have the same kind of macro explanatory autonomy. Consider a transit strike that causes a massive traffic jam that makes everyone in the office late to work. Each person's car is blocked by cars with particular and idiosyncratic causal histories. The traffic jam's ability to make people late is constituted by the ability of individual cars to block other cars. Aggregating the individual causal histories of each blocking car explains why everyone was late. However, the aggregate micro explanation obscures the fact that everyone would still have been late if the micro causal histories had been different. The transit strike raised the traffic density above a critical level. So, even if different individual cars had been on the highway, the traffic still would have been jammed and everyone still would have been late. The critical traffic density provides a macro explanation that is autonomous from any particular aggregate micro explanation.

My strategy for showing that macro explanations of some weak emergent phenomena are autonomous is analogous to well-known strategies for showing that explanations in special sciences can be autonomous from the explanations provided in underlying sciences.[18] One complementary strategy for defending special sciences emphasizes that macro explanations can contain causally relevant information that is missing from micro explanations (e.g., Jackson and Pettit 1992, Sterelny 1996). My arguments above have this form. The original defense of special sciences focussed on multiple realization and the resulting irreducibility of macro explanations (e.g., Fodor 1974 and 1997). My arguments can be recast in this form. Note that glider guns are multiply realizable in the Game of Life, as are random slimes in 2-2 Life, as are traffic jams, and none is reducible to any particular collection of aggregate micro phenomena.

Either way the argument is put, the conclusion is the same: Macro explanations of some weak emergent phenomena have a strong form of autonomy. Note that this is not the mere epistemological autonomy that comes with all weak emergence. The accidental glider stream discussed above is just an effect of micro contingencies. By contrast, the glider gun, the random slime, and the traffic jam are instances of larger macro regularities that support counterfactuals about what would happen in an indefinite variety of different micro situations. An indefinite variety of micro configurations constitute glider guns in the Game of Life, and they all shoot regular streams of gliders. An indefinite variety of micro configurations constitute random slime in 2-2 Life, and they all spread in the same way. An indefinite variety of micro configurations constitute traffic jams, and they all block traffic. Each macro-level glider gun, random slime, and traffic jam is nothing more than the micro-level elements that constitute it. But they participate in macro regularities that unify an otherwise heterogeneous collection of micro instances. Fodor argues that macro regularities in the Game of Life and similar systems have micro reductions because the macro regularities are "logical or mathematical constructions" out of micro regularities (1997, n. 5). But Fodor fails to appreciate that micro realizations of the macro regularities in cellular automata are as wildly disjunctive as any in the special sciences.

So, the explanatory autonomy of weak emergence can take two forms. When the emergent phenomena are mere effects of micro contingencies, then their explanatory autonomy is merely epistemological. The explanatory autonomy does not signal any distinctive macro structure in reality. But weak emergent phenomena that would be realized in an indefinite variety of different micro contingencies can instantiate robust macro regularities that can be described and explained only at the macro level. The point is not just that macro explanation and description is irreducible, but that this irreducibility signals the existence of an objective macro structure. This kind of robust weak emergence reveals something about reality, not just about how we describe or explain it. So the autonomy of this robust weak emergence is ontological, not merely epistemological.[19]

Not all weak emergence is metaphysically or scientifically significant. In some quarters emergence *per se* is treated as a metaphysically significant category that signals a qualitative difference in the world. This is not the perspective provided by weak emergence. Much weak emergence is due just to complicated micro-level context-sensitivity, the same context-sensitivity that is ubiquitous in nature. In some cases, though, these context-sensitive micro interactions fall into regularities that indicate an objective macro structure in reality. These macro regularities are important scientifically, for they explain the generic behavior of complex systems in nature. The Game of Life instantiates fantastically complicated macro structures like universal Turing machines only by exploiting the ability of glider guns to send signals arbitrary distances in time and space. The law of spreading random slime in 2-2 Life is a hallmark of one of the four fundamental classes of cellular automata rules identified by Wolfram (1994). Explaining robust traffic patterns necessitates identifying the critical role of traffic density. A significant activity in complexity science is sifting through the emergent behavior of complex systems, searching for weak emergent macro properties that figure in robust regularities with deep explanatory import.

## Conclusions

The problem of emergence arises out of attempting to make sense of the apparent macro/micro layers in the natural world. I have argued that what I call weak emergence substantially solves this problem. The weak emergence perspective is ontologically and causally reductionistic, and this enables it to avoid many of the traditional worries about emergence, such as those involving downward causation. But weak emergence is still rich enough for an ontology of objective macro-level structures. Indeed, the search for robust weak emergent macro-structures is one of the main activities in complexity science—exactly the science that attempts to explain the apparent emergent phenomena in nature. Could there be a better guide for understanding emergence in nature than complexity science?

Weak emergence is prevalent in nature, but it is unclear whether it is all the emergence we need. In particular, some aspects of the mind still strenuously resist ontological and causal reduction; examples include fine-grained intentionality, the qualitative aspects of consciousness, freedom, and certain normative states. Weak emergence can get no purchase on these phenomena until we have a (context-sensitive) reductionistic account of them. As long as this is in doubt, so is the final reach of weak emergence.

However this turns out, weak emergence should still illuminate a variety of debates and confusions about the relations between macro and micro. These range from long-standing controversies over the autonomy of the special sciences to newer debates about whether macro evolutionary patterns are mere effects of micro processes or reflect genuine species selection (Vrba 1984, Sterelny 1996).

Emergence is often viewed synchronically. An organism at a given time is thought to be more than the sum of its parts that exist at that time. Your mental states at a given time are thought to emerge from your neuro-physical states at that time. By contrast, the primary focus of weak emergence is diachronic. It concerns how the macro arises over time from the micro, i.e., the causal process (derivation) by which the micro constructs the macro. This is a bottom-up generative process, rooted in context-sensitive micro-level causal interactions.

The advent of modern philosophy is conventionally presented as the Cartesian triumph over Aristotelian scholasticism. An Aristotelian thesis that attributed natures on the basis of a rich dependence on generating context was supplanted by a Cartesian antithesis that attributed reductionistic essences independent of context. Computer simulations allow weak emergence to extend reductionism into new territory, but they do so by embodying the idea that something's nature can depend on its genesis. Thus, the macro can depend on the context-sensitive process from which it arises and by which it is maintained. In this way, weak emergence can be viewed as a new synthesis.

## Acknowledgments

Thanks for helpful comments to audiences at SCTPLS'99 in Berkeley CA (July 1999), at ISHPSSB'01 in Quinnipiac CT (July 2001), and at the philosophy department at the University of Oklahoma (October 2001), where some of the ideas in this paper were presented. Thanks also for helpful discussion to Carl Gillett, Paul Hovda, Brian Keeley, Dan McShea, Norman Packard, Steen Rasmussen, David Reeve, Edmund Ronald, Andre Skusa, Kelly Smith, and Pietro Speroni di Fenizio.

### Notes

This chapter originally appeared in *Principia Revista Internacional de Epistemologica* 6 (2003), 5–50.

1. These cellular automata include the Game of Life so this chapter illustrates the philosophical versatility of cellular automata, which Dennett (1991) recently emphasized.

2. There is no standard accepted terminology for referring to different kinds of emergence, so my terminology of "nominal," "weak" and "strong" might clash with the terminology used by some other authors. In particular, Gillett (unpublished) means something else by "strong" emergence.

3. Although my point in the text is unaffected by this, note that this example really involves multiple levels of emergence, for we could split these levels more finely into macro (micelles), meso (polymers), and micro (monomers). See Rasmussen et al. (2001) for an analysis and a model of this situation.

4. Since the two more restricted notions of emergence are proper subsets of nominal emergence, they of course exhibit the two hallmarks of emergence that characterize nominal emergence. However, they each also capture their own distinctive and specific forms of dependence and autonomy, as the subsequent discussion shows.

5. Supervenient properties, in this context, are macro properties that can differ only if their micro property bases differ; there can be no difference in supervenient properties without a difference in their micro bases.

6. The qualifier "weak" is intended to highlight the contrast with the "strong," irreducible macro causal powers characteristic of strong emergence. I need some qualifier, since weak emergence is just one among many kinds of emergence, but "weak" has the drawback of vagueness. I would prefer a more descriptive term, but I have not found an appropriate one. For example, "reductive" would emphasize weak emergence's ontological and causal reducibility, but it would obscure its explanatory irreducibility. One sometimes sees weak emergence described as "innocent" emergence (e.g., Chalmers 1996). This calls attention to our metaphysical evaluation of weak emergence, but it does not identify the source of this evaluation. This is unfortunate since different kinds of emergence are metaphysically innocent for different reasons (compare nominal and weak emergence). Unfortunate for a related reason is "statistical" emergence. "Statistical" does bring to mind a picture of macro phenomena arising out of the aggregation of micro phenomena, but it does not help distinguish the special kind of aggregation involved in weak emergence. Terms like "explainable" or "non-brute" emergence have the same problem. Thus, I will continue to use "weak" until I find a better alternative.

7. Thus, weak emergence can be exhibited by systems that also involve strong emergence. The fates of weak and strong emergence are independent.

8. A "backward looking" emergent object is one the existence of which is weakly emergent, and a "forward looking" emergent object is one with weak emergent behavior, causal powers, etc.

9. There are different kinds of simulations. My account of weak emergence fits best the agent-based simulations that explicitly represent micro causal interactions, but it can be extended to other simulation methods like those based on differential equations.

10. A variety of other kinds of systems studied in complexity science can be found by surveying conference proceedings, such as Farmer et al. 1986, Forrest 1989, Langton et al. 1992, Varela and Bourgine 1992, Gaussier and Nicoud 1994, and Bedau et al. 2000.

11. The rare exceptions arise when the initial configuration is too sparse to support any life.

12. I will speak of macro objects as causes, where referring to their macro properties or events involving them as causes might be more appropriate. I trust that no confusion will result.

13. This worry made Beckner (1974) conclude that emergent downward causation would require micro-level indeterminism, so that macro causes can have micro effects without violating micro

physical laws. Micro-level indeterminism is clearly an unsatisfactory way to save emergent downward causation, though. There is no guarantee that the indeterminism would be available exactly where and when it is needed, and brute downward causal determination of micro-indeterministic events would be mysterious.

14. This would be the analog of Vrba's effect hypothesis about macro evolutionary properties (Vrba 1994).

15. The opacity of the aggregate micro-level causal mechanisms in the agent-based models is a current source of unease about complexity science.

16. Note also that if weak emergence has mere epistemological autonomy, then weak emergent macro causation is spurious rather than genuine causation. For the apparent macro causation is really nothing more than an effect of micro causal processes. It would follow that weak downward causation is also spurious. So, the fate of genuine weak downward causation hinges on weak emergence having more than epistemological autonomy.

17. Context-sensitive micro interactions are necessary for weak emergence but they are not sufficient. All Life and Spreading Life have context-sensitive micro interactions but they are so trivial that the resulting macro properties are not weakly emergent.

18. Nonreductive physicalism in contemporary philosophy of mind is probably most plausible to cast as an instance of weak emergence. However, my defense of weak emergence here is not tied to the fate of nonreductive physicalism.

19. Some, such as Silberstein and McGeever (1999) and perhaps Gillett (unpublished) will still classify this robust weak emergence as mere epistemological emergence, on the grounds that it embraces ontological and causal reduction (mereological supervenience). However, I think this is an excessively liberal view of epistemological emergence. Consider an analogy: Is the difference between the (presumably hypothetical) world in which all special sciences are reducible to fundamental physics and the (presumably actual) world in which they are autonomous merely epistemological? Is there nothing in the ontological structure of the second world that *makes* the special sciences autonomous? Presumably not.

## References

Assad, A. M., and N. H. Packard. 1992. Emergent Colonization in an Artificial Ecology. In Varela and Bourgine, eds., *Towards a practice of autonomous systems* (Cambridge: MIT Press), pp. 143–152.

Baas, N. A. 1994. Emergence, hierarchies, and hyperstructures. In C. G. Langton, ed., *Artificial life III* (Redwood City: Addison-Wesley), pp. 515–537.

Beckner, Morton. 1974. Reduction, hierarchies and organicism. In F. J. Ayala and T. Dobzhansky, eds., *Studies in the philosophy of biology: Reduction and related problems* (Berkeley: University of California Press), pp. 163–176.

Bedau, M. A. 1997. Weak emergence. *Philosophical Perspectives* 11, 375–399.

Bedau, M. A., J. McCaskill, N. Packard, S. Rasmussen, eds. 2000. *Artificial life VII*. Cambridge: MIT Press.

Berlekamp, E. R., J. H. Conway, and R. K. Guy. 1982. *Winning ways for your mathematical plays.* Vol. 2. New York: Academic Press.

Campbell, Donald T. 1974. "Downward causation" in hierarchically organised biological systems. In F. J. Ayala and T. Dobzhansky, eds., *Studies in the philosophy of biology: Reduction and related problems* (Berkeley: University of California Press), pp. 179–186.

Chaitin, G. J. 1966. On the length of programs for computing finite binary sequences. *Journal of the Association of Computing Machinery* 13: 547–569.

Chaitin, G. J. 1975. A theory of program size formally identical to information theory. *Journal of the Association of Computing Machinery* 22: 329–340.

Chalmers, D. J. 1996. *The conscious mind: In search of a fundamental theory*. New York: Oxford University Press.

Dennett, Daniel. 1991. Real patterns. *Journal of Philosophy* 87: 27–51.

Farmer, J. D., Lapedes, A., Packard, N., and Wendroff, B., eds. 1986. *Evolution, games, and learning: Models for adaptation for machines and nature.* Amsterdam: North Holland.

Fodor, Jerry. 1974. Special sciences. *Synthese* 28: 97–115.

Fodor, Jerry. 1997. Special sciences: Still autonomous after all these years. *Philosophical Perspectives* 11: 149–163.

Forrest, S., ed. 1989. *Emergent computation: Self-organizing, collective, and cooperative phenomena in natural and artificial computing networks.* Amsterdam: North-Holland.

Gardner, M. 1983. *Wheels, life, and other mathematical amusements.* New York: Freeman.

Gaussier, P., and Nicoud, J.-D., eds. 1994. *From perception to action.* Los Alamitos, Calif.: IEEE Computer Society Press.

Gillett, Carl. Unpublished. Strong emergence as a defense of non-reductive physicalism: A physicalist metaphysics for "downward" determination.

Harré, Rom. 1985. *The philosophies of science.* Oxford: Oxford University Press.

Holland, John. 1998. *Emergence: From chaos to order.* Reading, MA: Helix Books.

Jackson, F., and P. Pettit. 1992. In defense of explanatory ecumenism. *Economics and Philosophy* 8: 1–21.

Kauffman, Stuart. 1995. *At home in the universe: The search for the laws of self-organization and complexity*. New York: Oxford University Press.

Kim, Jaegwon. 1992. "Downward-causation" in emergentism and nonreductive physicalism. In A. Beckerman, H. Flohr, and J. Kim, eds., *Emergence or reduction? Essays on the prospects of nonreductive physicalism* (Berlin: Walter de Gruyter), pp. 119–138.

Kim, Jaegwon. 1997. The mind-body problem: taking stock after forty years. *Philosophical Perspectives* 11: 185–207.

Kim, Jaegwon. (1999). Making sense of emergence. *Philosophical Studies* 95, 3–36.

Langton, C., C. E. Taylor, J. D. Farmer, S. Rasmussen, eds. 1992. *Artificial life II.* SFI Studies in the Sciences of Complexity, Vol. X. Reading, Calif.: Addison-Wesley.

Newman, David V. 1996. Emergence and strange attractors. *Philosophy of Science* 63, 245–261.

O'Connor, T. 1994. Emergent properties. *American Philosophical Quarterly* 31: 91–104.

Poundstone, W. 1985. *The recursive universe.* Chicago: Contemporary Books.

Rasmussen, S., N. A. Baas, B. Mayer, M. Nilson, and M. W. Olesen. 2001. Ansatz for dynamical hierarchies. *Artificial Life* 7: 329–353.

Rasmussen, S., and C. L. Barrett. 1995. Elements of a theory of simulation. In F. Morán, A. Moreno, J. J. Merelo, and P. Chacón, eds., *Advances in artificial life* (Berlin: Springer), pp. 515–529.

Rueger, Alexander. 2000. Physical emergence, diachronic and synchronic. *Synthese* 124: 297–322.

Ronald, E. M. A., M. Sipper, and M. S. Capcarrère. 1999. Design, observation, surprise! A test of emergence. *Artificial Life* 5: 225–239.

Silberstein, M., and J. McGeever. 1999. The search for ontological emergence. *The Philosophical Quarterly* 49: 182–200.

Simon, Herbert A. 1996. *The sciences of the artificial.* Cambridge; MIT Press.

Smart, J. J. C. 1963. *Philosophy and scientific realism.* London: Routledge and Keagan Paul.

Sperry, R. W. (1969). A modified concept of consciousness. *Psychological Review* 76, 532–536.

Sterelny, Kim. 1996. Explanatory pluralism in evolutionary biology. *Biology and Philosophy* 11: 193–214.

Strawson, P. F. 1963. *Individuals.* Garden City, N.Y.: Doubleday.

Varela, F., and P. Bourgine. 1992. *Towards a practice of autonomous systems.* Cambridge, Mass.: MIT Press.

Vrba, E. S. 1984. What is species selection? *Systematic Zoology* 33: 318–329.

Wimsatt, William. 1986. Forms of aggregativity. In A. Donagan, A. N. Perovich, Jr., and M. V. Wedin, eds., *Human nature and natural knowledge* (Dordrecht: Reidel), pp. 259–291.

Wimsatt, William. 1997. Aggregativity: reductive heuristics for finding emergence. *Philosophy of Science* 64 (Proceedings), S372–S384.

Wolfram, S. 1994. *Cellular automata and complexity.* Reading, Mass.: Addison-Wesley.

# 9 Real Patterns

**Daniel C. Dennett**

Are there really beliefs? Or are we learning (from neuroscience and psychology, presumably) that, strictly speaking, beliefs are figments of our imagination, items in a superseded ontology? Philosophers generally regard such ontological questions as admitting just two possible answers: either beliefs exist or they do not. There is no such state as quasi existence; there are no stable doctrines of semirealism. Beliefs must either be vindicated along with the viruses or banished along with the banshees. A bracing conviction prevails, then, to the effect that when it comes to beliefs (and other mental items) one must be either a realist or an eliminative materialist.

## 9.1 Realism about Beliefs

This conviction prevails in spite of my best efforts over the years to undermine it with various analogies: are *voices* in your ontology?[1] Are *centers of gravity* in your ontology?[2]

It is amusing to note that my analogizing beliefs to centers of gravity has been attacked from both sides of the ontological dichotomy, by philosophers who think it is simply obvious that centers of gravity are useful fictions, and by philosophers who think it is simply obvious that centers of gravity are perfectly real:

> The trouble with these supposed parallels . . . is that they are all strictly speaking *false*, although they are no doubt useful simplifications for many purposes. It is false, for example, that the gravitational attraction between the Earth and the Moon involves two point masses; but it is a good enough first approximation for many calculations. However, this is not at all what Dennett really wants to say about intentional states. For he insists that to adopt the intentional stance and interpret an agent as acting on certain beliefs and desires is to discern a pattern in his actions which is genuinely there (a pattern which is missed if we instead adopt a scientific stance): Dennett certainly does not hold that the role of intentional ascriptions is merely to give us a useful approximation to a truth that can be more accurately expressed in non-intentional terms.[3]

Compare this with Fred Dretske's[4] equally confident assertion of realism:

> I am a realist about centers of gravity. . . . The earth obviously exerts a gravitational attraction on *all* parts of the moon—not just its center of gravity. The *resultant* force, a vector sum, acts through

a point, but this is something quite different. One should be very clear about what centers of gravity are *before* deciding whether to be literal about them, *before* deciding whether or not to be a center-of-gravity realist. (ibid., p. 511)

Dretske's advice is well-taken. What are centers of gravity? They are mathematical points—abstract objects or what Hans Reichenbach called *abstracta*—definable in terms of physical forces and other properties. The question of whether abstract objects are real—the question of whether or not "one should be a realist about them"—can take two different paths, which we might call the metaphysical and the scientific. The metaphysical path simply concerns the reality or existence of abstract objects generally, and does not distinguish them in terms of their scientific utility. Consider, for instance, the *center of population* of the United States. I define this as the mathematical point at the intersection of the two lines such that there are as many inhabitants north as south of the latitude, and as many inhabitants east as west of the longitude. This point is (or can be) just as precisely defined as the center of gravity or center of mass of an object. (Since these median strips might turn out to be wide, take the midline of each strip as the line; count as inhabitants all those within the territorial waters and up to twenty miles in altitude—orbiting astronauts do not count—and take each inhabitant's navel to be the determining point, etc.) I do not know the center of population's current geographic location, but I am quite sure it is west of where it was ten years ago. It jiggles around constantly, as people move about, taking rides on planes, trains, and automobiles, etc. I doubt that this abstract object is of any value at all in any scientific theory, but just in case it is, here is an even more trivial abstract object: Dennett's lost sock center: the point defined as the center of the smallest sphere that can be inscribed around all the socks I have ever lost in my life.

These abstract objects have the same metaphysical status as centers of gravity. Is Dretske a realist about them all? Should we be? I do not intend to pursue this question, for I suspect that Dretske is—and we should be—more interested in the scientific path to realism: centers of gravity are real because they are (somehow) *good* abstract objects. They deserve to be taken seriously, learned about, used. If we go so far as to distinguish them as *real* (contrasting them, perhaps, with those abstract objects which are *bogus*), that is because we think they serve in perspicuous representations of real forces, "natural" properties, and the like. This path brings us closer, in any case, to the issues running in the debates about the reality of beliefs.

I have claimed that beliefs are best considered to be abstract objects rather like centers of gravity. Smith considers centers of gravity to be useful fictions which Dretske considers them to be useful (and hence?) real abstractions, and each takes his view to constitute a criticism of my position. The optimistic assessment of these opposite criticisms is that they cancel each other out; my analogy must have hit the nail on the head. The pessimistic assessment is that more needs to be said to convince philosophers that a mild and intermediate sort of realism is a positively attractive position, and not just the desperate dodge of ontological responsibility it has sometimes been taken to be. I have just such a case to present, a generalization and extension of my earlier attempts, via the concept of a *pattern*. My aim on this occasion is not so much to prove that my intermediate doctrine about the reality of psychological states is right, but just that it is quite possibly right, because a parallel doctrine is demonstrably right about some simpler cases.

We use folk psychology—interpretation of each other as believers, wanters, intenders, and the like—to predict what people will do next. Prediction is not the only thing we care about, of course. Folk psychology helps us understand and empathize with others, organize our memories, interpret our emotions, and flavor our vision in a thousand ways, but at the heart of all these is the enormous predictive leverage of folk psychology. Without its predictive power, we could have no interpersonal projects or relations at all; human activity would be just so much Brownian motion; we would be baffling ciphers to each other and to ourselves—we could not even conceptualize our own failings. In what follows, I shall concentrate always on folk-psychological prediction, not because I make the mistake of ignoring all the other interests we have in people aside from making bets on what they will do next, but because I claim that our power to *interpret* the actions of others depends on our power—seldom explicitly exercised—to predict them.[5]

Where utter patternlessness or randomness prevails, nothing is predictable. The success of folk-psychological prediction, like the success of any prediction, depends on there being some order or pattern in the world to exploit. Exactly where in the world does this pattern exist? What is the pattern a pattern *of*?[6] Some have thought, with Fodor, that the pattern of belief must in the end be a pattern of structures in the brain, formulae written in the language of thought. Where else could it be? Gibsonians might say the pattern is "in the light"—and Quinians (such as Donald Davidson and I) could almost agree: the pattern is discernible in agents' (observable) behavior when we subject it to "radical interpretation" (Davidson) "from the intentional stance" (Dennett).

When are the elements of a pattern real and not merely apparent? Answering this question will help us resolve the misconceptions that have led to the proliferation of "ontological positions" about beliefs, the different grades or kinds of realism. I shall concentrate on five salient exemplars arrayed in the space of possibilities: Fodor's industrial-strength Realism (he writes it with a capital "R"); Davidson's regular strength realism; my mild realism; Richard Rorty's milder-than-mild irrealism, according to which the pattern is *only* in the eyes of the beholders, and Paul Churchland's eliminative materialism, which denies the reality of beliefs altogether.

In what follows, I shall assume that these disagreements all take place within an arena of common acceptance of what Arthur Fine[7] calls NOA, the natural ontological attitude. That is, I take the interest in these disagreements to lie not in differences of opinion about the ultimate metaphysical status of physical things or abstract things

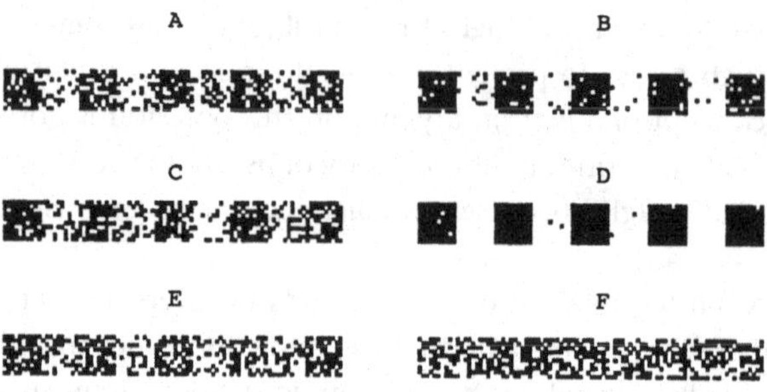


Figure 9.1

(e.g., electrons or centers of gravity), but in differences of opinion about whether beliefs and other mental states are, shall we say, *as real as* electrons or centers of gravity. I want to show that mild realism is the doctrine that makes the most sense when what we are talking about is real patterns, such as the real patterns discernible from the intentional stance.[8]

In order to make clear the attractions and difficulties of these different positions about patterns, I shall apply them first to a much simpler, more readily visualized, and uncontroversial sort of pattern.

## 9.2 The Reality of Patterns

Consider the six objects in figure 9.1 (which I shall call *frames*):

We can understand a frame to be a finite subset of data, a window on an indefinitely larger world of further data. In one sense *A–F* all display different patterns; if you look closely you will see that no two frames are exactly alike ("atom-for-atom replicas," if you like). In another sense, *A–F* all display the same pattern; they were all made by the same basic process, a printing of ten rows of ninety dots, ten black dots followed by ten white dots, etc. The overall effect is to create five equally spaced black squares or bars in the window. I take it that this pattern, which I shall dub *bar code*, is a real pattern if anything is. But some random (actually pseudo-random) "noise" has been allowed to interfere with the actual printing. The noise ratio is as follows:

*A*: 25% *B*: 10%

*C*: 25% *D*: 1%

*E*: 33% *F*: 50%

It is impossible to see that *F* is not purely (pseudo-) random noise; you will just have to take my word for it that it was actually generated by the same program that generated the other five patterns; all I changed was the noise ratio.

Now, what does it mean to say that a pattern in one of these frames is real, or that it is really there? Given our privileged information about how these frames were generated, we may be tempted to say that there is a single pattern in all six cases—even in *F*, where it is "indiscernible." But I propose that the self-contradictory air of "indiscernible pattern" should be taken seriously. We may be able to make some extended, or metaphorical, sense of the idea of indiscernible patterns (or invisible pictures or silent symphonies), but in the root case a pattern is "by definition" a candidate for pattern *recognition*. (It is this loose but unbreakable link to observers or perspectives, of course, that makes "pattern" an attractive term to someone perched between instrumentalism and industrial-strength realism.)

Fortunately, there is a standard way of making these intuitions about the discernibility-in-principle of patterns precise. Consider the task of transmitting information about one of the frames from one place to another. How many bits of information will it take to transmit each frame? The least efficient method is simply to send the "bit map," which identifies each dot *seriatim* ("dot one is black, dot two is white, dot three is white, . . . "). For a black-and-white frame of 900 dots (or pixels, as they are called), the transmission requires 900 bits. Sending the bit map is in effect verbatim quotation, accurate but inefficient. Its most important virtue is that it is equally capable of transmitting any pattern or any particular instance of utter patternlessness.

Gregory Chaitin's[9] valuable definition of mathematical randomness invokes this idea. A series (of dots or numbers or whatever) is random if and only if the information required to describe (transmit) the series accurately is *incompressible*: nothing shorter than the verbatim bit map will preserve the series. Then a series is not random—has a pattern—if and only if there is some more efficient way of describing it.[10] Frame *D*, for instance, can be described as "ten rows of ninety: ten black followed by ten white, etc., *with the following exceptions*: dots 57, 88, . . . ." This expression, suitably encoded, is much shorter than 900 bits long. The comparable expressions for the other frames will be proportionally longer, since they will have to mention, verbatim, more exceptions, and the degeneracy of the "pattern" in *F* is revealed by the fact that its description in this system will be no improvement over the bit map—in fact, it will tend on average to be trivially longer, since it takes some bits to describe the pattern that is then obliterated by all the exceptions.

Of course, there are bound to be other ways of describing the evident patterns in these frames, and some will be more efficient than others—in the precise sense of being systematically specifiable in fewer bits.[11] Any such description, if an improvement over the bit map, is the description of a real pattern in the data.[12]

Consider bar code, the particular pattern seen in *A–E*, and almost perfectly instantiated in *D*. *That* pattern is quite readily discernible to the naked human eye in these presentations of the data, because of the particular pattern-recognition machinery hard-wired in our visual systems—edge detectors, luminance detectors, and the like. But the very same data (the very same streams of bits) presented in some other format might well yield no hint of pattern to us, especially in the cases where bar code is contaminated by salt and pepper, as in frames *A* through *C*. For instance, if we broke the 900-bit series of frame *B* into 4-bit chunks, and then translated each of these into hexadecimal notation, one would be hard pressed indeed to tell the resulting series of hexadecimal digits from a random series, since the hexadecimal chunking would be seriously out of phase with the decimal pattern—and hence the "noise" would not "stand out" as noise. There are myriad ways of displaying any 900-bit series of data points, and not many of them would inspire us to concoct an efficient description of the series. Other creatures with different sense organs, or different interests, might readily perceive patterns that were imperceptible to us. The patterns would be *there* all along, but just invisible to *us*.

The idiosyncracy of perceivers' capacities to discern patterns is striking. Visual patterns with axes of vertical symmetry stick out like sore thumbs for us, but if one simply rotates the frame a few degrees, the symmetry is often utterly beyond noticing. And the "perspectives" from which patterns are "perceptible" are not restricted to variations on presentation to the sense modalities. Differences in knowledge yield striking differences in the capacity to pick up patterns. Expert chess players can instantly perceive (and subsequently recall with high accuracy) the total board position in a real game, but are much worse at recall if the same chess pieces are randomly placed on the board, even though to a novice both boards are equally hard to recall.[13] This should not surprise anyone who considers that an expert speaker of English would have much less difficulty perceiving and recalling

The frightened cat struggled to get loose.

than

Te ser.ioghehnde t srugfcalde go tgtt ohle

which contains the same pieces, now somewhat disordered. Expert chess players, unlike novices, not only know how to *play* chess; they know how to *read* chess—how to see the patterns at a glance.

A pattern exists in some data—is real—if *there is* a description of the data that is more efficient than the bit map, whether or not anyone can concoct it. Compression algorithms, as general-purpose pattern describers, are efficient ways of transmitting exact copies of frames, such as *A–F*, from one place to another, but our interests often favor a somewhat different goal: transmitting *inexact* copies that nevertheless preserve "the" pattern that is important to us. For some purposes, we need not list the exceptions to bar code, but only transmit the information that the pattern is bar code with *n*% noise. Following this strategy, frames *A* and *C*, though discernibly different under careful inspection, count as *the same pattern*, since what matters to us is that the pattern is bar code with 25% noise, and we do not care which particular noise occurs, only that it occurs.

Sometimes we are interested in not just ignoring the noise, but eliminating it, improving the pattern in transmission. Copy-editing is a good example. Consider the likely effoct thes santince wull hive hod on tha cupy adutor whu preparis thas monescrupt fur prunteng. *My* interest in this particular instance is that the "noise" be transmitted, not removed, though I actually do not care exactly *which* noise is there.

Here then are three different attitudes we take at various times toward patterns. Sometimes we care about exact description or reproduction of detail, at whatever cost. From this perspective, a real pattern in frame *A* is *bar code with the following exceptions: 7, 8, 11,....* At other times we care about the noise, but not where in particular it occurs. From this perspective, a real pattern in frame *A* is *bar code with 25% noise.* And sometimes, we simply tolerate or ignore the noise. From this perspective, a real pattern in frame *A* is simply: *bar code.* But is bar code really there in frame A? I am tempted to respond: Look! You can see it with your own eyes. But there is something more constructive to say as well.

When two individuals confront the same data, they may perceive different patterns in them, but since we can have varied interests and perspectives, these differences do not all count as disagreements. Or in any event they should not. If Jones sees pattern $\alpha$ (with $n$% noise) and Brown sees pattern $\beta$ (with $m$% noise) there may be no ground for determining that one of them is right and the other wrong. Suppose they are both using their patterns to bet on the next datum in the series. Jones bets according to the "pure" pattern $\alpha$, but budgets for $n$% errors when he looks for odds. Brown does likewise, using pattern $\beta$. If both patterns are real, they will both get rich. That is to say, so long as they use their expectation of deviations from the "ideal" to temper their odds policy, they will do better than chance—perhaps very much better.

Now suppose they compare notes. Suppose that $\alpha$ is a simple, easy-to-calculate pattern, but with a high noise rate—for instance, suppose $\alpha$ is bar code as it appears in frame *E*. And suppose that Brown has found some periodicity or progression in the "random" noise that Jones just tolerates, so that $\beta$ is a much more complicated description of pattern-superimposed-on-pattern. This permits Brown to do better than chance, we may suppose, at predicting when the "noise" will come. As a result, Brown budgets for a lower error rate—say only 5%. "What you call noise, Jones, is actually pattern," Brown might say. "Of course there is still *some* noise in my pattern, but my pattern is better—more real—than yours! Yours is actually just a mere appearance." Jones might well reply that it is all a matter of taste; he notes how hard Brown has to work to calculate predictions, and points to the fact that he is getting just as rich (or maybe richer) by using a simpler, sloppier system and making more bets at good odds than Brown can muster. "My pattern is perfectly real—look how rich I'm getting. If it were an illusion, I'd be broke."

This crass way of putting things—in terms of betting and getting rich—is simply a vivid way of drawing attention to a real, and far from crass, trade-off that is ubiquitous in nature, and hence in folk psychology. Would we prefer an extremely compact pattern description with a high noise ratio or a less compact pattern description with a lower noise ratio? Our decision may depend on how swiftly and reliably we can discern the simple pattern, how dangerous errors are, how much of our resources we can afford to allocate to detection and calculation. These "design decisions" are typically not left to us to make by individual and deliberate choices; they are incorporated into the design of our sense organs by genetic evolution, and into our culture by cultural evolution. The product of this design evolution process is what Wilfrid Sellars[14] calls our *manifest image*, and it is composed of folk physics, folk psychology, and the other pattern-making perspectives we have on the buzzing blooming confusion that bombards us with data. The ontology generated by the manifest image has thus a deeply pragmatic source.[15]

Do these same pragmatic considerations apply to the scientific image, widely regarded as the final arbiter of ontology? Science is supposed to carve nature at the joints—at its *real* joints, of course. Is it permissible in science to adopt a carving system so simple that it makes sense to tolerate occasional misdivisions and consequent mispredictions? It happens all the time. The ubiquitous practice of using idealized models is exactly a matter of trading off reliability and accuracy of prediction against computational tractability. A particularly elegant and handy oversimplification may under some circumstances be irresistible. The use of Newtonian rather than Einsteinian mechanics in most mundane scientific and engineering calculations is an obvious example. A tractable oversimplification may be attractive even in the face of a high error rate; considering inherited traits to be carried by single genes "for" those traits is an example; considering agents in the marketplace to be perfectly rational self-aggrandizers with perfect information is another.

## 9.3 Patterns in Life

The time has come to export these observations about patterns and reality to the controversial arena of belief attribution. The largish leap we must make is nicely expedited by pausing at a stepping-stone example midway between the world of the dot frames and the world of folk psychology: John Horton Conway's Game of Life. In my opinion, every philosophy student should be held responsible for an intimate acquaintance with the Game of Life. It should be considered an essential tool in every thought-experimenter's kit, a prodigiously versatile generator of philosophically important examples and thought experiments of admirable clarity and vividness. In *The Intentional Stance*, I briefly exploited it to make a point about the costs and benefits of risky prediction from the intentional stance,[16] but I have since learned that I presumed too much familiarity with the underlying ideas. Here, then, is a somewhat expanded basic introduction to Life.[17]

Life is played on a two-dimensional grid, such as a checkerboard or a computer screen; it is not a game one plays to win; if it is a game at all, it is solitaire. The grid divides space into square cells, and each cell is either ON or OFF at each moment. Each cell has eight neighbors: the four adjacent cells north, south, east, and west, and the four diagonals: northeast, southeast, southwest, and northwest. Time in the Life world is also discrete, not continuous; it advances in ticks, and the state of the world changes between each tick according to the following rule:

Each cell, in order to determine what to do in the next instant, counts how many of its eight neighbors is ON at the present instant. If the answer is exactly two, the cell stays in its present state (ON or OFF) in the next instant. If the answer is exactly three, the cell is ON in the next instant whatever its current state. Under all other conditions the cell is OFF.

The entire physics of the Life world is captured in that single, unexceptioned law. [While this is the fundamental law of the "physics" of the Life world, it helps at first to conceive this curious physics in biological terms: think of cells going ON as births, cells going OFF as deaths, and succeeding instants as generations. Either overcrowding (more than three inhabited neighbors) or isolation (less than two inhabited neighbors) leads to death.] By the scrupulous application of this single law, one can predict with perfect accuracy the next instant of any configuration of ON and OFF cells, and the instant after that, and so forth. In other words, the Life world is a toy world that perfectly instantiates Laplace's vision of determinism: given the state description of this world at an instant, we finite observers can perfectly predict the future instants by the simple application of our one law of physics. Or, in my terms, when we adopt the physical stance toward a configuration in the Life world, our powers of prediction are perfect: there is no noise, no uncertainty, no probability less than one. Moreover, it follows from the two-dimensionality of the Life world that nothing is hidden from view. There is no backstage; there are no hidden variables; the unfolding of the physics of objects in the Life world is directly and completely visible.

There are computer simulations of the Life world in which one can set up configurations on the screen and then watch them evolve according to the single rule. In the best simulations, one can change the scale of both time and space, alternating between close-up and bird's-eye view. A nice touch added to some color versions is that ON cells (often just called pixels) are color-coded by their age; they are born blue, let us say, and then change color each generation, moving through green to yellow to orange to red to brown to black and then staying black unless they die. This permits one to see at a glance how old certain patterns are, which cells are co-generational, where the birth action is, and so forth.[18]

One soon discovers that some simple configurations are more interesting than others. In addition to those configurations which never change—the "still lifes" such

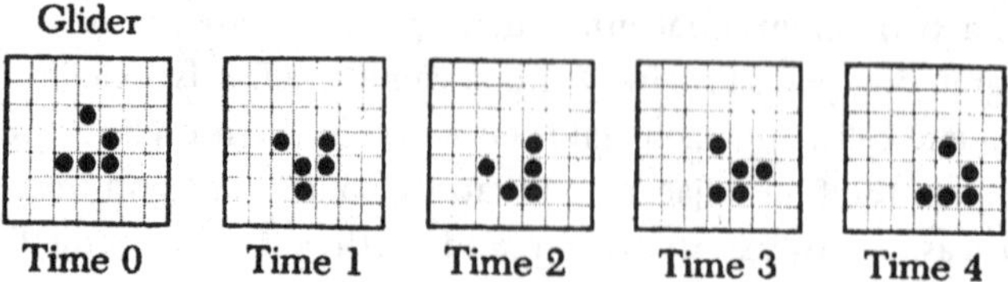


**Figure 9.2**
From Poundstone, *op. cit.*

as four pixels in a square—and those which evaporate entirely—such as any long diagonal line segment, whose two tail pixels die of isolation each instant until the line disappears entirely—there are configurations with all manner of periodicity. Three pixels in a line make a simple flasher, which becomes three pixels in a column in the next instant, and reverts to three in a line in the next, ad infinitum, unless some other configuration encroaches. Encroachment is what makes Life interesting: among the periodic configurations are some that swim, amoeba-like, across the plane. The simplest is the *glider*, the five-pixel configuration shown taking a single stroke to the southeast in figure 9.2. Then there are the eaters, the puffer trains, and space rakes, and a host of other aptly named denizens of the Life world that emerge in the ontology of a new level, analogous to what I have called the design level. This level has its own language, a transparent foreshortening of the tedious descriptions one could give at the physical level. For instance:

> An eater can eat a glider in four generations. Whatever is being consumed, the basic process is the same. A bridge forms between the eater and its prey. In the next generation, the bridge region dies from overpopulation, taking a bite out of both eater and prey. The eater then repairs itself. The prey usually cannot. If the remainder of the prey dies out as with the glider, the prey is consumed. (ibid., p. 38)

Note that there has been a distinct ontological shift as we move between levels; whereas at the physical level there is no motion, and the only individuals, cells, are defined by their fixed spatial location, at this design level we have the motion of persisting objects; it is one and the same glider that has moved southeast in figure 9.2, changing shape as it moves, and there is one less glider in the world after the eater has eaten it in figure 9.3. (Here is a warming-up exercise for what is to follow: should we say that there is *real* motion in the Life world, or only *apparent* motion? The flashing pixels on the computer screen are a paradigm case, after all, of what a psychologist would call apparent motion. Are there *really* gliders that move, or are there just patterns of cell state that move? And if we opt for the latter, should we say at least that these moving patterns are real?)

Notice, too, that at this level one proposes generalizations that require 'usually' or 'provided nothing encroaches' clauses. Stray bits of debris from earlier events can

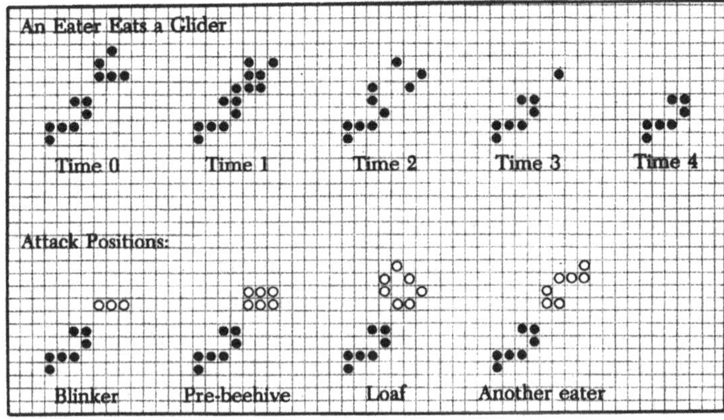


**Figure 9.3**
From Poundstone, *op. cit.*

"break" or "kill" one of the objects in the ontology at this level; their *salience as real things* is considerable, but not guaranteed. To say that their salience is considerable is to say that one can, with some small risk, ascend to this design level, adopt its ontology, and proceed to predict—sketchily and riskily—the behavior of larger configurations or systems of configurations, without bothering to compute the physical level. For instance, one can set oneself the task of designing some interesting supersystem out of the "parts" that the design level makes available. Surely the most impressive triumph of this design activity in the Life world is the proof that a working model of a universal Turing machine can in principle be constructed in the Life plane! Von Neumann had already shown that in principle a two-dimensional universal Turing machine could be constructed out of cellular automata, so it was "just" a matter of "engineering" to show how, in principle, it could be constructed out of the simpler cellular automata defined in the Life world. Glider streams can provide the tape, for instance, and the tape reader can be some huge assembly of eaters, gliders, and other bits and pieces. What does this huge Turing machine look like? Poundstone calculates that the whole construction, a self-reproducing machine incorporating a universal Turing machine, would be on the order of $10^{13}$ pixels.

Displaying a $10^{13}$-pixel pattern would require a video screen about 3 million pixels across at least. Assume the pixels are 1 millimeter square (which is very high resolution by the standards of home computers). Then the screen would have to be 3 kilometers (about two miles) across. It would have an area about six times that of Monaco.

Perspective would shrink the pixels of a self-reproducing pattern to invisibility. If you got far enough away from the screen so that the entire pattern was comfortably in view, the pixels (and even the gliders, eaters and guns) would be too tiny to make out. A self-reproducing pattern would be a hazy glow, like a galaxy. (ibid., pp. 227–228)

Now, since the universal Turing machine can compute any computable function, it can play chess—simply by mimicking the program of any chess-playing computer you like. Suppose, then, that such an entity occupies the Life plane, playing chess against itself. Looking at the configuration of dots that accomplishes this marvel would almost certainly be unilluminating to anyone who had no clue that a configuration with such powers could exist. But from the perspective of one who had the hypothesis that this huge array of black dots was a chess-playing computer, enormously efficient ways of predicting the future of that configuration are made available. As a first step one can shift from an ontology of gliders and eaters to an ontology of symbols and machine states, and, adopting this higher design stance toward the configuration, predict its future *as* a Turing machine. As a second and still more efficient step, one can shift to an ontology of chess-board positions, possible chess moves, and the grounds for evaluating them; then, adopting the intentional stance toward the configuration, one can predict its future *as* a chess player performing intentional actions—making chess moves and trying to achieve checkmate. Once one has fixed on an interpretation scheme, permitting one to say which configurations of pixels count as which symbols (either, at the Turing machine level, the symbols "0" or "1," say, or at the intentional level, "*QxBch*" and the other symbols for chess moves), one can use the interpretation scheme to predict, for instance, that the next configuration to emerge from the galaxy will be such-and-such a glider stream (the symbols for "*RxQ*," say). There is risk involved in either case, because the chess program being run on the Turing machine may be far from perfectly rational, and, at a different level, debris may wander onto the scene and "break" the Turing machine configuration before it finishes the game.

In other words, real but (potentially) noisy patterns abound in such a configuration of the Life world, there for the picking up if only we are lucky or clever enough to hit on the right perspective. They are not *visual* patterns but, one might say, *intellectual* patterns. Squinting or twisting the page is not apt to help, while posing fanciful interpretations (or what W. V. Quine would call "analytical hypotheses") may uncover a goldmine. The opportunity confronting the observer of such a Life world is analogous to the opportunity confronting the cryptographer staring at a new patch of cipher text, or the opportunity confronting the Martian, peering through a telescope at the Superbowl Game. If the Martian hits on the intentional stance—or folk psychology—as the right level to look for pattern, shapes will readily emerge through the noise.

## 9.4 The Reality of Intentional Patterns

The scale of compression when one adopts the intentional stance toward the two-dimensional chess-playing computer galaxy is stupendous: it is the difference between figuring out in your head what white's most likely (best) move is versus calculating the state of a few trillion pixels through a few hundred thousand generations. But the scale of the savings is really no greater in the Life world than in our own. Predicting that someone will duck if you throw a brick at him is easy from the folk-psychological stance; it is and will always be intractable if you have to trace the photons from brick to eyeball, the neurotransmitters from optic nerve to motor nerve, and so forth.

For such vast computational leverage one might be prepared to pay quite a steep price in errors, but in fact one belief that is shared by all of the representatives on the spectrum I am discussing is that "folk psychology" provides a description system that permits highly reliable prediction of human (and much nonhuman) behavior.[19] They differ in the explanations they offer of this predictive prowess, and the implications they see in it about "realism."

For Fodor, an industrial-strength Realist, beliefs and their kin would not be real unless the pattern dimly discernible from the perspective of folk psychology could also be discerned (more clearly, with less noise) as a pattern of structures in the brain. The pattern would have to be discernible from the different perspective provided by a properly tuned *syntactoscope* aimed at the purely formal (non-semantic) features of Mentalese terms written in the brain. For Fodor, the pattern seen through the noise by everyday folk psychologists would tell us nothing about reality, unless it, and the noise, had the following sort of explanation: what we discern from the perspective of folk psychology is the net effect of two processes: an ulterior, hidden process wherein the pattern exists quite pure, overlaid, and partially obscured by various intervening sources of noise: performance errors, observation errors, and other more or less random obstructions. He might add that the interior belief-producing process was in this respect *just* like the process responsible for the creation of frames *A–F*. If you were permitted to peer behind the scenes at the program I devised to create the frames, you would see, clear as a bell, the perfect bar-code periodicity, with the noise thrown on afterward like so much salt and pepper.

This is often the explanation for the look of a data set in science, and Fodor may think that it is either the only explanation that can ever be given, or at any rate the only one that makes any sense of the success of folk psychology. But the rest of us disagree. As G. E. M. Anscombe[20] put it in her pioneering exploration of intentional explanation, "if Aristotle's account [of reasoning using the practical syllogism] were supposed to describe actual mental processes, it would in general be quite absurd. The interest of the account is that it describes an order which is there whenever actions are done with intentions . . ." (ibid., p. 80).

But how *could* the order be there, so visible amidst the noise, if it were not the direct outline of a concrete orderly process in the background? Well, it *could* be there thanks to the statistical effect of very many concrete minutiae producing, as if by a hidden hand, an approximation of the "ideal" order. Philosophers have tended to ignore a variety of regularity intermediate between the regularities of planets and other objects "obeying" the laws of physics and the regularities of rule-following (that is,

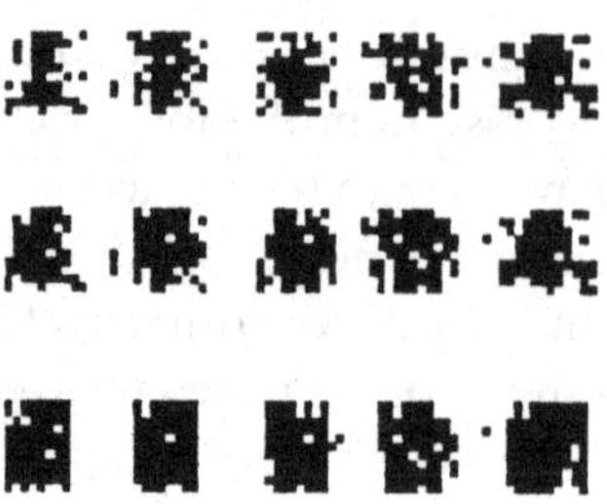

Figure 9.4

rule-*consulting*) systems.[21] These intermediate regularities are those which are preserved under selection pressure: the regularities dictated by principles of good design and hence homed in on by self-designing systems. That is, a "rule of thought" may be much more than a mere regularity; it may be a *wise* rule, a rule one would design a system by if one were a system designer, and hence a rule one would expect self-designing systems to "discover" in the course of settling into their patterns of activity. Such rules no more need be explicitly represented than do the principles of aerodynamics that are honored in the design of birds' wings.[22]

The contrast between these different sorts of pattern-generation processes can be illustrated. The frames in figure 9.1 were created by a hard-edged process (ten black, ten white, ten black, ...) obscured by noise, while the frames in figure 9.4 were created by a process almost the reverse of that: the top frame shows a pattern created by a normal distribution of black dots around means at $x = 10, 30, 50, 70$, and 90 (rather like Mach bands or interference fringes); the middle and bottom frames were created by successive applications of a very simple contrast enhancer applied to the top frame: a vertical slit "window" three pixels high is thrown randomly onto the frame; the pixels in the window vote, and majority rules. This gradually removes the salt from the pepper and the pepper from the salt, creating "artifact" edges such as those discernible in the bottom frame. The effect would be more striking at a finer pixel scale, where the black merges imperceptibly through grays to white but I chose to keep the scale at the ten-pixel period of bar code. I do not mean to suggest that it is impossible to tell the patterns in figure 9.4 from the patterns in figure 9.1. Of course it is possible; for one thing, the process that produced the frames in figure 9.1 will almost always show edges at exactly 10, 20, 30, ... and almost never at 9, 11, 19, 21, ... while there is a higher probability of these "displaced" edges being created by the process of figure 9.4 (as a close inspection of figure 9.4 reveals). Fine tuning could of course reduce these probabilities, but that is not my point. My point is that *even if* the evidence is substantial that the discernible pattern is produced by one process rather than another, it can be rational to ignore those differences and use the simplest pattern description (e.g., *bar code*) as one's way of organizing the data. ...

## Acknowledgment

Thanks to Kathleen Akins, Akeel Bilgrami, Donald Davidson, Barbara Hannan, Douglas Hofstadter, Norton Nelkin, W. V. O. Quine, Richard Rorty, George Smith, Peter Suber, Stephen White, and the MIT/Tufts philosophy of psychology discussion group for the discussions that provoked and shaped this chapter.

### Notes

This chapter originally appeared in *Journal of Philosophy* 87 (1991), 27–45. This reprinting omits pp. 45–51 from the original source.

1. *Content and Consciousness* (Boston: Routledge & Kegan Paul, 1969), ch. 1.

2. "Three Kinds of Intentional Psychology," in R. Healey, ed., *Reduction, Time and Reality* (New York: Cambridge, 1981); and *The Intentional Stance* (Cambridge: MIT, 1987).

3. Peter Smith, "Wit and Chutzpah," review of *The Intentional Stance* and Jerry A. Fodor's *Psychosemantics, Times Higher Education Supplement* (August 7, 1988), p. 22.

4. "The Stance Stance," commentary on *The Intentional Stance*, in *Behavioral and Brain Sciences*, XI (1988): 511–512.

5. R. A. Sharpe, in "Dennett's Journey Towards Panpsychism," *Inquiry*, XXXII (1989): 233–240, takes me to task on this point, using examples from Proust to drive home the point that "Proust draws our attention to possible lives and these possible lives are various. But in none of them is prediction of paramount importance" (240). I agree. I also agree that what makes people interesting (in novels and in real life) is precisely their unpredictability. But that unpredictability is only interesting against the backdrop of routine predictability on which all interpretation depends. As I note in *The Intentional Stance* (p. 79) in response to a similar objection of Fodor's, the same is true of chess: the game is interesting only because of the unpredictability of one's opponent, but that is to say: the intentional stance can usually eliminate *only* ninety percent of the legal moves.

6. Norton Nelkin, "Patterns," forthcoming.

7. *The Shaky Game: Einstein Realism and the Quantum Theory* (Chicago: University Press, 1986); see esp. p. 153n, and his comments there on Rorty, which I take to be consonant with mine here.

8. See *The Intentional Stance*, pp. 38–42, "Real patterns, deeper facts, and empty questions."

9. "Randomness and Mathematical Proof," *Scientific American*, CCXXXII (1975): 47–52.

10. More precisely: "A series of numbers is random if the smallest algorithm capable of specifying it to a computer has about the same number of bits of information as the series itself" (Chaitin, p. 48). This is what explains the fact that the "random number generator" built into most computers is not really properly named, since it is some function describable in a few bits (a little subroutine that is called for some output whenever a program requires a "random" number or series). If I send you the description of the pseudo-random number generator on my computer, you can use it to generate exactly the same infinite series of random-seeming digits.

11. Such schemes for efficient description, called compression algorithms, are widely used in computer graphics for saving storage space. They break the screen into uniformly colored regions, for instance, and specify region boundaries (rather like the "paint by numbers" line drawings sold in craft shops). The more complicated the picture on the screen, the longer the compressed description will be; in the worst case (a picture of confetti randomly sprinkled over the screen) the compression algorithm will be stumped, and can do no better than a verbatim bit map.

12. What about the "system" of pattern description that simply baptizes frames with proper names (*A* through *F*, in this case) and tells the receiver which frame is up by simply sending "*F*"? This looks much shorter than the bit map until we consider that such a description must be part of an entirely general system. How many proper names will we need to name all possible 900-dot frames? Trivially, the 900-bit binary number, 11111111.... To send the "worst-case" proper name will take exactly as many bits as sending the bit map. This confirms our intuition that proper names are maximally inefficient ways of couching generalizations ("Alf is tall and Bill is tall and...").

13. A. D. de Groot, *Thought and Choice in Chess* (The Hague: Mouton, 1965).

14. *Science, Perception and Reality* (Boston: Routledge & Kegan Paul, 1963).

15. In "Randomness and Perceived Randomness in Evolutionary Biology," *Synthese*, XLIII (1980): 287–329, William Wimsatt offers a nice example (296): while the insectivorous bird tracks individual insects, the anteater just averages over the ant-infested area; one might say that, while the bird's manifest image quantifies over insects, "ant" is a mass term for anteaters. See the discussion of this and related examples in my *Elbow Room* (Cambridge: MIT, 1984), pp. 108–110.

16. *The Intentional Stance*, pp. 37–39.

17. Martin Gardner introduced the Game of Life to a wide audience in two columns in *Scientific American* in October, 1970, and February, 1971. William Poundstone, *The Recursive Universe: Cosmic Complexity and the Limits of Scientific Knowledge* (New York: Morrow, 1985), is an excellent exploration of the game and its philosophical implications. Two figures from Poundstone's book are reproduced, with kind permission from the author and publisher, on pp. 198 and 199.

18. Poundstone, *op. cit.*, provides simple BASIC and IBM-PC assembly language simulations you can copy for your own home computer, and describes some of the interesting variations.

19. To see that the opposite poles share this view, see Fodor, *Psychosemantics* (Cambridge: MIT, 1987), ch. 1, "Introduction: the Persistence of the Attitudes"; and Paul Churchland, *Scientific Realism and the Plasticity of Mind* (New York: Cambridge, 1979), esp. p. 100: "For the P-theory [folk psychology] is in fact a marvelous intellectual achievement. It gives its possessor an explicit and systematic insight into the behaviour, verbal and otherwise, of some of the most complex agents in the environment, and its overall prowess in that respect remains unsurpassed by anything else our considerable theoretical efforts have produced."

20. *Intention* (New York: Blackwell, 1957).

21. A notable early exception is Sellars, who discussed the importance of just this sort of regularity in "Some Reflections on Language Games," *Philosophy of Science*, XXI (1954): 204–228. See especially the subsection of this classic paper, entitled "Pattern Governed and Rule Obeying Behavior," reprinted in Sellars's *Science, Perception and Reality*, pp. 324–327.

22. Several interpreters of a draft of this article have supposed that the conclusion I am urging here is that beliefs (or their contents) are *epiphenomena* having no causal powers, but this is a misinterpretation traceable to a simplistic notion of causation. If one finds a predictive pattern of the sort just described one has *ipso facto* discovered a causal power—a difference in the world that makes a subsequent difference testable by standard empirical methods of variable manipulation. Consider the crowd-drawing power of a sign reading "Free Lunch" placed in the window of a restaurant, and compare its power in a restaurant in New York to its power in a restaurant in Tokyo. The intentional level is obviously the right level at which to predict and explain such causal powers; the sign more reliably produces a particular belief in one population of perceivers than in the other, and variations in the color of typography of the sign are not as predictive of variations in crowd-drawing power as are variations in (perceivable) meaning. The fact that the regularities on which these successful predictions are based are efficiently capturable (only) in intentional terms and are not derived from "covering laws" does not show that the regularities are not "causal"; it just shows that philosophers have often relied on pinched notions of causality derived from exclusive attention to a few examples drawn from physics and chemistry. Smith has pointed out to me that here I am echoing Aristotle's claim that his predecessors had ignored final causes.

# II Scientific Perspectives on Emergence

# Introduction to Scientific Perspectives on Emergence

The chapters in this section are contemporary scientific perspectives on emergence. They cover a wide range of approaches to emergence and include examples of each of the leading ideas on emergence: irreducibility, unpredictability, unexplainability, conceptual novelty, and holism. Full of concrete cases, the chapters illustrate the flourishing scientific literature that invokes emergence today and are essential reading for anyone attempting to understand how science deals with emergence. The chapters in this section also provide rich raw material for empirically grounded generalizations about emergence in contemporary science. Comparing the philosophical analyses of emergence in part I with the empirical examples of purported emergent phenomena found in part II is a useful exercise.

## Complexity as a Source of Emergence

The paradigm cases of emergence in science today show much variety, but all the examples discussed in part II notably involve the behavior of a certain sort of complex system. The rise in scientific discussions of emergence coincides with the recent surge in scientific studies of complexity, and the selections included in this section reflect the central role of complex systems. Of course, some scientific discussions of emergence are not obviously tied to complex systems, and thus readers must make up their own minds about whether complex systems provide the only genuine examples of emergence in contemporary science. For example, a number of recent scientific articles discuss emergence specifically in the context of quantum mechanics, but those are not included here because they tend to be quite technical. For material in that domain, see chapter 6 in part I and the annotated bibliography at the end of this anthology.

Complex systems involve the collective behavior of a large group of relatively simple elements or agents, in which the interactions among these elements typically are local and nonlinear. Such systems support a distinction between what we might call *base* and *derivative* entities, properties, and phenomena. The base entities are the individual elements or agents in the system; the derivative entities are certain groups or collections or patterns or aggregations of the base entities. The base entities come in different forms. They might be individual molecules, individual organisms, perhaps human beings; they might be species of molecules, or populations or species of organisms. Note that not all base elements are microscopic, and not all derivative elements are macroscopic. *Base* and *derivative* are relative terms with a precise spatial scale only in a specific context. And in a given context, they could point to a temporal rather than spatial distinction.

The behaviors of such collective systems discussed in the chapters in this section include phase transitions in physical materials, such as melting ice; chaotic patterns such as the intermittent dripping of a faucet; robust nonchaotic patterns in collective physical phenomena, such as the rigidity of solids or the self-assembly, growth, and fission of vesicles; the flexible and fluid flocking of birds and schooling of fish; various kinds of ecological dynamics, such as predation, arms races, mutualism, and cheating; and unintended patterns in the way that humans distribute themselves in social groups.

None of these examples involve self-consciousness or the qualitative aspects of individual human experience, though. For such examples of emergent phenomena are perhaps the most typical examples of emergence found in contemporary philosophy of mind (this common philosophical approach is represented, for example, in chapters 3 and 7 of part I). Bringing the chapters in this section to the attention of philosophers is an attempt to redress this skew between the philosophy and the science of emergence. Valuable as that philosophical work has been, it has not yet paid sufficient attention to a rich and distinctively different set of examples within physics, chemistry, biology, and many of the social sciences. Philosophers should be able to help illuminate this new kind of emergence in science by applying expertise in philosophical analyses of complex and abstract concepts.

Many simple systems can be analyzed in such a way that, given initial conditions and boundary conditions, their typical behavior can be derived mathematically with complete certainty arbitrarily far into the future. By contrast, complex systems resist such mathematical analysis and divulge their typical long-term behavior only through patient observation of their behavior itself or perhaps the behavior of an appropriate computer simulation. Part II repeatedly emphasizes this lesson, and an especially vivid expression of the core problem occurs in chapter 14 (see also chapter 10). Conventional nonrelativistic quantum mechanics is a simple, well-confirmed theory that in principle explains the behavior of all matter, including the behavior of the familiar objects that we experience and interact with every day, so Laughlin and Pines term it a *theory of everything*. However, in practice this theory cannot be applied to systems containing more than a very few elementary particles. Laughlin and Pines put it this way: "We have succeeded in reducing all of ordinary physical behavior to a simple, correct Theory of Everything only to discover that it has revealed exactly nothing about many things of great importance"—in particular, it is not at all useful for explaining most of the behavior of everyday objects. That is to say, although in principle the structure and temporal development of most physical systems can be reduced to this theory of everything, reconstructing the details of that structure and development from first principles is impossible in practice. It is the failure of this reconstruction project, which is necessary for fully understanding derivative phenomena in terms of their basal features, that forms the essence of Anderson's argument for emergence in chapter 10.

If the theory of everything fails to provide practical explanations of the behavior of complex systems, where can successful explanations be found? Contemporary scientific approaches to complex systems offers the best place to look. In chapter 13 Simon briefly mentions many recent approaches to explaining the behavior of complex systems, including cybernetics, general systems theory, catastrophe theory, dynamical systems theory, genetic algorithms, and cellular automata. These approaches have various similarities, but also various differences, and, although references to the "science of complexity" occur occasionally, whether there is a shared subject matter unifying these approaches is unclear. Nevertheless, a characteristic feature of many of these explanatory frameworks is that macro-level collective behavior results from the iterated aggregation of basal-level interactions. This is well illustrated in this section by chapters 11, 12, 16, and 17, and in other sections by chapters 8, 9, and 21.

## Computational Models of Complexity

The recent boom in the scientific study of complex systems is partly a result of advances in the capability of inexpensive computers. Most scientific explanations of emergent behavior in part II take the form of individual- or agent-based computer systems (see also chapters 8 and 21). In individual- or agent-based models, the model's explicit dynamical rules apply explicitly only to basal individuals or to agents. If an individual-based model is simple enough, its behavior can be studied by working through examples on paper. Schelling in chapter 12 gives an especially simple and concrete example of such an individual-based model that explains human social group structure as the unintended and unanticipated result of individual human decisions. Schelling's model shows how derivative social structures can emerge out of the interaction of many elementary personal preferences of many individual people. The model vastly simplifies the myriad factors that affect the behavior of actual people, but the explanation's overall shape is general and reasonably persuasive. This leads to the potentially surprising, and perhaps unsettling, realization that social patterns involving a person could be merely the unplanned and unanticipated result of aggregating that person's own local individual preferences with those of many other people.

Most individual-based models are much too complex to study without the aid of computer simulations. This is illustrated in chapter 17, which presents a detailed and rather realistic individual-based model of prebiotic conditions in a complex fluid. In this model, the individuals are single molecules of water and nonaqueous monomers. The monomers are of two types: hydrophobic (having a thermodynamic tendency to avoid close contact with water molecules) and hydrophilic (having a thermodynamic tendency to accept close contact with water). Covalent bonds can bind these monomers into amphiphilic polymers with both hydrophobic and hydrophilic ends. In an aqueous medium, such amphiphiles spontaneously self-assemble into molecular aggregations, such as micelles. (Micelles are roughly spherical clusters of amphiphiles in which hydrophilic "heads" are all on the outside of the sphere and in close proximity with water molecules, and the hydrophobic "tails" are packed into the inside of the sphere as far away from the water as possible). Micelles can show continuously spontaneous growth through the addition of free-floating amphiphiles, until they get so large that they undergo fission and split into two "daughter" micelles. This model is intended to help explain how prebiotic conditions could give rise to the formation of spontaneously self-reproducing molecular aggregations that could function as primitive containers in which primitive energy-harvesting and information-carrying chemical systems could become localized and chemically integrated, thus providing a model of the origin of primitive lifelike molecular aggregations. The origin and functioning of living systems always has been one of the main sources of apparent emergent phenomena. It is no surprise, then, that a computer model concerned with the origin of life should exemplify palpable forms of emergence.

The computational embedding of the science of complex systems itself raises certain issues. In part II, readers should ask themselves whether the notion of emergence in complex systems applies only or especially to computational systems or models that are created by scientists, or whether it applies also to the real complex systems in nature that those models are intended to describe and explain. The computer models themselves exhibit dramatic apparent emergent behavior, so some are inclined to focus the application of emergence to the computer models themselves. Others, such as Bedau in chapter 8, are inclined to generalize the notion of emergence in computer models to apply to emergent phenomena in nature itself.

## Reduction in Principle and in Practice

A central consequence of the computational methodology of the scientific study of emergent behavior is that the investigation of emergent phenomena is inextricably empirical. Individual-based models are purely formal mathematical systems, so their behavior at any point in the future, given boundary conditions for the individuals in the system, in principle can be derived from the system's rules governing the basic entities. This in-principle derivability of a complex system's derivative state from its earlier basal states might be termed *dynamic reduction in principle.* Dynamic reduction in principle entails the base laws in an element's basal context (typically, just its local basal context) determining the subsequent basal state of the base element. *Dynamic* refers to the fact that the determination of the derivative elements by the base elements occurs over time. Dynamic reduction also entails that a system's derivative state at a given instant is determined by the combination of its base states at that instant—a *static* form of reduction in principle, one in which the base state determines the derivative state at a moment in time. The theory of everything in chapter 14 is so called because it provides both a static and a dynamic reduction in principle of the properties of everything, including familiar everyday objects.

However, as shown above, the theory of everything cannot be used to derive the behavior of everyday objects in practice. So, dynamic and static reduction in principle must be distinguished from what might be called dynamic and static *reduction in practice,* which involves the *practical feasibility* of calculating a system's derivative behavior from complete knowledge of its basal features. Reduction in practice is much stronger than reduction in principle because it entails being able to adequately understand or explain the behavior of derivative objects once the behavior of all its base elements can be fully understood or explained. Reduction in principle is more or less universally accepted, and Weinberg (chapter 18) perhaps comes closest to championing reduction in practice, but the vast majority of the scientists in this book repudiate reduction in practice. This point is expressed in different but recognizable forms in chapters 10, 12, 13, 14, and 17. For example, the failure of reduction in practice is evident in the failure of any attempt to reconstruct the current state or temporal development of many common physical principles from first principles, emphasized in different ways by Anderson in chapter 10 and Laughlin and Pines in chapter 14.

The failure of reduction in practice is the common ground for most appeals to emergence in the practice of contemporary science. In turn, this failure stems from the complexity of the systems in question. This explains why contemporary scientific discussions of emergence typically arise in attempts to describe and explain complex systems. It also accounts for a striking difference between philosophical and scientific discussions of emergence. The failure of reduction in principle is a dominant thread in traditional philosophical discussions of emergence. But this thread is largely absent in scientific explanations of complex systems because scientists typically view reduction in principle as tantamount to scientific common sense. Conversely, philosophers tend to be uninterested in the failure of reduction in practice, not simply because it would follow from the failure of reduction in principle, but also because focusing on what can be done in practice rather than in principle runs counter to a deeply entrenched philosophical culture. Even a nonreductionist apparatus such as supervenience concerns what is possible in principle. Nothing can be supervenient in principle but not in practice.

Scientists studying complex systems assume that their scientific activities are part of normal science. Novel methods might be required, with the heavy reliance on computer simulations rather than traditional analytically tractable theories being the most obvious example, but these methods are viewed simply as good scientific practice. Thus, scientific treatments of complex systems embrace emergence as a simple consequence of normal science and also as consistent with reduction, at least in principle. This scientific perspective is consistent with some philosophical perspectives on emergence, such as the views found in chapters 5 and 8. But it contrasts with the traditional philosophical equation of emergence with the failure of reduction even in principle, and the placement of emergent phenomena outside the scope of normal science. Echoes of this traditional perspective are evident in some of the chapters in part I, while other chapters emphasize various contrasting views. One main lesson to appreciate from this is that the physicalistic philosopher's insistence on the truth of reduction in principle simply leaves open the question of the existence of scientifically salient forms of emergence based on the failure of reduction in practice, even if these involve types of emergence that clearly are about real systems.

Although the distinction between reduction in principle and reduction in practice appears in virtually every chapter in part II, the distinction takes different forms and is expressed with different terminology. Often the distinction concerns dynamic reduction, that is, the derivation of system behavior over time, but sometimes static reduction is the issue. This complication is compounded by the lack of a standard terminology for these distinctions. Another terminological complication is that the distinction between dynamic and static reduction in principle and reduction in practice is orthogonal to the different analyses of reduction proposed by Nagel in chapter 19 and by Kim in chapters 7 and 24. Readers should be prepared to think through the consequences of the different combinations of these two distinctions.

If reduction in practice fails for a given system, then the system's derivative properties cannot be understood or explained from a complete understanding of its base properties and the basal laws governing them. In other words, the denial of reduction in practice entails fundamental epistemological limitations on the ability to understand or explain certain derivative behavior. Some people view the form of emergence associated with the denial of reduction in practice as essentially limited to this epistemological import, reflecting just the need for an epistemological crutch when explaining derivative phenomena. Others view this epistemological feature of scientific emergence as at least sometimes the consequence of a distinctive ontological structure in derivative phenomena, and therefore as reflecting a distinctive ontology for those phenomena. An important open question, addressed in chapters 8, 9, and 15, is whether the various antipathies to reduction in practice and corresponding sympathies to certain forms of emergence should be interpreted merely epistemologically, or whether they reflect some real ontological structure in the world.

A central point emphasized throughout this section (see also chapter 20) is that reduction in practice fails for complex systems. These systems are so complicated that their future global behavior can be derived from their underlying basal rules, given initial and boundary conditions, only by painstakingly working through all the basal effects of all their basic individuals and iterating this process sequentially through time. In other words, no alternative exists to simulating the systems on a computer and empirically observing the results. For this reason, the behavior of complex systems must be studied empirically.

This empirical quality of the study of emergent phenomena is readily apparent throughout the chapters in part II. What makes Schelling's model in chapter 12 surprising and unexpected, at least upon first encounter, is that what derivative structures will arise are not apparent without observing the global effect of a sequence of basic interactions. Similarly, even the authors of the model prebiotic chemistry in chapter 17 had no way to know under what conditions, if any, the model would yield growing and dividing micelles, short of actually observing a host of computer simulations. The same holds for the diverse models mentioned in chapters 13 and 15. This inability to anticipate global properties inclines some people to characterize emergence itself by whether the system's derivative behavior is surprising (see chapter 16, discussed below).

## Definitions and Distinctions Concerning Emergence

Empirical study of emergent phenomena is possible only if they can be defined precisely and practically. So it is no surprise that the scientific literature includes attempts to formulate critical distinctions needed for practical conceptions of emergence. As noted above, the scientific literature on emergence repeatedly is drawn to distinguish reduction in practice from reduction in principle, and a similar impulse is to distinguish different forms of emergence. Some of the chapters in part II employ informal verbal definitions of emergence (e.g., chapters 10, 13, and 14), while others attempt to make more precise and formal definitions (e.g., chapters 11, 15, 16, and 17). In both cases, comparison of these conceptions with those offered by philosophers in part I is useful.

An operational and empirically applicable definition of emergence is the motivation behind the differentiation of dimensions of emergent phenomena in chapter 11. Assad and Packard focus on two dimensions that distinguish kinds of emergence. One has to do with the nature of the emergent entity or process. The emergence literature contains discussions about whether emergence should be interpreted as fundamentally applied to properties, entities, states of affairs, processes, or something else. Assad and Packard, by contrast, call attention to the distinction between the emergence of structures, of functions, and of computational capacities. They assume that all these forms of emergence are possible, and that this dimension marks a useful distinction between different kinds of emergent phenomena.

As shown above, much of the scientific literature on emergence concerns the failure of reduction in practice, and the complexity of the systems in question is the cause. A little reflection reveals that the failure of reduction in practice comes in degrees; reduction can be more or less impractical because systems can be more or less complex. This second dimension of emergent phenomena emphasized by Assad and Packard involves distinguishing grades of underivability. (The notion of computational incompressibility emphasized in chapter 21 is part of the necessary background for understanding this dimension.) Bedau explicitly adopts Assad and Packard's conceptual framework of degrees of underivability in chapter 8. Readers might decide that they prefer to array kinds of emergence along some alternative scale of impracticality of reduction, but it is likely that some form of this distinction will prove useful.

The existence of degrees of derivability implies a lack of a sharp distinction between emergent and nonemergent phenomena for those views that define emergence in terms of underivability. Instead, different phenomena are simply more or less emergent, depending on their degree of resistance to derivation. From this perspective, the behavior of natural systems falls on a scale of more or less emergent. Viewing emergence as a matter of degree contrasts sharply with the dichotomous definitions typically favored by philosophers such as McLaughlin (chapter 4) and Kim (chapter 7) but conforms to the approach suggested by Wimsatt (chapter 5).

Scientific examples of emergent phenomena are sometimes epiphenomenal; that is, they are effects that do not themselves cause anything else. This raises the question of whether such examples of scientific emergence are objective scientific facts or merely exist in the eye of the beholder. One response is that emergent derivative structure is objective precisely when it is *not* epiphenomenal but has its own effects on the system in which it arises. This kind of emergent structure can have a function of its own inside the system; it can play a causal role in determining the system's behavior. In chapter 15 Crutchfield sketches a theory of "epsilon" machines that explain when a derivative structure has autonomous explanatory power. For example, a glider in a cellular automaton (see chapters 8, 9, and 21) is a dynamic pattern regularly moving across a lattice. Since gliders can change other derivative structures they encounter, they meet Crutchfield's criterion for not being merely in the eye of the beholder. (Dennett discusses the implications of related issues in chapter 9.)

Artificial life is the attempt to understand the essential nature of living systems by creating artificial systems that exhibit lifelike behavior. Those systems could be synthesized in software (computer models), in hardware (robots and autonomous agents), or in wetware (in a biochemical laboratory). The literature on artificial life is especially full of references to emergence, perhaps partly because life itself is one of the paradigm sources of emergent phenomena, and the debates about exactly what emergence means in the context of artificial life are seemingly interminable. Chapter 16 is an attempt to cut through these debates. The authors propose an operational test for emergence, somewhat analogous to the famous operational test for thinking machines proposed by Alan Turing at the dawn of artificial intelligence. The authors then apply their proposed test to eight examples of individual-based artificial life models. This test checks whether a scientist who knows the local basic rules that govern the system nevertheless is surprised by the global behavior that the system exhibits. On this view, a system with persistently surprising derivative behavior in the face of substantial examination and reflection about its basic rules exhibits a robust form of emergence. Some will object to the human subjectivity employed by this test, and it is an open question whether any subjective notion like surprise captures the apparently objective form of emergence that is rampant in artificial life systems. But the scientific literature on emergence contains widespread consensus about the need for some operational and empirical method of identifying emergent phenomena.

## Emergent Dynamical Hierarchies

Typical scientific examples of emergence, as noted above, involve systems that support a distinction between derivative and basic entities, properties, or phenomena. This distinction is not absolute but relative. Something is derivative (or basal) only relative to something else that is comparatively basic (or derivative). Furthermore, something can be derivative relative to one thing and basic relative to something else. For example, an individual grain of sand is derivative relative to the collection of many molecules out of which it is composed, and it is basal relative to the sand pile composed of that grain of sand along with many others. So, structural hierarchies can consist of successive layers of derivative phenomena that serve as base phenomena for successive derivative phenomena, and one kind of emergent phenomena can serve as the base for further, successive emergent phenomena.

This leads to the familiar picture of a hierarchy of potentially emergent phenomena, and of a parallel hierarchy of explanatory frameworks. The details of these hierarchies can be debated, and their resolution can vary, but the idea is very roughly reflected by the platitudes that chemical phenomena arise out of the phenomena of atomic physics, molecular biological phenomena arise out of chemical phenomena, cellular biological phenomena arise out of molecular biological phenomena, and so on. The precise contents of correct formulations of such platitudes are quite controversial, but many believe that they contain an important element of truth, and versions of this view are expressed in all three parts of this book. (See, among others, chapters 7, 10, and 22. Chapter 6 expresses a skeptical view about the usefulness of such levels.)

Simon draws attention to how emergent scientific hierarchies lead to the development of autonomous sciences of derivative phenomena. (Compare Fodor on special sciences in chapter 22.) Because of the practical impossibility in many cases of arriving at derivative properties from base properties, Simon emphasizes that "we can build nearly independent theories of each successive level of complexity." At the same time, in-principle reductionism enables scientists to "build bridging theories that show how each higher level can be accounted for in terms of the elements and relations of the next level below." (Compare Nagel on bridge laws in chapter 19.) In this way, certain sciences are both autonomous from basic sciences and dependent on them. This simultaneous autonomy and dependence is the most general signature of emergence.

Earlier, "dynamic" and "static" reduction were distinguished, both in principle and in practice. Dynamic reduction concerns how current derivative properties depend on and arise from previous base properties. Static reduction concerns how current derivative properties can be defined in terms of current base properties. Many scientific examples of emergence in this section amount to the practical difficulty of obtaining derivative features from base features over time. This dynamic form of emergence, although evident in chapter 8, contrasts with the static forms of emergence typically discussed by most philosophers. Dynamic emergence essentially involves how derivative states arise over time. A given derivative state is dynamically emergent not merely because of its intrinsic instantaneous properties, such as the global configuration of all its base properties, but as a result of how that global state arises over time.

Dynamical hierarchies are an especially dramatic form of dynamic emergence. These are scientific hierarchies of the type discussed above, but they are dynamical in two ways. First, dynamic hierarchies arise over time and develop new features over time; in particular, new levels are produced over time, perhaps in an open-ended manner. In this way, the creation and development of hierarchies is dynamic. Second, once a dynamical hierarchy arises and has a particular structure, it exhibits characteristic temporal patterns. These patterns are essentially dynamic, since they concern regular change over time. These characteristic dynamic patterns constitute a kind of robustness of derivative state, for they involve the continual flexible restoration of the system's derivative state. Robust dynamical patterns are characteristic of living systems, as with how living systems continually assimilate food from their environment and use it to maintain and repair themselves.

Dynamical hierarchies are a central focus of chapter 17. Rasmussen et al. formalize their conception of dynamical hierarchies and emergence, then simulate a dynamical hierarchy of growing and dividing micelles that emerge from an aqueous prebiotic environment containing simple monomers (see above). They are careful to explain what kind of downward causation arises in their dynamical hierarchy (compare the discussions of downward causation in chapters 6–8 and 24). An especially notable feature of this model is that it creates a dynamical hierarchy with multiple levels of robust emergence. One level of emergence is the production of amphiphilic polymers from hydrophilic and hydrophobic monomers. The next level of emergence involves the production of growing and dividing micelles from those same amphiphiles. The key point is that these two levels of emergent phenomena ultimately both emerge from the same fundamental base level: the monomers in water. This three-layer dynamical hierarchy reflects only a tiny fragment of the full dynamical hierarchies involved in complex real emergent phenomena, such as real forms of life, but it is a concrete step in exactly the right direction, and the path of the next steps is easy to see.

The chapters in this section are only the tip of a vast iceberg of discussions of emergence in contemporary science. The bibliography at the end of this anthology is a guide to some of the further literature. Flourishing innovations in the scientific study of complex systems and other areas of science make now an exciting time to be thinking about the variety of emergent phenomena in our accounts of nature and in nature itself. The chapters in this section provide a point of embarkation for this intellectual odyssey.

# 10 More is Different: Broken Symmetry and the Nature of the Hierarchical Structure of Science

**P. W. Anderson**

The reductionist hypothesis may still be a topic for controversy among philosophers, but among the great majority of active scientists I think it is accepted without question. The workings of our minds and bodies, and of all the animate or inanimate matter of which we have any detailed knowledge, are assumed to be controlled by the same set of fundamental laws, which except under certain extreme conditions we feel we know pretty well.

It seems inevitable to go on uncritically to what appears at first sight to be an obvious corollary of reductionism: that if everything obeys the same fundamental laws, then the only scientists who are studying anything really fundamental are those who are working on those laws. In practice, that amounts to some astrophysicists, some elementary particle physicists, some logicians and other mathematicians, and few others. This point of view, which it is the main purpose of this article to oppose, is expressed in a rather well-known passage by Weisskopf (*1*):

> Looking at the development of science in the Twentieth Century one can distinguish two trends, which I will call "intensive" and "extensive" research, lacking a better terminology. In short: intensive research goes for the fundamental laws, extensive research goes for the explanation of phenomena in terms of known fundamental laws. As always, distinctions of this kind are not unambiguous, but they are clear in most cases. Solid state physics, plasma physics, and perhaps also biology are extensive. High energy physics and a good part of nuclear physics are intensive. There is always much less intensive research going on than extensive. Once new fundamental laws are discovered, a large and ever increasing activity begins in order to apply the discoveries to hitherto unexplained phenomena. Thus, there are two dimensions to basic research. The frontier of science extends all along a long line from the newest and most modern intensive research, over the extensive research recently spawned by the intensive research of yesterday, to the broad and well developed web of extensive research activities based on intensive research of past decades.

The effectiveness of this message may be indicated by the fact that I heard it quoted recently by a leader in the field of materials science, who urged the participants at a meeting dedicated to "fundamental problems in condensed matter physics" to accept that there were few or no such problems and that nothing was left but extensive science, which he seemed to equate with device engineering.

The main fallacy in this kind of thinking is that the reductionist hypothesis does not by any means imply a "constructionist" one: The ability to reduce everything to simple fundamental laws does not imply the ability to start from those laws and reconstruct the universe. In fact, the more the elementary particle physicists tell us about the nature of the fundamental laws, the less relevance they seem to have to the very real problems of the rest of science, much less to those of society.

The constructionist hypothesis breaks down when confronted with the twin difficulties of scale and complexity. The behavior of large and complex aggregates of elementary particles, it turns out, is not to be understood in terms of a simple extrapolation of the properties of a few particles. Instead, at each level of complexity entirely new properties appear, and the understanding of the new behaviors requires research which I think is as fundamental in its nature as any other. That is, it seems to me that one may array the sciences roughly linearly in a hierarchy, according to the idea: The elementary entities of science X obey the laws of science Y.

| X | Y |
|---|---|
| solid state or many-body physics | elementary particle physics |
| chemistry | many-body physics |
| molecular biology | chemistry |
| cell biology | molecular biology |
| ⋮ | ⋮ |
| psychology | physiology |
| social sciences | psychology |

But this hierarchy does not imply that science X is "just applied Y." At each stage entirely new laws, concepts, and generalizations are necessary, requiring inspiration and creativity to just as great a degree as in the previous one. Psychology is not applied biology, nor is biology applied chemistry.

In my own field of many-body physics, we are, perhaps, closer to our fundamental, intensive underpinnings than in any other science in which non-trivial complexities occur, and as a result we have begun to formulate a general theory of just how this shift from quantitative to qualitative differentiation takes place. This formulation, called the theory of "broken symmetry," may be of help in making more generally clear the breakdown of the constructionist converse of reductionism. I will give an elementary and incomplete explanation of these ideas, and then go on to some more general speculative comments about analogies at other levels and about similar phenomena.

Before beginning this I wish to sort out two possible sources of misunderstanding. First, when I speak of scale change causing fundamental change I do not mean the rather well-understood idea that phenomena at a new scale may obey actually different fundamental laws—as, for example, general relativity is required on the cosmological scale and quantum mechanics on the atomic. I think it will be accepted that all ordinary matter obeys simple electrodynamics and quantum theory, and that really covers most of what I shall discuss. (As I said, we must all start with reductionism, which I fully accept.) A second source of confusion may be the fact that the concept of broken symmetry has been borrowed by the elementary particle physicists, but their use of the term is strictly an analogy, whether a deep or a specious one remaining to be understood.

Let me then start my discussion with an example on the simplest possible level, a natural one for me because I worked with it when I was a graduate student: the ammonia molecule. At that time everyone knew about ammonia and used it to calibrate his theory or his apparatus, and I was no exception. The chemists will tell you than ammonia "is" a triangular pyramid

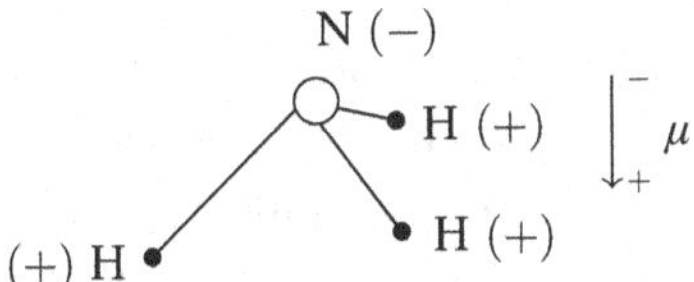


with the nitrogen negatively charged and the hydrogens positively charged, so that it has an electric dipole moment ($\mu$), negative toward the apex of the pyramid. Now this seemed very strange to me, because I was just being taught that nothing has an electric dipole moment. The professor was really proving that no nucleus has a dipole moment, because he was teaching nuclear physics, but as his arguments were based on the symmetry of space and time they should have been correct in general.

I soon learned that, in fact, they were correct (or perhaps it would be more accurate to say not incorrect) because he had been careful to say that no stationary state of a system (that is, one which does not change in time) has an electric dipole moment. If ammonia starts out from the above unsymmetrical state, it will not stay in it very long. By means of quantum mechanical tunneling, the nitrogen can leak through the triangle of hydrogens to the other side, turning the pyramid inside out, and, in fact, it can do so very rapidly. This is the so-called "inversion," which occurs at a frequency of about $3 \times 10^{10}$ per second. A truly stationary state can only be an equal superposition of the unsymmetrical pyramid and its inverse. That mixture does not have a dipole moment. (I warn the reader again that I am greatly oversimplifying and refer him to the textbooks for details.)

I will not go through the proof, but the result is that the state of the system, if it is to be stationary, must always have the same symmetry as the laws of motion which govern it. A reason may be put very simply: In quantum mechanics there is always a way, unless symmetry forbids, to get from one state to another. Thus, if we start from any one unsymmetrical state, the system will make transitions to others, so only by adding up all the possible unsymmetrical states in a symmetrical way can we get a stationary state. The symmetry involved in the case of ammonia is parity, the equivalence of left- and right-handed ways of looking at things. (The elementary particle experimentalists' discovery of certain violations of parity is not relevant to this question; those effects are too weak to affect ordinary matter.)

Having seen how the ammonia molecule satisfies our theorem that there is no dipole moment, we may look into other cases and, in particular, study progressively bigger systems to see whether the state and the symmetry are always related. There are other similar pyramidal molecules, made of heavier atoms. Hydrogen phosphide, $PH_3$, which is twice as heavy as ammonia, inverts, but at one-tenth the ammonia frequency. Phosphorus trifluoride, $PF_3$, in which the much heavier fluorine is substituted for hydrogen, is not observed to invert at a measurable rate, although theoretically one can be sure that a state prepared in one orientation would invert in a reasonable time.

We may then go on to more complicated molecules, such as sugar, with about 40 atoms. For these it no longer makes any sense to expect the molecule to invert itself. Every sugar molecule made by a living organism is spiral in the same sense, and they never invert, either by quantum mechanical tunneling or even under thermal agitation at normal temperatures. At this point we must forget about the possibility of inversion and ignore the parity symmetry: the symmetry laws have been, not repealed, but broken.

If, on the other hand, we synthesize our sugar molecules by a chemical reaction more or less in thermal equilibrium, we will find that there are not, on the average, more left- than right-handed ones or vice versa. In the absence of anything more complicated than a collection of free molecules, the symmetry laws are never broken, on the average. We needed living matter to produce an actual unsymmetry in the populations.

In really large, but still inanimate, aggregates of atoms, quite a different kind of broken symmetry can occur, again leading to a net dipole moment or to a net optical rotating power, or both. Many crystals have a net dipole moment in each elementary unit cell (pyroelectricity), and in some this moment can be reversed by an electric field (ferroelectricity). This asymmetry is a spontaneous effect of the crystal's seeking its lowest energy state. Of course, the state with the opposite moment also exists and has, by symmetry, just the same energy, but the system is so large that no thermal or quantum mechanical force can cause a conversion of one to the other in a finite time compared to, say, the age of the universe.

There are at least three inferences to be drawn from this. One is that symmetry is of great importance in physics. By symmetry we mean the existence of different viewpoints from which the system appears the same. It is only slightly overstating the case to say that physics is the study of symmetry. The first demonstration of the power of this idea may have been by Newton, who may have asked himself the question: What if the matter here in my hand obeys the same laws as that up in the sky—that is, what if space and matter are homogeneous and isotropic?

The second inference is that the internal structural of a piece of matter need not be symmetrical even if the total state of it is. I would challenge you to start from the fundamental laws of quantum mechanics and predict the ammonia inversion and its easily observable properties without going through the stage of using the unsymmetrical pyramidal structure, even though no "state" ever has that structure. It is fascinating that it was not until a couple of decades ago (*2*) that nuclear physicists stopped thinking of the nucleus as a featureless, symmetrical little ball and realized that while it really never has a dipole moment, it can become football-shaped or plate-shaped. This has observable consequences in the reactions and excitation spectra that are studied in nuclear physics, even though it is much more difficult to demonstrate directly than the ammonia inversion. In my opinion, whether or not one calls this intensive research, it is as fundamental in nature as many things one might so label. But it needed no new knowledge of fundamental laws and would have been extremely difficult to derive synthetically from those laws; it was simply an inspiration, based, to be sure, on everyday intuition, which suddenly fitted everything together.

The basic reason why this result would have been difficult to derive is an important one for our further thinking. If the nucleus is sufficiently small there is no real way to define its shape rigorously: Three or four or ten particles whirling about each other do not define a rotating "plate" or "football." It is only as the nucleus is considered to be a many-body system—in what is often called the $N \to \infty$ limit—that such behavior is rigorously definable. We say to ourselves: A macroscopic body of that shape would have such-and-such a spectrum of rotational and vibrational excitations, completely different in nature from those which would characterize a featureless system. When we see such a spectrum, even not so separated, and somewhat imperfect, we recognize that the nucleus is, after all, not macroscopic; it is merely approaching macroscopic behavior. Starting with the fundamental laws and a computer, we would have to do two impossible things—solve a problem with infinitely many bodies, and then apply the result to a finite system—before we synthesized this behavior.

A third insight is that the state of a really big system does not at all have to have the symmetry of the laws which govern it; in fact, it usually has less symmetry. The outstanding example of this is the crystal: Built from a substrate of atoms and space according to laws which express the perfect homogeneity of space, the crystal suddenly and unpredictably displays an entirely new and very beautiful symmetry. The general rule, however, even in the case of the crystal, is that the large system is less symmetrical than the underlying structure would suggest: Symmetrical as it is, a crystal is less symmetrical than perfect homogeneity.

Perhaps in the case of crystals this appears to be merely an exercise in confusion. The regularity of crystals could be deduced semiempirically in the mid-19th century without any complicated reasoning at all. But sometimes, as in the case of superconductivity, the new symmetry—now called broken symmetry because the original symmetry is no longer evident—may be of an entirely unexpected kind and extremely difficult to visualize. In the case of superconductivity, 30 years elapsed between the time when physicists were in possession of every fundamental law necessary for explaining it and the time when it was actually done.

The phenomenon of superconductivity is the most spectacular example of the broken symmetries which ordinary macroscopic bodies undergo, but it is of course not the only one. Antiferromagnets, ferroelectrics, liquid crystals, and matter in many other states obey a certain rather general scheme of rules and ideas, which some many-body theorists refer to under the general heading of broken symmetry. I shall not further discuss the history, but give a bibliography at the end of this article (*3*).

The essential idea is that in the so-called $N \to \infty$ limit of large systems (on our own, macroscopic scale) it is not only convenient but essential to realize that matter will undergo mathematically sharp, singular "phase transitions" to states in which the microscopic symmetries, and even the microscopic equations of motion, are in a sense violated. The symmetry leaves behind as its expression only certain characteristic behaviors, for instance, long-wavelength vibrations, of which the familiar example is sound waves; or the unusual macroscopic conduction phenomena of the superconductor; or, in a very deep analogy, the very rigidity of crystal lattices, and thus of most solid matter. There is, of course, no question of the system's really violating, as opposed to breaking, the symmetry of space and time, but because its parts find it energetically more favorable to maintain certain fixed relationships with each other, the symmetry allows only the body as a whole to respond to external forces.

This leads to a "rigidity," which is also an apt description of superconductivity and superfluidity in spite of their apparent "fluid" behavior. [In the former case, London noted this aspect very early (*4*).] Actually, for a hypothetical gaseous but intelligent citizen of Jupiter or of a hydrogen cloud somewhere in the galactic center, the properties of ordinary crystals might well be a more baffling and intriguing puzzle than those of superfluid helium.

I do not mean to give the impression that all is settled. For instance, I think there are still fascinating questions of principle about glasses and other amorphous phases, which may reveal even more complex types of behavior. Nevertheless, the role of this type of broken symmetry in the properties of inert but macroscopic material bodies is now understood, at least in principle. In this case we can see how the whole becomes not only more than but very different from the sum of its parts.

The next order of business logically is to ask whether an even more complete destruction of the fundamental symmetries of space and time is possible and whether new phenomena then arise, intrinsically different from the "simple" phase transition representing a condensation into a less symmetric state.

We have already excluded the apparently unsymmetric cases of liquids, gases, and glasses. (In any real sense they are more symmetric.) It seems to me that the next stage is to consider the system which is regular but contains information. That is, it is regular in space in some sense so that it can be "read out," but it contains elements which can be varied from one "cell" to the next. An obvious example is DNA; in everyday life, a line of type or a movie film have the same structure. This type of "information-bearing crystallinity" seems to be essential to life. Whether the development of life requires any further breaking of symmetry is by no means clear.

Keeping on with the attempt to characterize types of broken symmetry which occur in living things, I find that at least one further phenomenon seems to be identifiable and either universal or remarkably common, namely, ordering (regularity or periodicity) in the time dimension. A number of theories of life processes have appeared in which regular pulsing in time plays an important role: theories of development, of growth and growth limitation, and of the memory. Temporal regularity is very commonly observed in living objects. It plays at least two kinds of roles. First, most methods of extracting energy from the environment in order to set up a continuing, quasi-stable process involve time-periodic machines, such as oscillators and generators, and the processes of life work in the same way. Second, temporal regularity is a means of handling information, similar to information-bearing spatial regularity. Human spoken language is an example, and it is noteworthy that all computing machines use temporal pulsing. A possible third role is suggested in some of the theories mentioned above: the use of phase relationships of temporal pulses to handle information and control the growth and development of cells and organisms (5).

In some sense, structure—functional structure in a teleological sense, as opposed to mere crystalline shape—must also be considered a stage, possibly intermediate between crystallinity and information strings, in the hierarchy of broken symmetries.

To pile speculation on speculation, I would say that the next stage could be hierarchy or specialization of function, or both. At some point we have to stop talking about decreasing symmetry and start calling it increasing complication. Thus, with increasing complication at each stage, we go on up the hierarchy of the sciences. We expect to encounter fascinating and, I believe, very fundamental questions at each stage in fitting together less complicated pieces into the more complicated system and understanding the basically new types of behavior which can result.

There may well be no useful parallel to be drawn between the way in which complexity appears in the simplest cases of many-body theory and chemistry and the way it appears in the truly complex cultural and biological ones, except perhaps to say that, in general, the relationship between the system and its parts is intellectually a one-way street. Synthesis is expected to be all but impossible; analysis, on the other hand, may be not only possible but fruitful in all kinds of ways: Without an understanding of the broken symmetry in superconductivity, for instance, Josephson would probably not have discovered his effect. [Another name for the Josephson effect is "macroscopic quantum-interference phenomena": interference effects observed between macroscopic wave functions of electrons in superconductors, or of helium atoms in superfluid liquid helium. These phenomena have already enormously extended the accuracy of electromagnetic measurements, and can be expected to play a great role in future computers, among other possibilities, so that in the long run they may lead to some of the major technological achievements of this decade (*6*).] For another example, biology has certainly taken on a whole new aspect from the reduction of genetics to biochemistry and biophysics, which will have untold consequences. So it is not true, as a recent article would have it (*7*), that we each should "cultivate our own valley, and not attempt to build roads over the mountain ranges . . . between the sciences." Rather, we should recognize that such roads, while often the quickest shortcut to another part of our own science, are not visible from the viewpoint of one science alone.

The arrogance of the particle physicist and his intensive research may be behind us (the discoverer of the positron said "the rest is chemistry"), but we have yet to recover from that of some molecular biologists, who seem determined to try to reduce everything about the human organism to "only" chemistry, from the common cold and all mental disease to the religious instinct. Surely there are more levels of organization between human ethology and DNA than there are between DNA and quantum electrodynamics, and each level can require a whole new conceptual structure.

In closing, I offer two examples from economics of what I hope to have said. Marx said that quantitative differences become qualitative ones, but a dialogue in Paris in the 1920's sums it up even more clearly:

Fitzgerald: The rich are different from us.

Hemingway: Yes, they have more money.

### Note

This chapter originally appeared in *Science* 177 (1972), 393–396. The original article was an expanded version of a Regents' Lecture given in 1967 at the University of California, La Jolla.

### References

1. V. F. Weisskopf, in *Brookhaven Nat. Lab. Publ. 888T360* (1965). Also see *Nuovo Cimento Suppl. Ser I* 4, 465 (1966); *Phys. Today* 20 (No. 5), 23 (1967).

2. A. Bohr and B. R. Mottelson, *Kgl. Dan. Vidensk. Selsk. Mat. Fys. Medd.* 27, 16 (1953).

3. Broken symmetry and phase transitions: L. D. Landau, *Phys. Z. Sowjetunion* 11, 26, 542 (1937). Broken symmetry and collective motion, general: J. Goldstone, A. Salam, S. Weinberg, *Phys. Rev.* 127, 965 (1962); P. W. Anderson, *Concepts in Solids* (Benjamin, New York, 1963), pp. 175–182;

B. D. Josephson, thesis, Trinity College, Cambridge University (1962). Special cases: antiferromagnetism, P. W. Anderson, *Phys. Rev.* 86, 694 (1952); superconductivity, ———, *ibid.* 110, 827 (1958); *ibid.* 112, 1900 (1958); Y. Nambu, *ibid.* 117, 648 (1960).

4. F. London, *Superfluids* (Wiley, New York, 1950), vol. 1.

5. M. H. Cohen, *J. Theor. Biol.* 31, 101 (1971).

6. J. Clarke, *Amer. J. Phys.* 38, 1075 (1969); P. W. Anderson, *Phys. Today* 23 (No. 11), 23 (1970).

7. A. B. Pippard, *Reconciling Physics with Reality* (Cambridge Univ. Press, London, 1972).

# 11 Emergence

**Andrew Assad and Norman H. Packard**

Since its inception, the field of Artificial Life has consistently referred to the property of "emergent" phenomena as one of its primary distinguishing features [1]. However, despite its frequent use in the literature, a clear, concise, and widely accepted definition of the term "emergent" has not yet been realized. Rather than attempting to propose such a standard, this chapter offers a definition that is relevant to the study at hand, and perhaps, will contribute to what may ultimately be an accepted definition.

In his opening essay to the *Proceedings of the First Workshop on Artificial Life*, Langton [2] includes *emergent behavior* as one of the fundamental characteristics of an ALife system. At first glance, his contrasting of emergent behavior with the *prespecified behavior* characteristic of AI systems seems to provide a sharp, intuitive distinction between the fields. However, upon further inspection this distinction may not be so clear.

We will be studying emergence in the realm of an ALife model, but it is worth noting that the concept has a rich history, which the ALife approach exemplified by this work will hopefully expand. The idea of emergence could perhaps be traced to Heraclitus and his theory of flux, and Anaxagoras, with his theory of *perichoresis*, which held that all discernable structure in the world is a result of a dynamical unmixing process (his version of emergence) that began with a homogeneous chaos.

In modern thought, the scientific paradigm beginning with Newton has been somewhat antithetical to the idea of emergence, as the power of the paradigm has often come from the exact derivability of phenomena (which, as we argue below, make a phenomena non-emergent). The first ideas of emergence in biological evolution were implicit in Darwin [3], though he was perhaps too conservative to voice the concept very loudly. Subsequently, Bergson, a rather nonconservative philosopher, voiced the concept rather more loudly [4].

Since these beginnings, discussion of emergence may have developed in two broad overlapping areas, theoretical biology and cybernetics [5, 6, 7, 8, 9, 10], and more recently and more or less independently in the area of computational models, starting with the work of McCullouch and Pitts [11], Kauffman [12], and Holland [13], though this early work on computational models does not always explicitly emphasize emergence. Computational emergence has only recently been named and studied in its own right [14], and is now the basis of most ALife studies [2]. Extended discussions of emergence in the context of ALife have been taking place on the network [15], and an extended bibliography may be found also be found there [16].

There is some controversy about whether true, life-like emergence can occur within a purely computational domain [5, 8, 17, 18]; we take the view here that some sort of nontrivial emergence can indeed occur in a purely computational domain, and we aim to demonstrate its occurrence by example.

Quite generally, emergence usually refers to two or more levels of description. In the simplest case of two levels (which is all we consider in this chapter), we will call the level on which the model is defined in terms of interactions between components the *microscopic level*. The level on which phenomena emerge, as a global property of the collection of components, is the *macroscopic level*.

Typically, Alife models which produce unexpected macroscopic behavior that is not immediately predictable upon inspection of the specification of the system are said to exhibit some form of emergent behavior. Although certainly not a rigorous definition, this notion of emergence seems to be fairly pervasive in the literature. Thus, in many respects, "emergence" seems to be in the eye of the beholder. What is a wholly unexpected behavior from one perspective may be immediately obvious from another, though it is clear that some types of emergence (any emergence resulting in universal computation) must be able to produce truly unexpected behavior.

It is in this spirit of relativity that we propose a *scale* in which to measure emergence rather than a singular definition. While this scale does not circumvent the problem of lack of precision, it hopefully provides a common yardstick from which the level of emergence of a behavior can be estimated. The scale ranges from a "weak" or non-emergent level at the top to a "strong" or maximally emergent level at the bottom:

Non-emergent Behavior is immediately deducible upon inspection of the specification or rules generating it.

Weakly emergent Behavior is deducible in hindsight from the specification *after* observing the behavior.

⋮

Strongly emergent Behavior is deducible in theory, but its elucidation is prohibitively difficult.

Maximally emergent Behavior is impossible to deduce from the specification.

Note that the levels in this scale are not intended to be absolutes at this point, since we have refrained from attempting to give a definition of terms like "deducible," general enough to be independent of the context of different behaviors and models.

However, the scale should at least allow for an increased accuracy or resolution in discussing a potentially emergent aspect of a complex system's behavior.

What is it that emerges? We distinguish between three different answers to this question:

Structure The emergence of patterned structure, either in space time configurations, or in more abstract symbolic spaces. Examples include Benard cells in fluid convection, flocking behavior of Boids, and gliders in the game of life.

Computation The emergence of computational processing, over and above the computation automatically implemented in the formation of a structure. Examples include Holland's classifier system, and artificial life models that include the possibility for evolution of computation performed by organisms [1, 19], or, in the unique model of Fontana, computation performed directly by the microscopic elements [20].

Functionality The emergence of functionality, where functionality is defined in terms of actions that are functional, or beneficial to the microscopic components. So far, the primary examples of emergence of functionality are a subset of the examples of computational emergence, where the computation performs a function for the microscopic organisms.

The relationship between these different types of emergence is unclear. We hypothesize that there is a hierarchy of necessity: emergence of functionality requires the emergence of computation, which requires the emergence of structure. On the other hand, emergence of structure does not necessarily imply the emergence of computation, which does not necessarily imply the emergence of functionality.

## Notes

This chapter originally appeared as section 2 of "Emergent Colonization in an Artificial Ecology," in F. Varela and P. Bourgine (eds.), *Towards a Practice of Autonomous Systems: Proceedings of the First European Conference on Artifical Life*, pp. 143–152, Cambridge, MA: The MIT Press, 1992. The notes have been renumbered to reflect the text reprinted here.

1. Packard, N. H., and M. Bedau. 1991. Measurement of evolutionary activity and teleology. In *Artificial Life II*, edited by C. Langton, J. Farmer, and S. Rasmussen. Reading, MA: Addison-Wesley.

2. Langton, C. 1989. Artifical Life. In *Artificial Life*, edited by C. Langton. Reading, MA: Addison-Wesley.

3. Darwin, C. 1859. *On the Origin of Species by Means of Natural Selection.* London: reprinted Cambridge, MA: Harvard University Press, 1964.

4. Bergson, H. 1911. *Creative Evolution.* Translated by A. Mitchell. New York.

5. Rosen, R. 1973. On the generation of metabolic novelties in evolution. In *Biogenesis, Evolution, Homeostasis*, edited by A. Locker. New York: Pergamon Press.

6. Rosen, R. 1985. *Anticipatory Systems.* New York: Pergamon Press.

7. Pattee, H. H. 1972. The nature of hierarchical controls in living matter. In *Foundations of Mathematical Biology, Vol. I,* edited by R. Rosen. New York: Academic Press.

8. Pattee, H. H. 1989. Simulations, realizations, and theories of life. In *Artificial Life,* edited by C. Langton. Reading, MA: Addison-Wesley.

9. Rossler, O. 1984. Deductive prebiology. In *Molecular Evolution and the Prebiological Paradigm,* edited by K. Matsuno, K. Dose, K. Harada, and D. L. Rohlfing. Plenum.

10. Wiener, N. 1948. *Cybernetics, or Control and Communication in the Animal and the Machine.* Cambridge, MA: MIT Press.

11. McCulloch, W., and W. Pitts. 1943. A logical calculus of the ideas immanent in nervous activity. *Bull. Math. Biophysiscs* 5:115.

12. Kauffman, S. A. 1969. *J. Theoret. Biol.* 22:437.

13. Holland, J. 1975. *Adaptation in Natural and Artificial Systems.* Ann Arbor: The University of Michigan Press.

14. Forrest, S. 1989. Emergent computation: Self-organizing, collective, and cooperative phenomena in natural and artificial computing networks. *Proceedings of the Ninth Annual Center for Nonlinear Studies and Computing Division Conference, Physica D.*

15. Freeman, E. T., and M. W. Lugowsky. 1989–1991. *Artificial Life Digest,* alife-request@iuvax.cs.indiana.edu.

16. Cariani, P. 1991. In *Artificial Life Digest* 49, maintained by E. T. Freeman and M. W. Lugowsky, alife-request@iuvax.cs.indiana.edu.

17. Cariani, P. 1991. Emergence and Artificial Life. In *Artificial Life II,* edited by C. Langton, D. Farmer, and S. Rasmussen. Reading, MA: Addison-Wesley.

18. Cariani, P. 1990. Adaptation and emergence in organisms and devices. In press, *Journal of General Evolution.*

19. Ackley, D. H., and M. L. Littman. 1991. Interaction between learning and evolution. In *Artificial Life II,* edited by C. Langton, D. Farmer, and S. Rasmussen. Reading, MA: Addison-Wesley.

20. Fontana, W. 1991. Algorithmic Chemistry. In *Artificial Life II,* edited by C. Langton, D. Farmer, and S. Rasmussen. Reading, MA: Addison-Wesley.

# 12 Sorting and Mixing: Race and Sex

**Thomas Schelling**

People get separated along many lines and in many ways. There is segregation by sex, age, income, language, religion, color, personal taste, and the accidents of historical location. Some segregation results from the practices of organizations. Some is deliberately organized. Some results from the interplay of individual choices that discriminate. Some of it results from specialized communication systems, like languages. And some segregation is a corollary of other modes of segregation: residence is correlated with job location and transport.

If blacks exclude whites from their church, or whites exclude blacks, the segregation is organized; and it may be reciprocal or one-sided. If blacks just happen to be Baptists and whites Methodists, the two colors will be segregated Sunday morning whether they intend to be or not. If blacks join a black church because they are more comfortable among their own color, and whites a white church for the same reason, undirected individual choice can lead to segregation. And if the church bulletin board is where people advertise rooms for rent, blacks will rent rooms from blacks and whites from whites because of a communication system that is connected with churches that are correlated with color.

Some of the same mechanisms segregate college professors. The college may own some housing, from which all but college staff are excluded. Professors choose housing commensurate with their incomes, and houses are clustered by price while professors are clustered by income. Some professors prefer an academic neighborhood; any differential in professorial density will cause them to converge and increase the local density, and attract more professors. And house-hunting professors learn about available housing from colleagues and their spouses, and the houses they learn about are naturally the ones in neighborhood where professors already live.

The similarity ends there, and nobody is about to propose a commission to desegregate academics. Professors are not much missed by those they escape from in their residential choices. They are not much noticed by those they live among, and, though proportionately concentrated, are usually a minority in their neighborhood. While indeed they escape classes of people they would not care to live among, they are more conscious of where they do live than of where they don't, and the active choice is more like congregation than segregation, though the result may not be so different.

This chapter is about the kind of segregation—or separation, or sorting—that can result from discriminatory individual behavior. By "discriminatory" I mean reflecting an awareness, conscious or unconscious, of sex or age or religion or color or whatever the basis of segregation is, an awareness that influences decisions on where to live, whom to sit by, what occupation to join or to avoid, whom to play with, or whom to talk to. It examines some of the *individual* incentives and individual perceptions of difference that can lead *collectively* to segregation. It also examines the extent to which inferences can be drawn from actual collective segregation about the preferences of individuals, the strengths of those preferences, and the facilities for exercising them.

The main concern is segregation by "color" in the United States. The analysis, though, is so abstract that any twofold distinction could constitute an interpretation—whites and blacks, boys and girls, officers and enlisted men, students and faculty. The only requirement of the analysis is that the distinction be twofold, exhaustive, and recognizable. (Skin color, of course, is neither dichotomous nor even unidimensional, but by convention the distinction is nearly twofold, even in the United States census.)

At least two main processes of segregation are outside this analysis. One is organized action—legal or illegal, coercive or merely exclusionary, subtle or flagrant, open or covert, kindly or malicious, moralistic or pragmatic. The other is the process, largely but not entirely economic, by which the poor get separated from the rich, the less educated from the more educated, the unskilled from the skilled, the poorly dressed from the well dressed—in where they work and live and eat and play, in whom they know and whom they date and whom they go to school with. Evidently color is correlated with income, and income with residence; so even if residential choices were colorblind and unconstrained by organized discrimination, whites and blacks would not be randomly distributed among residences.

It is not easy to draw the lines separating "individually motivated" segregation from the more organized kind or from the economically induced kind. Habit and tradition are substitutes for organization. Fear of sanctions can coerce behavior whether or not the fear is justified, and whether the sanctions are consensual, conspiratorial, or dictated. Common expectations can lead to concerted behavior.

The economically induced separation is also intermixed with discrimination. To choose a neighborhood is to choose neighbors. To pick a neighborhood with good schools, for example, is to pick a neighborhood of *people* who want good schools. People may furthermore rely, even in making economic choices, on information that is color-discriminating; believing that darker-skinned people are on the average poorer than lighter-skinned, one may consciously or unconsciously rely on color as an index of poverty or, believing that others rely on color as an index, adopt their signals and indices accordingly.

For all these reasons, the lines dividing the individually motivated, the collectively enforced, and the economically induced segregation are not clear lines at all. They are furthermore not the only mechanisms of segregation. Separate or specialized communication systems—especially distinct languages—can have a strong segregating influence that, though interacting with the three processes mentioned, is nevertheless a different one.

### Individual Incentives and Collective Results

Economists are familiar with systems that lead to aggregate results that the individual neither intends nor needs to be aware of, results that sometimes have no recognizable counterpart at the level of the individual. The creation of money by a commercial banking system is one; the way savings decisions cause depressions or inflations is another.

Biological evolution is responsible for a lot of sorting and separating, but the little creatures that mate and reproduce and forage for food would be amazed to know that they were bringing about separation of species, territorial sorting, or the extinction of species. Among social examples, the coexistence or extinction of second languages is a phenomenon that, though affected by decrees and school curricula, corresponds to no conscious collective choice.

Romance and marriage... are exceedingly individual and private activities, at least in this country, but their genetic consequences are altogether aggregate. The law and the church may constrain us in our choices, and some traditions of segregation are enormously coercive; but, outside of royal families, there are few marriages that are part of a genetic plan. When a short boy marries a tall girl, or a blonde a brunette, it is no part of the individual's purpose to increase genetic randomness or to change some frequency distribution within the population.

Some of the phenomena of segregation may be similarly complex in relation to the dynamics of individual choice. One might even be tempted to suppose that some "unseen hand" separates people in a manner that, though foreseen and intended by no one, corresponds to some consensus or collective preference or popular will. But in economics we know a great many macro-phenomena, like depression and inflation, that do not reflect any universal desire for lower incomes or higher prices. The same applies to bank failures and market crashes. What goes on in the "hearts and minds" of small savers has little to do with whether or not they cause a depression. The hearts and minds and motives and habits of millions of people who participate in a segregated society may or may not bear close correspondence with the massive results that collectively they can generate.

A special reason for doubting any social efficiency in aggregate segregation is that the range of choice is often so meager. The demographic map of almost any American metropolitan area suggests that it is easy to find residential areas that are all white or nearly so and areas that are all black or nearly so but hard to find localities in which neither whites nor nonwhites are more than, say, three-quarters of the total. And, comparing decennial maps, it is nearly impossible to find an area that, if integrated within that range, will remain integrated long enough for a couple to get their house paid for or their children through school.

## Some Quantitative Constraints

Counting blacks and whites in a residential block or on a baseball team will not tell how they get along. But it tells something, especially if numbers and ratios matter to the people who are moving in or out of the block or being recruited for the team. With quantitative analysis there are a few logical constraints, analogous to the balance-sheet identities in economics. (Being logical constraints, they contain no news unless one just never thought of them before.)

The simplest constraint on dichotomous mixing is that, within a given set of boundaries, not both groups can enjoy numerical superiority. For the whole population the numerical ratio is determined at any given time; but locally, in a city or a neighborhood, a church or a school or a restaurant, either blacks or whites can be a majority. But if each insists on being a local majority, there is only one mixture that will satisfy them—complete segregation.

Relaxing the condition, if whites want to be at least three-fourths and blacks at least one-third, it won't work. If whites want to be at least two-thirds and blacks no fewer than one-fifth, there is a small range of mixtures that meet the conditions. And not everybody can be in the mixtures if the overall ratio is outside the range.

In spatial arrangements, like a neighborhood or a hospital ward, everybody is next to somebody. A neighborhood may be 10 percent black or white; but if you have a neighbor on either side, the minimum nonzero percentage of opposite color is 50. If people draw their boundaries differently we can have everybody in a minority: at dinner, with men and women seated alternately, everyone is outnumbered two to one locally by the opposite sex but can join a three-fifths majority if he extends his horizon to the next person on either side.

## Separating Mechanisms

The simple mathematics of ratios and mixtures tells us something about what outcomes are logically possible, but tells us little about the behavior that leads to, or that leads away from, particular outcomes. To understand what kinds of segregation or integration may result from individual choice, we have to look at the processes by which various mixtures and separations are brought about. We have to look at the incentives and the behavior that the incentives motivate, and particularly the way that different individuals comprising the society impinge on each other's choices and react to each other's presence.

There are many different incentives or criteria by which blacks and whites, or boys and girls, become separated. Whites may simply prefer to be among whites and blacks among blacks. Alternatively, whites may merely avoid or escape blacks and blacks avoid or escape whites. Whites may prefer the company of whites, while the blacks don't care. Whites may prefer to be among whites and blacks also prefer to be among whites, but if the whites can afford to live or to eat or to belong where the blacks cannot afford to follow, separation can occur.

Whites and blacks may not mind each other's presence, may even prefer integration, but may nevertheless wish to avoid minority status. Except for a mixture at exactly 50:50, no mixture will then be self-sustaining because there is none without a minority, and if the minority evacuates, complete segregation occurs. If both blacks and whites can tolerate minority status but place a limit on how small the minority is—for example, a 25 percent minority—initial mixtures ranging from 25 percent to 75 percent will survive but initial mixtures more extreme than that will lose their minority members and become all of one color. And if those who leave move to where they constitute a majority, they will increase the majority there and may cause the other color to evacuate.

Evidently if there are lower limits to the minority status that either color can tolerate, and if complete segregation obtains initially, no individual will move to an area dominated by the other color. Complete segregation is then a stable equilibrium.

### Sorting and Scrambling

Minor-league players at Dodgertown—the place where Dodger-affiliated clubs train in the spring—are served cafeteria-style. "A boy takes the first seat available," according to the general manager. "This has been done deliberately. If a white boy doesn't want to eat with a colored boy, he can go out and buy his own food. We haven't had any trouble."[8]

Major-league players are not assigned seats in their dining hall; and though mixed tables are not rare, they are not the rule either. If we suppose that major- and minor-league racial attitudes are not strikingly different, we may conclude that racial preference in the dining hall is positive but less than the price of the nearest meal.

Actually, though, there is an alternative: whites and blacks in like-colored clusters can enter the line together and, once they have their trays, innocently take the next seats alongside each other. Evidently they don't. If they did, some scrambling system would have had to be invented. Maybe we conclude, then, that the racial preferences, though enough to make separate eating the general rule, are not strong enough to induce the slight trouble of picking partners before getting food. Or perhaps we conclude that players lack the strategic foresight to beat the cafeteria line as a seat-scrambling device.

But even a minor-league player knows how to think ahead a couple of outs in deciding whether a sacrifice fly will advance the ball team. It is hard to believe that if a couple of players wanted to sit together it would not occur to them to meet at the beginning of the line; and the principle extends easily to segregation by color.

We are left with some alternative hypotheses. One is that players are relieved to have an excuse to sit without regard to color, and cafeteria-line–scrambling eliminates an embarrassing choice. Another is that players can ignore, accept, or even prefer mixed tables but become uncomfortable or self-conscious, or think that others are uncomfortable or self-conscious, when the mixture is lopsided. Joining a table with blacks and whites is a casual thing, but being the seventh at a table with six players of opposite color imposes a threshold of self-consciousness that spoils the easy atmosphere and can lead to complete and sustained separation.

Hostesses are familiar with the problem. Men and women mix nicely at stand-up parties until, partly at random and partly because a few men or women get stuck in a specialized conversation, some clusters form that are nearly all male or all female; selective migration then leads to the cocktail-party equivalent of the Dodgertown major-league dining hall. Hostesses, too, have their equivalent of the cafeteria-line rule: they alternate sexes at the dinner table, grasp people by the elbows and move them around the living room, or bring in coffee and make people serve themselves to disturb the pattern.

Sometimes the problem is the other way around. It is usually good to segregate smokers from non-smokers in planes and other enclosed public places; swimmers and surfers should be segregated in the interest of safety; and an attempt is made to keep slow-moving vehicles in the right-hand lane of traffic. Many of these dichotomous groupings are asymmetrical: cigar smokers are rarely bothered by people who merely breathe; the surfer dislikes having his board hit anybody in the head but there is somebody else who dislikes it much more; and the driver of a slow truck passing a slower one on a long grade is less conscious of who is behind him than the driver behind is of the truck in front. Styles of behavior differ: surfers like to be together and cluster somewhat in the absence of regulation; water-skiers prefer dispersal and are engaged in a mobile sport, and rarely reach accommodation with swimmers on how to share the water.

These several processes of separation, segregation, sharing, mixing, dispersal—sometimes even pursuit—have a feature in common. The consequences are aggregate but the decisions are exceedingly individual. The swimmer who avoids the part of the beach where the surfers are clustered, and the surfer who congregates where the surfboards are, are reacting individually to an environment that consists mainly of other individuals who are reacting likewise. The results can be unintended, even unnoticed.

Non-smokers may concentrate in the least smoky railroad car; as that car becomes crowded, smokers, choosing less crowded cars, find themselves among smokers, whether they notice it or not, and less densely crowded, whether they appreciate it or not.

The more crucial phenomena are of course residential decisions and others, like occupational choice, inter-city migration, school- and church-population, where the separating and mixing involve lasting associations that matter. The minor-league players who eat lunch at Dodgertown have no cafeteria-line–mechanism to scramble their home addresses; and even if they were located at random, they would usually not be casually integrated, because mixed residential areas are few and the choice, for a black or for a white, is between living among blacks or living among whites—unless even that choice is restricted.

It is not easy to tell from the aggregate phenomenon just what the motives are behind the individual decisions, or how strong they are. The smoker on an airplane may not know that the person in front of him is sensitive to tobacco smoke; the water-skier might be willing to stay four hundred yards offshore if doing so didn't just leave a preferred strip to other skiers. The clustered men and women at that cocktail party may be bored and wish the hostess could shake things up, but without organization no one can do any good by himself. And people who are happy to work where English and French are both spoken may find it uncomfortable if their own language falls to extreme minority status; and by withdrawing they only aggravate the situation that induced them to withdraw.

People who have to choose between polarized extremes—a white neighborhood or a black, a French-speaking club or one where English alone is spoken, a school with few whites or one with few blacks—will often choose in the way that reinforces the polarization. Doing so is no evidence that they prefer segregation, only that, if segregation exists and they have to choose between exclusive association, people elect like rather than unlike environments.

The dynamics are not always transparent. There are chain reactions, exaggerated perceptions, lagged responses, speculation on the future, and organized efforts that may succeed or fail. Three people of a particular group may break leases and move out of an apartment without being noticed, but if they do it the same week somebody will notice and comment. Other residents are then alerted to whether the whites or the blacks or the elderly, or the families with children or the families without, are moving away, thereby generating the situation of minority status they thought they foresaw.

Some of the processes may be passive, systemic, unmotivated but nevertheless biased. If job vacancies are filled by word of mouth or apartments go to people who have acquaintances in the building, or if boys can marry only girls they know and can know only girls who speak their language, a biased communication system will preserve and enhance the prevailing homogeneities.

## A Self-Forming Neighborhood Model

Some vivid dynamics can be generated by any reader with a half-hour to spare, a roll of pennies and a roll of dimes, a tabletop, a large sheet of paper, a spirit of scientific inquiry, or, lacking that spirit, a fondness for games.

Get a roll of pennies, a roll of dimes, a ruled sheet of paper divided into one-inch squares, preferably at least the size of a checkerboard (sixty-four squares in eight rows and eight columns) and find some device for selecting squares at random. We place dimes and pennies on some of the squares, and suppose them to represent the members of two homogeneous groups—men and women, blacks and whites, French-speaking and English-speaking, officers and enlisted men, students and faculty, surfers and swimmers, the well dressed and the poorly dressed, or any other dichotomy that is exhaustive and recognizable. We can spread them at random or put them in contrived patterns. We can use equal numbers of dimes and pennies or let one be a minority. And we can stipulate various rules for individual decision.

For example, we can postulate that every dime wants at least half its neighbors to be dimes, every penny wants a third of its neighbors to be pennies, and any dime or penny whose immediate neighborhood does not meet these conditions gets up and moves. Then by inspection we locate the ones that are due to move, move them, keep on moving them if necessary and, when everybody on the board has settled down, look to see what pattern has emerged. (If the situation never "settles down," we look to see what kind of endless turbulence or cyclical activity our postulates have generated.)

Define each individual's neighborhood as the eight squares surrounding him; he is the center of a 3-by-3 neighborhood. He is content or discontent with his neighborhood according to the colors of the occupants of those eight surrounding squares, some of which may be empty. We furthermore suppose that, if he is discontent with the color of his own neighborhood, he moves to the nearest empty square that meets his demands.

As to the order of moves, we can begin with the discontents nearest the center of the board and let them move first, or start in the upper left and sweep downward to the right, or let the dimes move first and then the pennies; it usually turns out that the precise order is not crucial to the outcome.

Then we choose an overall ratio of pennies to dimes, the two colors being about equal or one of them being a "minority." There are two different ways we can distribute the dimes and the pennies. We can put them in some prescribed pattern that we want to test, or we can spread them at random.

Start with equal numbers of dimes and pennies and suppose that the demands of both are "moderate"—each wants something more than one-third of his neighbors to

```
  # O # O # O
# O # O # O # O
O # O # O # O #
# O # O # O # O
O # O # O # O #
# O # O # O # O
O # O # O # O #
  O # O # O #
```

Figure 12.3

```
– # – # O # – O      – – – # – # – –
# # # O – O # O      – – – – – – – –
– # O – – # O #      – – – – – – – –
– O # O # O # O      – – # – # – # –
O O O # O O O –      – – – – – – – –
# – # # # – – O      # – – – – – – –
– # O # O # O –      – – O – O – O –
– O – O – – # –      – – – – – – – –
```

Figure 12.4

be like himself. The number of neighbors that a coin can have will be anywhere from zero to eight. We make the following specifications. If a person has one neighbor, he must be the same color; of two neighbors, one must be his color; of three, four, or five neighbors, two must be his color; and of six, seven, or eight neighbors, he wants at least three.

It is possible to form a pattern that is regularly "integrated" that satisfies everybody. An alternating pattern does it (figure 12.3), on condition that we take care of the corners.

No one can move, except to a corner, because there are no other vacant cells; but no one wants to move. We now mix them up a little, and in the process empty some cells to make movement feasible.

There are 60 coins on the board. We remove 20, using a table of random digits; we then pick 5 empty squares at random and replace a dime or a penny with a 50-50 chance. The result is a board with 64 cells, 45 occupied and 19 blank. Forty individuals are just where they were before we removed 20 neighbors and added 5 new ones. The left side of figure 12.4 shows one such result, generated by exactly this process. The #'s are dimes and the *O*'s are pennies; alternatively, the #'s speak French and the *O*'s speak English, the #'s are black and the *O*'s are white, the #'s are boys and the *O*'s are girls, or whatever you please.

The right side of figure 12.4 identifies the individuals who are not content with their neighborhoods. Six #'s and three *O*'s want to move; the rest are content as things stand. The pattern is still "integrated"; even the discontent are not without some neighbors like themselves, and few among the content are without neighbors of opposite color. The general pattern is not strongly segregated in appearance. One is hard-put to block out #-neighborhoods or *O*-neighborhoods at this stage. The problem is to satisfy a fraction, 9 of 45, among the #'s and *O*'s by letting them move somewhere among the 19 blank cells.

Anybody who moves leaves a blank cell that somebody can move into. Also, anybody who moves leaves behind a neighbor or two of his own color; and when he leaves a neighbor, his neighbor loses a neighbor and may become discontent. Anyone who moves gains neighbors like himself, adding a neighbor like them to their neighborhood but also adding one of opposite color to the unlike neighbors he acquires.

I cannot too strongly urge you to get the dimes and pennies and do it yourself. I can show you an outcome or two. A computer can do it for you a hundred times, testing variations in neighborhood demands, overall ratios, sizes of neighborhoods, and so forth. But there is nothing like tracing it through for yourself and seeing the thing work itself out. In an hour you can do it several times and experiment with different rules of behavior, sizes and shapes of boards, and (if you turn some of the coins heads and some tails) subgroups of dimes and pennies that make different demands on the color compositions of their neighborhoods.

### Chain Reaction

What is instructive about the experiment is the "unraveling" process. Everybody who selects a new environment affects the environments of those he leaves and those he moves among. There is a chain reaction. It may be quickly damped, with little motion, or it may go on and on and on with striking results. (The results of course are only suggestive, because few of us live in square cells on a checkerboard.)

One outcome for the situation depicted in figure 12.4 is shown in figure 12.5. It is "one outcome" because I have not explained exactly the order in which individuals moved. If the reader reproduces the experiment himself, he will get a slightly different configuration, but the general pattern will not be much different. Figure 12.6 is a replay from figure 12.4, the only difference from figure 12.5 being in the order of moves. It takes a few minutes to do the experiment again, and one quickly gets an impression of the kind of outcome to expect. Changing the neighborhood demands, or using twice as many dimes as pennies, will drastically affect the results; but for any given set of numbers and demands, the results are fairly stable.

All the people are content in figures 12.5 and 12.6. And they are more segregated. This is more than just a visual impression: we can make a few comparisons. In figure

```
  # #   O # #
# # # O O O # #
# # O O     O #
# O   O   O O O
O O O # O O O
  O # # # O O O
    # # # #
O O         #
```

Figure 12.5

```
  # # # O     O
# # # O   O   O
# # O       O
  O   O   O   O
O O O # O O O
    # # # O O O
O # # # # # # #
O O       # # #
```

Figure 12.6

12.4 the *O*'s altogether had as many *O*'s for neighbors as they had #'s; some had more or less than the average, and 3 were discontent. For the #'s the ratio of #-neighbors to *O*-neighbors was 1:1, with a little colony of #'s in the upper left corner and 6 widely distributed discontents. After sorting themselves out in figure 12.5, the average ratio of like to unlike neighbors for #'s and *O*'s together was 2.3:1, more than double the original ratio. And it is about triple the ratio that any individual demanded! Figure 12.6 is even more extreme. The ratio of like to unlike neighbors is 2.8:1, nearly triple the starting ratio and four times the minimum demanded.

Another comparison is the number who had no opposite neighbors in figure 12.4. Three were in that condition before people started moving; in figure 12.5 there are 8 without neighbors of opposite color, and in figure 12.6 there are 14.

What can we conclude from an exercise like this? We may at least be able to disprove a few notions that are themselves based on reasoning no more complicated than the checker-board. Propositions beginning with "It stands to reason that..." can sometimes be discredited by exceedingly simple demonstrations that, though perhaps true, they do not exactly "stand to reason." We can at least persuade ourselves that certain mechanisms could work, and that observable aggregate phenomena could be compatible with types of "molecular movement" that do not closely resemble the aggregate outcomes that they determine.

```
# # # # O     O
# # # # O O   O
# # # #     O
  O # O O O   O
O O O # O O O
      # # O
    O # # # O
  O   O # # #
```

Figure 12.7

```
  # #     # #
# # #   # # #
# # O O O # O
  O O O O O O O
O O O # O O O
  O # # # O O O
  # # #   O O
  # #
```

Figure 12.8

There may be a few surprises. What happens if we raise the demands of one color and lower the demands of the other? Figure 12.7 shows typical results. Here we increased by one the number of like neighbors that a # demanded and decreased by one the number that an *O* demanded, as compared with figures 12.5 and 12.6. By most measures, "segregation" is about the same as in figures 12.5 and 12.6. The difference is in population densities: the *O*'s are spread out all over their territory, while the #'s are packed in tight. The reader will discover, if he actually gets those pennies and dimes and tries it for himself, that something similar would happen if the demands of the two colors were equal but one color outnumbered the other by two or three to one. The minority then tends to be noticeably more tightly packed. Perhaps from figure 12.7 we could conclude that if surfers mind the presence of swimmers less than swimmers mind the presence of surfers, they will become almost completely separated, but the surfers will enjoy a greater expanse of water.

### Is it "Segregated"?

The reader might try guessing what set of individual preferences led from figure 12.4 to the pattern in figure 12.8.

The ratio of like to unlike neighbors for all the #'s and *O*'s together is slightly more than three to one; and there are 6 *O*'s and 8 #'s that have no neighbors of opposite color. The result is evidently segregation; but, following a suggestion of my dictionary, we might say that the process is one of *aggregation*, because the rules of behavior ascribed both to #'s and to *O*'s in figure 12.8 were simply that each would move to acquire three neighbors of like color irrespective of the presence or absence of neighbors of opposite color. As an individual motivation, this is quite different from the one that formed the patterns in figures 12.5 and 12.6. But in the aggregate it may be hard to discern which motivation underlies the pattern, and the process, of segregated residence. And it matters!

The first impact of a display like this on a reader may be—unless he finds it irrelevant—discouragement. A moderate urge to avoid small-minority status may cause a nearly integrated pattern to unravel, and highly segregated neighborhoods to form. Even a deliberately arranged viable pattern, as in figure 12.3, when buffeted by a little random motion, proves unstable and gives way to the separate neighborhoods of figures 12.5 through 12.8. These then prove to be fairly immune to continued random turnover.

For those who deplore segregation, however, and especially for those who deplore more segregation than people were seeking when they collectively segregated themselves, there may be a note of hope. The underlying motivation can be far less extreme than the observable patterns of separation. What it takes to keep things from unraveling is to be learned from figure 12.4; the later figures indicate only how hard it may be to restore such "integration" as would satisfy the individuals, once the process of separation has stabilized. In figure 12.4 only 9 of the 45 individuals are motivated to move, and if we could persuade them to stay everybody else would be all right. Indeed, the reader might exercise his own ingenuity to discover how few individuals would need to be invited into figure 12.4 from outside, or how few individuals would need to be relocated in figure 12.4, to keep anybody from wanting to move. If two lonely #'s join a third lonely #, none of them is lonely anymore, but the first will not move to the second unless assured that the third will arrive, and without some concert or regulation, each will go join some larger cluster, perhaps abandoning some nearby lonely neighbor in the process and surely helping to outnumber the opposite color at their points of arrival.

## The Bounded-Neighborhood Model

Turn now to a different model, and change the definition of "neighborhood." Instead of everyone's defining his neighborhood by reference to his own location, there is a common definition of the neighborhood and its boundaries. A person is either inside it or outside. Everyone is concerned about the color *ratio* within the neighborhood but not with the arrangement of colors within the neighborhood. "Residence" can therefore just as well be interpreted as membership or participation in a job, office, university, church, voting bloc, restaurant, or hospital.

In this model there is one particular area that everybody, black or white, prefers to its alternatives. He will live in it unless the percentage of residents of opposite color exceeds some limit. Each person, black or white, has his own limit. ("Tolerance," I shall occasionally call it.) If a person's limit is exceeded in this area he will go someplace else—a place, presumably, where his own color predominates or where color does not matter.

"Tolerance," it should be noticed, is a *comparative* measure. And it is specific to this location. Whites who appear, in this location, to be less tolerant of *blacks* than other whites may be merely more tolerant of the alternative *locations*.

Evidently the limiting ratios must be compatible for some blacks and some whites—as percentages they must add to at least 100—or no contented mixture of any whites and blacks is possible. Evidently, too, if nobody can tolerate extreme ratios, an area initially occupied by one color alone would remain so.

### Notes

This chapter originally appeared as part of chapter 4 in *Micromotives and Macrobehavior*, pp. 137–155, New York, W. W. Norton & Co., 1978. The original numbering of figures and notes has been retained in this reprint.

8. Charles Maher, "The Negro Athlete in America," *The Los Angeles Times* Sports Section, March 29, 1968.

# 13 Alternative Views of Complexity

**Herbert Simon**

## Conceptions of Complexity

This century has seen recurrent bursts of interest in complexity and complex systems. An early eruption, after World War I, gave birth to the term "holism," and to interest in "Gestalts" and "creative evolution." In a second major eruption, after World War II, the favorite terms were "information," "feedback," "cybernetics," and "general systems." In the current eruption, complexity is often associated with "chaos," "adaptive systems," "genetic algorithms," and "cellular automata."

While sharing a concern with complexity, the three eruptions selected different aspects of the complex for special attention. The post-WWII interest in complexity, focusing on the claim that the whole transcends the sum of the parts, was strongly anti-reductionist in flavor. The post-WWII outburst was rather neutral on the issue of reductionism, focusing on the roles of feedback and homeostasis (self-stabilization) in maintaining complex systems. The current interest in complexity focuses mainly on mechanisms that create and sustain complexity and on analytic tools for describing and analyzing it.

### Holism and Reductionism

"Holism" is a modern name for a very old idea. In the words of its author, the South African statesman and philosopher, J. C. Smuts:

> [Holism] regards natural objects as wholes.... It looks upon nature as consisting of discrete, concrete bodies and things... [which] are not entirely resolvable into parts; and.... which are more then the sums of their parts, and the mechanical putting together of their parts will not produce them or account for their characters and behavior.[1]

Holism can be given weaker or stronger interpretations. Applied to living systems, the strong claim that "the putting together of their parts will not produce them or account for their characters and behaviors" implies a vitalism that is wholly antithetical to modern molecular biology. Applied to minds in particular, it is used to support both the claim that machines cannot think and the claim that thinking involves more than the arrangement and behavior of neurons. Applied to complex systems in general, it postulates new system properties and relations among subsystems that had no place in the system components; hence it calls for emergence, a "creative" principle. Mechanistic explanations of emergence are rejected.

In a weaker interpretation, emergence simply means that the parts of a complex system have mutual relations that do not exist for the parts in isolation. Thus, there can be gravitational attractions among bodies only when two or more bodies interact with each other. We can learn something about the (relative) gravitational accelerations of binary stars, but not of isolated stars.

In the same way, if we study the structures only of individual proteins, nothing presages the way in which one protein molecule, serving as an enzyme, can provide a template into which two other molecules can insert themselves and be held while they undergo a reaction linking them. The template, a real enough physical property of the enzyme, has no function until it is placed in an environment of other molecules of a certain kind.

Even though the template's function is "emergent," having no meaning for the isolated enzyme molecule, the binding process, and the forces employed in it, can be given a wholly reductionist explanation in terms of the known physico-chemical properties of the molecules that participate in it. Consequently, this weak form of emergence poses no problems for even the most ardent reductionist.

"Weak emergence" shows up in a variety of ways. In describing a complex system we often find it convenient to introduce new theoretical terms, like inertial mass in mechanics, or voltage in the theory of circuits, for quantities that are not directly observable but are defined by relations among the observables.[2] We can often use such terms to avoid reference to details of the component subsystems, referring only to their aggregate properties.

Ohm, for example, established his law of electrical resistance by constructing a circuit containing a battery that drove current through a wire, and an ammeter that measured the magnetic force induced by the current. By changing the length of the wire, he altered the current. The equation relating the length of the wire (resistance) to the force registered by the ammeter (current) contained two constants, which were independent of the length of the wire but changed if he replaced the battery by another. These constants were labeled the *voltage* and *internal resistance* of the battery, which was otherwise unanalyzed and treated as a "black box." Voltage and internal resistance are not measured directly but are theoretical terms, inferred from the measured resistance and current with the aid of Ohm's Law.

Whereas the details of components can often be ignored while studying their interactions in the whole system, the short-run behavior of the individual subsystems can often be described in detail while ignoring the (slower) interactions among subsystems.

In economics, we often study the interaction of closely related markets—for example, the markets for iron ore, pig iron, sheet steel and steel products—under the assumption that all other supply and demand relations remain constant. In the next chapter, we will discuss at length this near independence of hierarchical systems from the detail of their component subsystems, as well as the short-run independence of the subsystems from the slower movements of the total system.

By adopting this weak interpretation of emergence, we can adhere (and I will adhere) to reductionism in principle even though it is not easy (often not even computationally feasible) to infer rigorously the properties of the whole from knowledge of the properties of the parts. In this pragmatic way, we can build nearly independent theories for each successive level of complexity, but at the same time, build bridging theories that show how each higher level can be accounted for in terms of the elements and relations of the next level below.

This is, of course, the usual conception of the sciences as building upward from elementary particles, through atoms and molecules to cells, organs and organisms. The actual history, however, has unfolded, as often as not, in the opposite direction—from top down. . . .

### Cybernetics and General Systems Theory

The period during and just after World War II saw the emergence of what Norbert Wiener dubbed "cybernetics": a combination of servomechanism theory (feedback control systems), information theory, and modern stored-program computers, all of which afford bold new insights into complexity. Information theory explains organized complexity in terms of the reduction of entropy (disorder) that is achieved when systems (organisms, for example) absorb energy from external sources and convert it to pattern or structure. In information theory, energy, information, and pattern all correspond to negative entropy.

Feedback control shows how a system can work toward goals and adapt to a changing environment,[3] thereby removing the mystery from teleology. What is required is ability to recognize the goal, to detect differences between the current situation and the goal, and actions that can reduce such differences: precisely the capabilities embodied in a system like the General Problem Solver. Soon this insight was being applied to constructing small robots that could maneuver around a room autonomously.[4] As computers became available, systems could be built at levels of complexity that had never before been contemplated; and by virtue of their capability for interpreting and executing their own internally stored programs, computers initiated the study of artificial intelligence.

These developments encouraged both the study of complex systems, especially adaptive goal-oriented systems, "as wholes," and simultaneously, the reductive explanation of system properties in terms of mechanisms. Holism was brought into confrontation with reductionism in a way that had never been possible before, and that confrontation continues today in philosophical discussion of artificial systems.

During these postwar years, a number of proposals were advanced for the development of "general systems theory," that, abstracting from the special properties of physical, biological, or social systems, would apply to all of them.[5] We might well feel that, while the goal is laudable, systems of these diverse kinds could hardly be expected to have any nontrivial properties in common. Metaphor and analogy can be helpful or they can be misleading. All depends on whether the similarities the metaphor captures are significant or superficial.

If a general systems theory is too ambitious a goal, it might still not be vain to search for common properties among broad *classes* of complex systems. The ideas that go by the name of cybernetics constitute, if not a theory, at least a point of view that has proved fruitful over a wide range of applications.[6] It has been highly useful to look at the behavior of adaptive systems in terms of feedback and homeostasis and to apply to these concepts the theory of selective information.[7] The concepts of feedback and information provide a frame of reference for viewing a wide range of situations, just as do the ideas of evolution, of relativism, of axiomatic method, and of operationalism.

The principal contribution of this second wave of inquiry into complexity lay precisely in the more specific concepts it brought to attention rather than in the broad idea of a general systems theory. This view is illustrated in the next chapter, which focuses on the properties of those particular complex systems that are hierarchical in structure, and draws out the consequences for system behavior of the strong assumption of hierarchy (or near-decomposability, as I shall call it).

## Current Interest in Complexity

The current, third, burst of interest in complexity shares many of the characteristics of the second. Much of the motivation for it is the growing need to understand and cope with some of the world's large-scale systems—the environment, for one, the worldwide society that our species has created, for another, and organisms, for a third. But this motivation could not, by itself, tie attention to complexity for very long if novel ways of thinking about it were not also provided. Going beyond the tools and concepts that appeared in the second wave, other new ideas have emerged, together with relevant mathematics and computational algorithms. The ideas have such labels as "catastrophe," "chaos," "genetic algorithms," and "cellular automata."

As always, the labels have some tendency to assume a life of their own. The foreboding tone of "catastrophe" and "chaos" says something about the age of anxiety in which these concepts were named. Their value as concepts, however, depends not on the rhetoric they evoke, but on their power to produce concrete answers to questions of complexity. For the particular concepts listed above, much of the verdict is not yet in. I want to comment briefly on each of them, for they are both alternatives and complements to the approach to hierarchical complexity that I will develop in the next chapter.

## Catastrophe Theory

Catastrophe theory appeared on the scene around 1968,[8] made an audible splash, and nearly faded from public sight within a few years. It is a solid body of mathematics dealing with the classification of nonlinear dynamic systems according to their modes of behavior. Catastrophic events occur in a special kind of system. Such a system can assume two (or more) distinct steady states (static equilibria, for example, or periodic cycles); but when the system is in one of these states, a moderate change in a system parameter may cause it to shift suddenly to the other—or into an unstable state in which variables increase without limit. The mathematician R. Thom constructed a topological classification of two-variable and three-variable systems according to the kinds of catastrophes they could or couldn't experience.

It is not hard to think of natural systems that exhibit behavior of this kind—stable behavior followed by a sudden shift to disequilibrium or to another, quite different, equilibrium. A commonly cited example is the threatened dog that either suddenly moves to the attack—or panics and flees. More complex examples have been studied: for instance, a budworm population infesting a spruce forest. The rapidly reproducing budworms quickly reach an equilibrium of maximum density; but the slow continuing growth of the spruce forest gradually alters the limit on the budworm population until, when a critical forest density is exceeded, the population explodes.[9] One can conjure up models of human revolutions embodying similar mechanisms.

In the circumstances that create it, the catastrophe mechanism is effective and the metaphor evocative, but in practice, only a limited number of situations have been found where it leads to any further analysis. Most of the initial applications that struck public fancy (like the attacking/fleeing dog) were after-the-fact explanations of phenomenon that were already familiar. For this reason, catastrophe theory is much less prominent in the public eye and in the literature of complexity today than it was twenty-five years ago.

## Complexity and Chaos

The theory of chaos also represents solid mathematics, which in this case has a long history reaching back to Poincaré.[10] Chaotic systems are deterministic dynamic systems that, if their initial conditions are disturbed even infinitesimally, may alter their paths radically. Thus, although they are deterministic, their detailed behavior over time is unpredictable, for small perturbations cause large changes in path. Chaotic systems were sufficiently intractable to mathematical treatment that, although the subject was kept alive by a few French mathematicians working in the tradition of Poincaré, only modest progress was made with them until well beyond the middle of this century. A major source of new progress has been the ability to use computers to display and explore their chaotic behavior.

Gradually, researchers in a number of sciences began to suspect that important phenomena they wished to understand were, in this technical sense, chaotic. One of the first was the meteorologist E. N. Lorenz, who started to explore in the early 1960s the possibility that weather was a chaotic phenomenon—the possibility that the butterfly in Singapore, by flapping its wings, could cause a thunderstorm in New York. Soon, fluid turbulence in general was being discussed in terms of chaos; and the possible inculpation of chaos in the complex behavior of a wide range of physical and biological systems was being studied. Solid experimental evidence that specific physical systems do, in fact, behave chaotically began to appear in the late 1970s.[11]

The growth in attention to chaos must be viewed against the background of our general understanding of dynamic systems. For a long time we have had a quite general theory of systems of *linear* differential equations and their solution in closed form. With systems of *nonlinear* equations, matters were less satisfactory. Under particular simple boundary conditions, solutions were known for a number of important systems of nonlinear partial differential equations that capture the laws of various kinds of wave motion. But beyond these special cases, knowledge was limited to methods for analyzing local behavior qualitatively—its stability or instability—in order to divide the space of achievable states into discrete regions. In each such region, specific kinds of behavior (e.g., movement to equilibrium, escape from unstable equilibrium, steady-state motion in limit cycles) would occur.[12]

This was the bread-and-butter content of the standard textbook treatments of nonlinear analysis, and beyond these qualitative generalizations, complex nonlinear systems had to be studied mainly by numerical simulation with computers. Most of the large computers and super-computers of the past half century have been kept busy simulating numerically the behavior of the systems of partial differential equations that describe the dynamics of airplanes, atomic piles, the atmosphere, and turbulent systems generally. As chaotic systems were not typically discussed in the textbooks, the then-current theory of nonlinear systems provided little help in treating such phenomena as turbulence except at an aggregate and very approximate level.

Under these circumstances, new computer-generated discoveries about chaos in the late 1970s and early 1980s created enormous interest and excitement in a variety of fields where phenomena were already suspected of being chaotic, and hence could perhaps be understood better with the new theory. Numerical computations on simple nonlinear systems revealed unsuspected invariants ("universal numbers") that predicted, for broad classes of such systems, at what point they would change from orderly to chaotic behavior.[13] Until high-speed computers were available to reveal them, such regularities were invisible.

Deep understanding has now been achieved of many aspects of chaos, but to say that we "understand" does not imply that we can predict. Chaos led to the recognition of a new, generalized, notion of equilibrium—the so-called "strange attractor." In classical nonlinear theory a system might come to a stable equilibrium, or it might oscillate permanently in a limit cycle, like the orbit of a planet. A chaotic system, however, might also enter a region of its state space, the strange attractor, in which it would remain permanently.

Within the strange attractor, motion would not cease, nor would it be predictable, but although deterministic, would appear to be random. That is, slightly different directions of entrance into the strange attractor, or slight perturbations when in it, would lead the system into quite different paths. A billiard ball aimed exactly at a 45° angle across a square "ideal" billiard table, will reflect off successive sides and, returning to the starting point, repeat its rectangular path indefinitely. But if you decrease or increase the 45° angle by an epsilon, the ball will never return to the starting point but will pursue a path that will eventually take it as close as you please to any spot on the table. The table's entire surface has become the strange attractor for the chaotic behavior and almost equal but different initial angles will produce continually diverging paths.

The theory of chaos has perhaps not maintained the hectic pace of development it experienced from the early 1960s to the late 1980s, but during this period it established itself as an essential conceptual framework and mathematical tool for the study of a class of systems that have major real-world importance in a number of scientific domains. The mechanisms of chaos are more general, but also of wider application, than those of catastrophe theory. Hence, we can expect chaos to continue to play a larger role than catastrophe in the continuing study of complex systems.

### Rationality in a Catastrophic or Chaotic World

What implications do catastrophe and chaos have for the systems—economies, the human mind, and designed complex systems—that we have been discussing in the previous six chapters? Although there have been some attempts to discover chaos in economic time series, the results thus far have been inconclusive. I am aware of no clear demonstration of chaos in the brain, but there is increasing evidence that chaos plays a role, although still a rather unclear one, in the functioning of the normal and defective heart. Designers frequently construct systems (e.g., airplanes and ships) that produce, and cope successfully with, turbulence and perhaps other kinds of chaos.

On the basis of the evidence, we should suppose neither that all of the complex systems we encounter in the world are chaotic, nor that few of them are. Moreover, as the airplane example shows, the ominous term "chaotic" should not be read as "unmanageable." Turbulence is frequently present in hydraulic and aerodynamic situations and artifacts. In such situations, although the future is not predictable in any detail, it is manageable as an aggregate phenomenon. And the paths of tornadoes and hurricanes are notoriously unstable but stable enough in the short run that we can usually be warned and reach shelter before they hit us.

Since Newton, astronomers have been able to compute the motion of a system of two bodies that exercise mutual gravitational attraction on each other. With three or more bodies, they never obtained more than approximations to the motion, and indeed, there is now good reason to believe that, in general, gravitational systems of three or more bodies, including the solar system, are chaotic. But we have no reason to anticipate untoward consequences from that chaos—its presence simply implies that astronomers will be frustrated in their attempts to predict the exact positions of the planets in the rather long run—a perplexity as frustrating as, but perhaps less damaging than, the difficulties meteorologists experience in predicting the weather.

Finally, there has been substantial progress in devising feedback devices that "tame" chaos by restricting chaotic systems, moving within their strange attractors, to small neighborhoods having desired properties, so that the chaos becomes merely tolerable noise. Such devices provide an example, consonant with the discussion in earlier chapters, of the substitution of control for prediction.

## Complexity and Evolution

Much current research on complex systems focuses upon the emergence of complexity, that is, system evolution. Two computational approaches to evolution that have attracted particular attention are the genetic algorithms first explored by Holland[14] and computer algorithms for cellular automata that simulate the multiplication and competition of organisms, playing the so-called "game of life."

**Genetic Algorithms** From an evolutionary standpoint, an organism can be represented by a list or vector of features (its genes). Evolution evaluates this vector in terms of fitness for survival. From generation to generation, the frequency distribution of features and their combinations over the members of a species change through sexual reproduction, crossover, inversion, and mutation. Natural selection causes features and combinations of features contributing to high fitness to multiply more rapidly than, and ultimately to replace, features and combinations conducive to low fitness.

By programming this abstraction on a modern computer we can build a computational model of the process of evolution. The simulation, in turn, can be used to study the relative rates at which fitness will grow under different assumptions about the model, including assumptions about rates of mutation and crossover. In the next chapter, we will consider the special case of evolution in hierarchical systems, which appears to be the kind of system that predominates in the natural world.

**Cellular Automata and The Game of Life** The computer is used not only to estimate the statistics of evolution but to carry out simulations, at an abstract level, of evolutionary processes. This research goes back, in fact, to the second eruption of interest in complexity, after World War II, when John von Neumann, building on some ideas of Stanislaw Ulam, defined abstractly (but did not implement) a system that was capable of reproducing itself. The idea was kept alive by Arthur Burks and others, but it was not until well into the current period of activity that Christopher Langton created a computer program that simulated a self-reproducing cellular automaton.[15] Computer programs can create symbolic objects of various kinds and apply rules for their replication or destruction as a function of their environments (which include other nearby objects). With appropriate selection of the system parameters, such simulations can provide vivid demonstrations of evolving self-reproducing systems. This line of exploration is still at a very early stage of development, is largely dependent on computer simulation, and lacks any large body of formal theory. It will be some time before we can assess its potential, but it has already presented us with a fundamental and exciting result: self-reproducing systems are a reality.

## Conclusion

Complexity is more and more acknowledged to be a key characteristic of the world we live in and of the systems that cohabit our world. It is not new for science to attempt to understand complex systems: astronomers have been at it for millennia, and biologists, economists, psychologists, and others joined them some generations ago. What is new about the present activity is not the study of particular complex systems but the study of the phenomenon of complexity in its own right.

If, as appears to be the case, complexity (like systems science) is too general a subject to have much content, then particular classes of complex systems possessing strong properties that provide a fulcrum for theorizing and generalizing can serve as the foci of attention. More and more, this appears to be just what is happening, with chaos, genetic algorithms, cellular automata, catastrophe, and hierarchical systems serving as some of the currently visible focal points.

### Notes

This chapter originally appeared as chapter 7 in *The Sciences of the Artificial*, third edition, Cambridge, MA: The MIT Press, 1996.

1. J. C. Smuts, "Holism," *Encyclopaedia Britannica*, 14th ed., vol. 11 (1929), p. 640.

2. H. A. Simon, "The Axiomatization of Physical Theories," *Philosophy of Science*, *37*(1970), 16–26.

3. A. Rosenblueth, N. Wiener and J. Bigelow, "Behavior, Purpose, and Teleology," *Philosophy of Science, 10*(1943), 18–24.

4. W. Grey Walter, "An Imitation of Life," *Scientific American, 182*(5) (1950):42.

5. See especially the yearbooks of the Society for General Systems Research. Prominent exponents of general systems theory were L. von Bertalanffy, K. Boulding, R. W. Gerard and, still active in this endeavor, J. G. Miller.

6. N. Wiener, *Cybernetics* (New York: Wiley, 1948). For an imaginative forerunner, see A. J. Lotka, *Elements of Mathematical Biology* (New York: Dover Publications, 1951), first published in 1924 as *Elements of Physical Biology*.

7. C. Shannon and W. Weaver, *The Mathematical Theory of Communication* (Urbana: University of Illinois Press, 1949); W. R. Ashby, *Design for a Brain* (New York: Wiley, 1952).

8. See R. Thom, *An Outline of a General Theory of Models* (Reading, MA: Benjamin, 1975).

9. For an account of the spruce/budworm model, see T. F. H. Allen and T. B. Starr, *Ecology: Perspectives for Ecological Complexity* (Chicago, IL: University of Chicago Press, 1982), and references cited there....

10. H. Poincaré, *Les Methodes Nouvelle de la Méchanique Céleste.* (Paris: Gauthier-Villars, 1892).

11. An excellent selection of the literature of chaos, both mathematical and experimental, up to the middle 1980s can be found in P. Cvitanović (ed.), *Universality in Chaos* (Bristol: Adam Hilger, 1986).

12. A standard source is A. A. Andronov, E. A. Leontovich, I. I. Gordon and A. G. Maier, *Qualitative Theory of Second-Order Dynamic Systems* (Wiley, NY: 1973).

13. M. J. Feigenbaum, "Universal Behavior in Nonlinear Systems," *Los Alamos Science, 1*(1980):4–27. This and other "classic" papers on chaos from the 1970s and 1980s are reprinted in P. Cvitanović, ed., *op. cit.*

14. J. H. Holland, *Adaptation in Natural and Artificial Systems* (Ann Arbor: University of Michigan Press, 1975).

15. A. W. Burks (ed.), *Essays on Cellular Automata* (Champaign-Urbana: University of Illinois Press, 1970); C. G. Langton (ed.), *Artificial Life.* Santa Fe Institute Studies in the Sciences of Complexity, Proceedings, vol. 6 (Redwood City, CA: Addison-Wesley, 1989); C. G. Langton, C. Taylor, J. D. Farmer and S. Rassmussen (eds.), *Artificial Life II.* Santa Fe Institute Studies in the Sciences of Complexity, Proceedings, vol. 10 (Redwood City, CA: Addison-Wesley, 1992).

## 14 The Theory of Everything

**Robert B. Laughlin and David Pines**

The Theory of Everything is a term for the ultimate theory of the universe—a set of equations capable of describing all phenomena that have been observed, or that will ever be observed (1). It is the modern incarnation of the reductionist ideal of the ancient Greeks, an approach to the natural world that has been fabulously successful in bettering the lot of mankind and continues in many people's minds to be the central paradigm of physics. A special case of this idea, and also a beautiful instance of it, is the equation of conventional nonrelativistic quantum mechanics, which describes the everyday world of human beings—air, water, rocks, fire, people, and so forth. The details of this equation are less important than the fact that it can be written down simply and is completely specified by a handful of known quantities: the charge and mass of the electron, the charges and masses of the atomic nuclei, and Planck's constant. For experts we write

$$i\hbar\frac{\partial}{\partial t}|\Psi\rangle = \mathcal{H}|\Psi\rangle \tag{14.1}$$

where

$$\mathcal{H} = -\sum_{j}^{N_e}\frac{\hbar^2}{2m}\nabla_j^2 - \sum_{\alpha}^{N_i}\frac{\hbar^2}{2M_\alpha}\nabla_\alpha^2 - \sum_{j}^{N_e}\sum_{\alpha}^{N_i}\frac{Z_\alpha e^2}{|\vec{r}_j - \vec{R}_\alpha|} + \sum_{j\ll k}^{N_e}\frac{e^2}{|\vec{r}_j - \vec{r}_k|} + \sum_{\alpha\ll\beta}^{N_j}\frac{Z_\alpha Z_\beta e^2}{|\vec{R}_\alpha - \vec{r}_\beta|}. \tag{14.2}$$

The symbols $Z_\alpha$ and $M_\alpha$ are the atomic number and mass of the $\alpha^{\text{th}}$ nucleus, $R_\alpha$ is the location of this nucleus, $e$ and $m$ are the electron charge and mass, $r_j$ is the location of the $j^{\text{th}}$ electron, and $\hbar$ is Planck's constant.

Less immediate things in the universe, such as the planet Jupiter, nuclear fission, the sun, or isotopic abundances of elements in space are not described by this equation, because important elements such as gravity and nuclear interactions are missing. But except for light, which is easily included, and possibly gravity, these missing parts are irrelevant to people-scale phenomena. Eqs. 14.1 and 14.2 are, for all practical purposes, the Theory of Everything for our everyday world.

However, it is obvious glancing through this list that the Theory of Everything is not even remotely a theory of every thing (2). We know this equation is correct because it has been solved accurately for small numbers of particles (isolated atoms and small molecules) and found to agree in minute detail with experiment (3–5). However, it cannot be solved accurately when the number of particles exceeds about 10. No computer existing, or that will ever exist, can break this barrier because it is a catastrophe of dimension. If the amount of computer memory required to represent the quantum wavefunction of one particle is $N$ then the amount required to represent the wavefunction of $k$ particles is $N^k$. It is possible to perform approximate calculations for larger systems, and it is through such calculations that we have learned why atoms have the size they do, why chemical bonds have the length and strength they do, why solid matter has the elastic properties it does, why some things are transparent while others reflect or absorb light (6). With a little more experimental input for guidance it is even possible to predict atomic conformations of small molecules, simple chemical reaction rates, structural phase transitions, ferromagnetism, and sometimes even superconducting transition temperatures (7). But the schemes for approximating are not first-principles deductions but are rather art keyed to experiment, and thus tend to be the least reliable precisely when reliability is most needed, i.e., when experimental information is scarce, the physical behavior has no precedent, and the key questions have not yet been identified. There are many notorious failures of alleged *ab initio* computation methods, including the phase diagram of liquid $^3$He and the entire phenomenology of high-temperature superconductors (8–10). Predicting protein functionality or the behavior of the human brain from these equations is patently absurd. So the triumph of the reductionism of the Greeks is a pyrrhic victory: We have succeeded in reducing all of ordinary physical behavior to a simple, correct Theory of Everything only to discover that it has revealed exactly nothing about many things of great importance.

In light of this fact it strikes a thinking person as odd that the parameters $e$, $\hbar$, and $m$ appearing in these equations may be measured accurately in laboratory experiments involving large numbers of particles. The electron charge, for example, may be accurately measured by passing current through an electrochemical cell, plating out metal atoms, and measuring the mass deposited, the separation of the atoms in the crystal being known from x-ray diffraction (11). Simple electrical measurements performed on superconducting rings determine to high accuracy the quantity the quantum of magnetic flux $hc/2e$ (11). A version of this phenomenon also is seen in superfluid helium, where coupling to electromagnetism is irrelevant (12). Four-point conductance measurements on semiconductors in the quantum Hall regime accurately determine the quantity $e^2/h$ (13). The magnetic field generated by a superconductor that is mechanically rotated measures $e/mc$ (14, 15). These things are clearly true, yet they cannot be deduced by direct calculation from the Theory of Everything, for exact results cannot be predicted by approximate calculations. This point is still not understood by many professional physicists, who find it easier to believe that a deductive link exists and has only to be discovered than to face the truth that there is no link. But it is true nonetheless. Experiments of this kind work because there are higher organizing principles in nature that make them work. The Josephson quantum is exact because of the principle of continuous symmetry breaking (16). The quantum Hall effect is exact because of localization (17). Neither of these things can be deduced from microscopics, and both are transcendent, in that they would continue to be true and to lead to exact results even if the Theory of Everything were changed. Thus the existence of these effects is profoundly important, for it shows us that for at least some fundamental things in nature the Theory of Everything is irrelevant. P. W. Anderson's famous and apt description of this state of affairs is "more is different" (2).

The emergent physical phenomena regulated by higher organizing principles have a property, namely their insensitivity to microscopics, that is directly relevant to the broad question of what is knowable in the deepest sense of the term. The low-energy excitation spectrum of a conventional superconductor, for example, is completely generic and is characterized by a handful of parameters that may be determined experimentally but cannot, in general, be computed from first principles. An even more trivial example is the low-energy excitation spectrum of a conventional crystalline insulator, which consists of transverse and longitudinal sound and nothing else, regardless of details. It is rather obvious that one does not need to prove the existence of sound in a solid, for it follows from the existence of elastic moduli at long length scales, which in turn follows from the spontaneous breaking of translational and rotational symmetry characteristic of the crystalline state (16). Conversely, one therefore learns little about the atomic structure of a crystalline solid by measuring its acoustics.

The crystalline state is the simplest known example of a quantum protectorate, a stable state of matter whose generic low-energy properties are determined by a higher organizing principle and nothing else. There are many of these, the classic prototype being the Landau fermi liquid, the state of matter represented by conventional metals and normal $^3$He (18). Landau realized that the existence of well-defined fermionic quasiparticles at a fermi surface was a universal property of such systems independent of microscopic details, and he eventually abstracted this to the more general idea that low-energy elementary excitation spectra were generic and characteristic of distinct stable states of matter. Other important quantum protectorates include superfluidity in Bose liquids such as $^4$He and the newly discovered atomic condensates (19–21), superconductivity (22, 23), band insulation (24), ferromagnetism (25), antiferromagnetism (26), and the quantum Hall states (27). The low-energy excited quantum states of these systems are particles in exactly the same sense that the electron in the vacuum of quantum electrodynamics is a particle: They carry momentum, energy, spin, and charge, scatter off one another according to simple rules, obey fermi or bose statistics depending on their nature, and in some cases are even "relativistic," in the sense of being described quantitively by Dirac or Klein-Gordon equations at low energy scales. Yet they are not elementary, and, as in the case of sound, simply do not exist outside the context of the stable state of matter in which they live. These quantum protectorates, with their associated emergent behavior, provide us with explicit demonstrations that the underlying microscopic theory can easily have no measurable consequences whatsoever at low energies. The nature of the underlying theory is unknowable until one raises the energy scale sufficiently to escape protection.

Thus far we have addressed the behavior of matter at comparatively low energies. But why should the universe be any different? The vacuum of space-time has a number of properties (relativity, renormalizability, gauge forces, fractional quantum numbers) that ordinary matter does not possess, and this state of affairs is alleged to be something extraordinary distinguishing the matter making up the universe from the matter we see in the laboratory (28). But this is incorrect. It has been known since the early 1970s that renormalizability is an emergent property of ordinary matter either in stable quantum phases, such as the superconducting state, or at particular zero-temperature phase transitions between such states called quantum critical points (29, 30). In either case the low-energy excitation spectrum becomes more and more generic and less and less sensitive to microscopic details as the energy scale of the measurement is lowered, until in the extreme limit of low energy all evidence of the microscopic equations vanishes away. The emergent renormalizability of quantum critical points is formally equivalent to that postulated in the standard model of elementary particles right down to the specific phrase "relevant direction" used to describe measurable quantities surviving renormalization. At least in some cases there is thought to be an emergent relativity principle in the bargain (29, 30). The rest of the strange agents in the standard model also have laboratory analogues. Particles carrying fractional quantum numbers and gauge forces between these particles occur as emergent phenomena in the fractional quantum Hall effect (17). The Higgs mechanism is nothing but superconductivity with a few technical modifications (31). Dirac fermions, spontaneous breaking of CP, and topological defects all occur in the low-energy spectrum of superfluid $^3$He (32–34).

Whether the universe is near a quantum critical point is not known one way or the other, for the physics of renormalization blinds one to the underlying microscopics as a matter of principle when only low-energy measurements are available. But that is exactly the point. The belief on the part of many that the renormalizability of the universe is a constraint on an underlying microscopic Theory of Everything rather than an emergent property is nothing but an unfalsifiable article of faith. But if proximity to a quantum critical point turns out to be responsible for this behavior, then just as it is impossible to infer the atomic structure of a solid by measuring long-wavelength sound, so might it be impossible to determine the true microscopic basis of the universe with the experimental tools presently at our disposal. The standard model and models based conceptually on it would be nothing but mathematically elegant phenomenological descriptions of low-energy behavior, from which, until experiments or observations could be carried out that fall outside the its region of validity, very little could be inferred about the underlying microscopic Theory of Everything. Big Bang cosmology is vulnerable to the same criticism. No one familiar with violent high-temperature phenomena would dare to infer anything about eqs. 14.1 and 14.2 by studying explosions, for they are unstable and quite unpredictable one experiment to the next (35, 36). The assumption that the early universe should be exempt from this problem is not justified by anything except wishful thinking. It could very well turn out that the Big Bang is the ultimate emergent phenomenon, for it is impossible to miss the similarity between the large-scale structure recently discovered in the density of galaxies and the structure of styrofoam, popcorn, or puffed cereals (37, 38).

Self-organization and protection are not inherently quantum phenomena. They occur equally well in systems with temperatures or frequency scales of measurement so high that quantum effects are unobservable. Indeed the first experimental measurements of critical exponents were made on classical fluids near their liquid-vapor critical points (39). Good examples would be the spontaneous crystallization exhibited by ball bearings placed in a shallow bowl, the emission of vortices by an airplane wing (40), finite-temperature ferromagnetism, ordering phenomena in liquid crystals (41), or the spontaneous formation of micelle membranes (42). To this day the best experimental confirmations of the renormalization group come from measurements of finite-temperature critical points (43). As is the case in quantum systems, these classical ones have low-frequency dynamic properties that are regulated by principles and independent of microscopic details (44, 45). The existence of classical protectorates raises the possibility that such principles might even be at work in biology (46).

What do we learn from a closer examination of quantum and classical protectorates? First, that these are governed by emergent rules. This means, in practice, that if you are locked in a room with the system Hamiltonian, you can't figure the rules out in the absence of experiment, and hand-shaking between theory and experiment. Second, one can follow each of the ideas that explain the behavior of the protectorates we have mentioned as it evolved historically. In solid-state physics, the experimental tools available were mainly long-wavelength, so that one needed to exploit the atomic perfection of crystal lattices to infer the rules. Imperfection is always present, but time and again it was found that fundamental understanding of the emergent rules had to wait until the materials became sufficiently free of imperfection. Conventional superconductors, for which nonmagnetic impurities do not interfere appreciably with superconductivity, provide an interesting counterexample. In general it took a long time to establish that there really were higher organizing principles leading to quantum protectorates. The reason was partly materials, but also the indirectness of the information provided by experiment and the difficulty in consolidating that information, including throwing out the results of experiments that have been perfectly executed, but provide information on minute details of a particular sample, rather than on global principles that apply to all samples.

Some protectorates have prototypes for which the logical path to microscopics is at least discernable. This helped in establishing the viability of their assignment as protectorates. But we now understand that this is not always the case. For example, superfluid $^3$He, heavy-fermion metals, and cuprate superconductors appear to be systems in which all vestiges of this link have disappeared, and one is left with nothing but the low-energy principle itself. This problem is exacerbated when the principles of self-organization responsible for emergent behavior compete. When more than one kind of ordering is possible the system decides what to do based on subtleties that are often beyond our ken. How can one distinguish between such competition, as exists for example, in the cuprate superconductors, and a "mess"? The history of physics has shown that higher organizing principles are best identified in the limiting case in which the competition is turned off, and the key breakthroughs are almost always associated with the serendipitous discovery of such limits. Indeed, one could ask whether the laws of quantum mechanics would ever have been discovered if there had been no hydrogen atom. The laws are just as true in the methane molecule and are equally simple, but their manifestations are complicated.

The fact that the essential role played by higher organizing principles in determining emergent behavior continues to be disavowed by so many physical scientists is a poignant comment on the nature of modern science. To solid-state physicists and chemists, who are schooled in quantum mechanics and deal with it every day in the context of unpredictable electronic phenomena such as organogels (47), Kondo insulators (48), or cuprate superconductivity, the existence of these principles is so obvious that it is a cliché not discussed in polite company. However, to other kinds of scientist the idea is considered dangerous and ludicrous, for it is fundamentally at odds with the reductionist beliefs central to much of physics. But the safety that comes from acknowledging only the facts one likes is fundamentally incompatible with science. Sooner or later it must be swept away by the forces of history.

For the biologist, evolution and emergence are part of daily life. For many physicists, on the other hand, the transition from a reductionist approach may not be easy, but should, in the long run, prove highly satisfying. Living with emergence means, among other things, focusing on what experiment tells us about candidate scenarios for the way a given system might behave before attempting to explore the consequences of any specific model. This contrasts sharply with the imperative of reductionism, which requires us never to use experiment, as its objective is to construct a deductive path from the ultimate equations to the experiment without cheating. But this is unreasonable when the behavior in question is emergent, for the higher organizing principles—

the core physical ideas on which the model is based—would have to be deduced from the underlying equations, and this is, in general, impossible. Repudiation of this physically unreasonable constraint is the first step down the road to fundamental discovery. No problem in physics in our time has received more attention, and with less in the way of concrete success, than that of the behavior of the cuprate superconductors, whose superconductivity was discovered serendipitously, and those properties, especially in the underdoped region, continue to surprise (49, 50). As the high-$T_c$ community has learned to its sorrow, deduction from microscopics has not explained, and probably cannot explain as a matter of principle, the wealth of crossover behavior discovered in the normal state of the underdoped systems, much less the remarkably high superconducting transition temperatures measured at optimal doping. Paradoxically high-$T_c$ continues to be the most important problem in solid-state physics, and perhaps physics generally, because this very richness of behavior strongly suggests the presence of a fundamentally new and unprecedented kind of quantum emergence.

In his book "The End of Science" John Horgan (51) argues that our civilization is now facing barriers to the acquisition of knowledge so fundamental that the Golden Age of Science must be thought of as over. It is an instructive and humbling experience to attempt explaining this idea to a child. The outcome is always the same. The child eventually stops listening, smiles politely, and then runs off to explore the countless infinities of new things in his or her world. Horgan's book might more properly have been called the End of Reductionism, for it is actually a call to those of us concerned with the health of physical science to face the truth that in most respects the reductionist ideal has reached its limits as a guiding principle. Rather than a Theory of Everything we appear to face a hierarchy of Theories of Things, each emerging from its parent and evolving into its children as the energy scale is lowered. The end of reductionism is, however, not the end of science, or even the end of theoretical physics. How do proteins work their wonders? Why do magnetic insulators superconduct? Why is $^3He$ a superfluid? Why is the electron mass in some metals stupendously large? Why do turbulent fluids display patterns? Why does black hole formation so resemble a quantum phase transition? Why do galaxies emit such enormous jets? The list is endless, and it does not include the most important questions of all, namely those raised by discoveries yet to come. The central task of theoretical physics in our time is no longer to write down the ultimate equations but rather to catalogue and understand emergent behavior in its many guises, including potentially life itself. We call this physics of the next century the study of complex adaptive matter. For better or worse we are now witnessing a transition from the science of the past, so intimately linked to reductionism, to the study of complex adaptive matter, firmly based in experiment, with its hope for providing a jumping-off point for new discoveries, new concepts, and new wisdom.

## Acknowledgments

We thank E. Abrahams, P. W. Anderson, G. Baym, S. Chakravarty, G. Volovik, and P. Nozières for thoughtful criticism of the manuscript. D. P. thanks the Aspen Institute for Physics, where part of the manuscript was written, for its hospitality. This work was supported primarily by the National Science Foundation under Grant DMR-9813899 and the Department of Energy. Additional support was provided by National Aeronautics and Space Administration Collaborative Agreement 974-9801 and the Ministry of International Trade and Industry of Japan through the New Energy and Industrial Technology Development Organization, and the Science and Technology Center for Superconductivity under National Science Foundation Grant No. DMR 91-2000.

### Note

This chapter originally appeared in *Proceedings of the National Academy of Sciences* 97 (2000), 28–31.

### References

1. Gribbin, G. R. (1999) *The Search for Superstrings, Symmetry, and the Theory of Everything* (Little Brown, New York).

2. Anderson, P. W. (1972) *Science* 177, 393–396.

3. Graham, R. L., Yeager, D. L., Olsen, J., Jorgensen, P., Harrison, R., Zarrabian, S., and Bartlett, R. (1986) *J. Chem. Phys.* 85, 6544–6549.

4. Wolnicwicz, L. (1995) *J. Chem. Phys.* 103, 1792.

5. Pople, J. (2000) *Rev. Mod. Phys.*, in press.

6. Slater, J. C. (1968) *Quantum Theory of Matter* (McGraw–Hill, New York).

7. Chang, K. J., Dacorogna, M. M., Cohen, M. L., Mignot, J. M., Chouteau, G., and Martinez, G. (1985) *Phys. Rev. Lett.* 54, 2375–2378.

8. Vollhardt, D., and Wölfle, P. (1990) *The Superfluid Phases of Helium 3* (Taylor and Francis, London).

9. Osheroff, D. D. (1997) *Rev. Mod. Phys.* 69, 667–682.

10. Bok, J., Deutscher, G., Pavuna, D., and Wolf, S. A., eds. (1998) *The Gap Symmetry and Fluctuations in High-$T_c$ Superconductors* (Plenum, New York).

11. Taylor, B. N., Parker, W. H., and Langenberg, D. N. (1969) *Rev. Mod. Phys.* 41, 375 and references therein.

12. Pereversev, S. V., Loshak, A., Backhaus, S., Davis, J. C., and Packard, R. E. (1997) *Nature (London)* 385, 449–451.

13. von Klitzing, K., Dorda, G., and Pepper, M. (1980) *Phys. Rev. Lett.* 45, 494.

14. Liu, M. (1998) *Phys. Rev. Lett.* 81, 3223–3226.

15. Tate, J., Felch, S. B., and Cabrera, B. (1990) *Phys. Rev. B* 42, 7885–7893.

16. Anderson, P. W. (1976) *Basic Notions of Condensed Matter Physics* (Benjamin, Menlo Park, CA).

17. Laughlin, R. B. (1999) *Rev. Mod. Phys.* 71, 863–874.

18. Pines, D., and Nozieres, P. (1966) *The Theory of Quantum Liquids* (Benjamin, New York).

19. Anderson, M. H., Ensher, J. R., Matthews, M. R., Wieman, C. E., and Cornell, E. A. (1995) *Science* 269, 198–201.

20. Bradley, C. C., Sackett, C. A., Tollett, J. J., and Hulet, R. G. (1995) *Phys. Rev. Lett.* 75, 1687–1690.

21. Davis, K. B., Mewes, M. O., Andrew, M. R., Vandruten, N. J., Durfee, D. S., Kurn, D. M., and Ketterle, W. (1995) *Phys. Rev. Lett.* 75, 3969–3973.

22. Schrieffer, J. R. (1983) *Theory of Superconductivity* (Benjamin, New York).

23. de Gennes, P. G. (1966) *Superconductivity of Metals and Alloys* (Benjamin, New York).

24. Sze, S. M. (1981) *Physics of Semiconductor Devices* (Wiley, New York).

25. Herring, C. (1966) in *Magnetism*, eds. Rado, G. T., and Suhl, H. (Academic, New York).

26. Kittel, C. (1963) *Quantum Theory of Solids* (Wiley, New York).

27. Prange, R. E., and Girvin, S. M. (1987) *The Quantum Hall Effect* (Springer, Heidelberg).

28. Peskin, M. E., and Schroeder, D. V. (1995) *Introduction to Quantum Field Theory* (Addison-Wesley, Reading, MA).

29. Wilson, K. G. (1983) *Rev. Mod. Phys.* 55, 583.

30. Fisher, M. E. (1974) *Rev. Mod. Phys.* 46, 597.

31. Anderson, P. W. (1963) *Phys. Rev.* 130, 439.

32. Volovik, G. E. (1992) *Exotic Properties of Superfluid* $^3$He (World Scientific, Singapore).

33. Volovik, G. E. (1998) *Physica B* 255, 86.

34. Volovik, G. E. (1999) *Proc. Natl. Acad. Sci. USA* 96, 6042–6047.

35. Zeldovich, Y. B., and Raizer, Y. P. (1966) *Physics of Shock Waves and High-Temperature Hydrodynamic Phenomena* (Academic, New York).

36. Sturtevant, B., Shepard, J. E., and Hornung, H. G., eds. (1995) *Shock Waves* (World Scientific, Singapore).

37. Huchra, J. P., Geller, M. J., Delapparent, V., and Corwin, H. G. (1990) *Astrophys. J.* Suppl., 72, 433–470.

38. Shechtman, S. A., Landy, S. D., Oemler, A., Tucker, D. L., Lin, H., Kirshner, R. P., and Schechter, P. L. (1996) *Astrophys. J.* 470, 172–188.

39. Levett Sengers, J. M. H. (1974) *Physica* 73, 73.

40. Saffman, P. G. (1992) *Vortex Dynamics* (Cambridge Univ. Press, Cambridge).

41. Lubensky, T. C., Harris, A. B., Kamien, R. D., and Yan, G. (1998) *Ferroelectrics* 212, 1–20.

42. Safran, S. (1994) *Statistical Thermodynamics of Surfaces, Interfaces, and Membranes* (Addison–Wesley, Reading, MA).

43. Stanley, H. E. (1987) *Introduction to Phase Transitions and Critical Phenomena* (Oxford, New York).

44. Riste, T., Samuelsen, E. J., Otnes, K., and Feder, J. (1971) *Solid State Comm.* 9, 1455–1458.

45. Courtery, E., and Gammon, R. W. (1979) *Phys. Rev. Lett.* 43, 1026.

46. Doniach, S. (1994) *Statistical Mechanics, Protein Structure, and Protein-Substrate Interactions* (Plenum, New York).

47. Geiger, C., Stanescu, M., Chen, L. H., and Whitten, D. G. (1999) *Langmuir* 15, 2241–2245.

48. Aeppli, G., and Fisk, Z. (1992) *Comments Condens. Matter Phys.* 16, 155.

49. Pines, D. (1997) *Z. Phys.* 103, 129.

50. Pines, D. (1998) in *The Gap Symmetry and Fluctuations in High-$T_c$ Superconductors*, eds. Bok, J., Deutscher, G., Pavuna, D. and Wolf, S. A. (Plenum, New York), pp. 111–142.

51. Horgan, J. (1997) *The End of Science: Facing the Limits of Knowledge in the Twilight of the Scientific Age* (Addison–Wesley, Reading, MA).

# 15 Is Anything Ever New? Considering Emergence

James P. Crutchfield

## 15.1 Emergent?

Some of the most engaging and perplexing natural phenomena are those in which highly-structured collective behavior emerges over time from the interaction of simple subsystems. Flocks of birds flying in lockstep formation and schools of fish swimming in coherent array abruptly turn together with no leader guiding the group.[2] Ants form complex societies whose survival derives from specialized laborers, unguided by a central director.[3] Optimal pricing of goods in an economy appears to arise from agents obeying the local rules of commerce.[4] Even in less manifestly complicated systems emergent global information processing plays a key role. The human perception of color in a small region of a scene, for example, can depend on the color composition of the entire scene, not just on the spectral response of spatially-localized retinal detectors.[5,6] Similarly, the perception of shape can be enhanced by global topological properties, such as whether or not curves are opened or closed.[7]

How does global coordination emerge in these processes? Are common mechanisms guiding the emergence across these diverse phenomena?

Emergence is generally understood to be a process that leads to the appearance of structure not directly described by the defining constraints and instantaneous forces that control a system. Over time "something new" appears at scales not directly specified by the equations of motion. An emergent feature also cannot be explicitly represented in the initial and boundary conditions. In short, a feature emerges when the underlying system puts some effort into its creation.

These observations form an intuitive definition of emergence. For it to be useful, however, one must specify what the "something" is and how it is "new". Otherwise, the notion has little or no content, since almost any time-dependent system would exhibit emergent features.

### 15.1.1 Pattern!

One recent and initially baffling example of emergence is deterministic chaos. In this, deterministic equations of motion lead over time to apparently unpredictable behavior. When confronted with chaos, one question immediately demands an answer—Where in the determinism did the randomness come from? The answer is that the effective dynamic, which maps from initial conditions to states at a later time, becomes so complicated that an observer can neither measure the system accurately enough nor compute with sufficient power to predict the future behavior when given an initial condition. The emergence of disorder here is the product of both the complicated behavior of nonlinear dynamical systems and the limitations of the observer.[8]

Consider instead an example in which order arises from disorder. In a self-avoiding random walk in two-dimensions the step-by-step behavior of a particle is specified directly in stochastic equations of motion: at each time it moves one step in a random direction, except the one it just came from. The result, after some period of time, is a path tracing out a self-similar set of positions in the plane. A "fractal" structure emerges from the largely disordered step-by-step motion.

Deterministic chaos and the self-avoiding random walk are two examples of the emergence of "pattern." The new feature in the first case is unpredictability; in the second, self-similarity. The "newness" in each case is only heightened by the fact that the emergent feature stands in direct opposition to the systems' defining character: complete determinism underlies chaos and near-complete stochasticity, the orderliness of self-similarity. But for whom has the emergence occurred? More particularly, to whom are the emergent features "new"? The state of a chaotic system always moves to a unique next state under the application of a deterministic function. Surely, the system state doesn't know its behavior is unpredictable. For the random walk, "fractalness" is not in the "eye" of the particle performing the local steps of the random walk, by definition. The newness in both cases is in the eye of an observer: the observer whose predictions fail or the analyst who notes that the feature of statistical self-similarity captures a commonality across length scales.

Such comments are rather straightforward, even trivial from one point of view, in these now-familiar cases. But there are many other phenomena that span a spectrum of novelty from "obvious" to "purposeful." The emergence of pattern is the primary theme, for example, in a wide range of phenomena that have come to be labeled "pattern formation." These include, to mention only a few, the convective rolls of Bénard and Couette fluid flows, the more complicated flow structures observed in weak turbulence,[9] the spiral waves and Turing patterns produced in oscillating chemical reactions,[10–12] the statistical order parameters describing phase transitions, the divergent correlations and long-lived fluctuations in critical phenomena,[13–15] and the forms appearing in biological morphogenesis.[10,16,17]

Although the behavior in these systems is readily described as "coherent," "self-organizing," and "emergent," the patterns which appear are detected by the observers and analysts themselves. The role of outside perception is evidenced by historical denials of patterns in the Belousov-Zhabotinsky reaction, of coherent structures in highly turbulent flows, and of the energy recurrence in anharmonic oscillator chains reported by Fermi, Pasta, and Ulam. Those experiments didn't suddenly start behaving differently once these key structures were appreciated by scientists. It is the observer or analyst who lends the teleological "self" to processes which otherwise simply "organize" according to the underlying dynamical constraints. Indeed, the detected patterns are often *assumed* implicitly by the analysts via the statistics selected to confirm the patterns' existence in experimental data. The obvious consequence is that "structure" goes unseen due to an observer's biases. In some fortunate cases, such as convection rolls, spiral waves, or solitons, the functional representations of "patterns" are shown to be consistent with mathematical models of the phenomena. But these models themselves rest on a host of theoretical assumptions. It is rarely, if ever, the case that the appropriate notion of pattern is extracted from the phenomenon itself using minimally-biased discovery procedures. Briefly stated, in the realm of pattern formation "patterns" are guessed and then verified.

### 15.1.2 Intrinsic Emergence

For these reasons, pattern formation is insufficient to capture the essential aspect of the emergence of coordinated behavior and global information processing in, for example, flocking birds, schooling fish, ant colonies, and in color and shape perception. At some basic level, though, pattern formation must play a role. The problem is that the "newness" in the emergence of pattern is always referred outside the system to some observer that anticipates the structures via a fixed palette of possible regularities. By way of analogy with a communication channel, the observer is a receiver that already has the codebook in hand. Any signal sent down the channel that is not already decodable using it is essentially noise, a pattern unrecognized by the observer.

When a new state of matter emerges from a phase transition, for example, initially no one knows the governing "order parameter." This is a recurrent conundrum in condensed matter physics, since the order parameter is the foundation for analysis and, even, further experimentation. After an indeterminant amount of creative thought and mathematical invention, one is sometimes found and then verified as appropriately capturing measurable statistics. The physicists' codebook is extended in just this way.

In the emergence of coordinated behavior, though, there is a closure in which the patterns that emerge are important *within* the system. That is, those patterns take on their "newness" with respect to other structures in the underlying system. Since there is no external referent for novelty or pattern, we can refer to this process as "intrinsic" emergence. Competitive agents in an efficient capital market control their individual production-investment and stock-ownership strategies based on the optimal pricing that has emerged from their collective behavior. It is essential to the agents' resource allocation decisions that, through the market's collective behavior, prices emerge that are accurate signals "fully reflecting" all available information.

What is distinctive about intrinsic emergence is that the patterns formed confer additional functionality which supports global information processing. Recently, examples of this sort have fallen under the rubric of "emergent computation."[18] The approach here differs in that it is based on explicit methods of detecting computation embedded in nonlinear processes. More to the point, the hypothesis in the following is that during intrinsic emergence there is an increase in intrinsic computational capability, which can be capitalized on and so lends additional functionality.

In summary, three notions will be distinguished:

1. The intuitive definition of emergence: "something new appears";
2. Pattern formation: an observer identifies "organization" in a dynamical system; and
3. Intrinsic emergence: the system itself capitalizes on patterns that appear.

## 15.2 What's in a Model?

In moving from the initial intuitive definition of emergence to the more concrete notion of pattern formation and ending with intrinsic emergence, it became clear that the essential novelty involved had to be referred to some evaluating entity. The relationship between novelty and its evaluation can be made explicit by thinking always of some observer that builds a model of a process from a series of measurements. At the level of the intuitive definition of emergence, the observer is that which recognizes the "something" and evaluates its "newness." In pattern formation, the observer is the scientist that uses prior concepts—e.g., "spiral" or "vortex"—to detect structure in experimental data and so to verify or falsify their applicability to the phenomenon. Of the three, this case is probably the most familiarly appreciated in terms of an "observer" and its "model." Intrinsic emergence is more subtle. The closure of "newness" evaluation pushes the observer inside the system. This requires in turn that intrinsic emergence be defined in terms of the "models" embedded in the observer. The observer in this view is a subprocess of the entire system. In particular, it is one that has the requisite information processing capability with which to take advantage of the emergent patterns.

"Model" is being used here in a sense that is somewhat more generous than found in daily scientific practice. There it often refers to an explicit representation—an analog—of a system under study. Here models will be seen in addition as existing implicitly in the dynamics and behavior of a process. Rather than being able to point to (say) an agent's model of its environment, one may have to excavate the "model." To do this one might infer that an agent's responses are in co-relation with its environment, that an agent has memory of the past, that the agent can make decisions, and so on. Thus, "model" here is more "behavioral" than "cognitive."

## 15.3 The Modeling Dilemma

The utility of this view of intrinsic emergence depends on answering a basic question: How does an observer understand the structure of natural processes? This includes both the scientist studying nature and an organism trying to predict aspects of its environment in order to survive. The answer requires stepping back to the level of pattern formation.

A key modeling dichotomy that runs throughout all of science is that between order and randomness. Imagine a scientist in the laboratory confronted after days of hard work with the results of a recent experiment—summarized prosaically as a simple numerical recording of instrument responses. The question arises, What fraction of the particular numerical value of each datum confirms or denies the hypothesis being tested and how much is essentially irrelevant information, just "noise" or "error"?

This dichotomy is probably clearest within science, but it is not restricted to it. In many ways, this caricature of scientific investigation gives a framework for understanding the necessary balance between order and randomness that appears whenever there is an "observer" trying to detect structure or pattern in its environment. The general puzzle of discovery then is: Which part of a measurement series does an observer ascribe to "randomness" and which part to "order" and "predictability"? Aren't we all in our daily activities to one extent or another "scientists" trying to ferret out the usable from the unusable information in our lives?

Given this basic dichotomy one can then ask: How does an observer actually make the distinction? The answer requires understanding how an observer models data—that is, the method by which elements in a representation, a "model," are justified in terms of given data.

A fundamental point is that *any* act of modeling makes a distinction between data that is accounted for—the ordered part—and data that is not described—the apparently random part. This distinction might be a null one: for example, for either completely predictable or ideally random (unstructured) sources the data is explained by one descriptive extreme or the other. Nature is seldom so simple. It appears that natural processes are an amalgam of randomness and order. In our view it is the organization of the interplay between order and randomness that makes nature "complex." A complex process then differs from a "complicated" process, a large system consisting of very many components, subsystems, degrees of freedom, and so on. A complicated system—such as an ideal gas—needn't be complex, in the sense used here. The ideal gas has no structure. Its microscopic dynamics are accounted for by randomness.

Experimental data is often described by a whole range of candidate models that are statistically and structurally consistent with the given data set. One important variation over this range of possible "explanations" is where each candidate draws the randomness-order distinction. That is, the models vary in the regularity captured and in the apparent error each induces.

It turns out that a balance between order and randomness can be reached and used to define a "best" model for a given data set. The balance is given by minimizing the model's size while minimizing the amount of apparent randomness. The first part is a version of Occam's dictum: causes should not be multiplied beyond necessity. The second part is a basic tenet of science: obtain the best prediction of nature. Neither component of this balance can be minimized alone, otherwise absurd "best" models would be selected. Minimizing the model size alone leads to huge error, since the smallest (null) model captures no regularities; minimizing the error alone produces a huge model, which is simply the data itself and manifestly not a useful encapsulation of what happened in the laboratory. So both model size and the induced error must be minimized together in selecting a "best" model. Typically, the sum of the model size and the error are minimized.[19–23]

From the viewpoint of scientific methodology the key element missing in this story of what to do with data is how to measure structure or regularity. (A particular notion of structure based on computation will be introduced shortly.) Just how structure is measured determines where the order-randomness dichotomy is set. This particular problem can be solved in principle: we take the size of the candidate model as the measure of structure. Then the size of the "best" model is a measure of the data's intrinsic structure. If we believe the data is a faithful representation of the raw behavior of the underlying process, this then translates into a measure of structure in the natural phenomenon originally studied.

Not surprisingly, this does not really solve the problem of quantifying structure. In fact, it simply elevates it to a higher level of abstraction. Measuring structure as the length of the description of the "best" model assumes one has chosen a language in which to describe models. The catch is that this representation choice builds in its own biases. In a given language some regularities can be compactly described, in others the same regularities can be quite baroquely expressed. Change the language and the same regularities could require more or less description. And so, lacking prior God-given knowledge of the appropriate language for nature, a measure of structure in terms of the description length would seem to be arbitrary.

And so we are left with a deep puzzle, one that precedes measuring structure: How is structure discovered in the first place? If the scientist knows beforehand the appropriate representation for an experiment's possible behaviors, then the amount of that kind of structure can be extracted from the data as outlined above. In this case, the prior knowledge about the structure is verified by the data if a compact, predictive model results. But what if it is not verified? What if the hypothesized structure is simply not appropriate? The "best" model could be huge or, worse, appear upon closer and closer analysis to diverge in size. The latter situation is clearly not tolerable. An infinite model is impractical to manipulate. These situations indicate that the behavior is so new as to not fit (finitely) into current understanding. Then what do we do?

This is the problem of "innovation." How can an observer ever break out of inadequate model classes and discover appropriate ones? How can incorrect assumptions be changed? How is anything new ever discovered, if it must always be expressed in the current language?

If the problem of innovation can be solved, then, as all of the preceding development indicated, there is a framework which specifies how to be quantitative in detecting and measuring structure.

## 15.4 Where Is Science Now?

Contemporary physics does not have the tools to address the problems of innovation, the discovery of patterns, or even the practice of modeling itself, since there are no physical principles that define and dictate how to measure natural structure. It is no surprise, though, that physics does have the tools for detecting and measuring complete order—equilibria and fixed point or periodic behavior—and ideal randomness—via temperature and thermodynamic entropy or, in dynamical contexts, via the Shannon entropy rate and Kolmogorov complexity.

For example, a physicist can analyze the dynamics of a box of gas and measure the degree of disorder in the molecular motion with temperature and the disorganization of the observed macroscopic state in terms of the multiplicity of associated microstates; that is, with the thermodynamic entropy. But the physicist has no analogous tools for deducing what mechanisms in the system maintain the disorder.

Then again, the raw production of information is just one aspect of a natural system's behavior. There are other important contributors to how nature produces patterns, such as how much memory of past behavior is required and how that memory is organized to support the production of information. Information processing in natural systems is a key attribute of their behavior and also how science comes to understand the underlying mechanisms.

The situation is a bit worse than a lack of attention to structure. Physics does not yet have a systematic approach to analyzing the complex information architectures embedded in patterns and processes that occur between order and randomness. This is, however, what is most needed to detect and quantify structure in nature.

The theories of phase transitions and, in particular, critical phenomena do provide mathematical hints at how natural processes balance order and randomness in that they study systems balancing different thermodynamic phases. Roughly speaking, one can think of crystalline ice as the ordered regime and of liquid water as the (relatively) disordered regime of the same type of matter ($H_2O$). At the phase transition, when both phases coexist, the overall state is more complex than either pure phase. What these theories provide is a set of coarse tools that describe large-scale statistical properties. What they lack are the additional, more detailed probes that would reveal, for example, the architecture of information processing embedded in those states; namely, the structure of those complex thermodynamic states. In fact, modern nonequilibrium thermodynamics can now describe the dominance of collective "modes" that give rise to the complex states found close to certain phase transitions.[24,25] What is still needed, though, is a definition of structure and way to detect and to measure it. This would then allow us to analyze, model, and predict complex systems at the "emergent" scales.

## 15.5 A Computational View of Nature

One recent approach is to adapt ideas from the theory of discrete computation, which has developed measures of information processing structure.[26] Computation theory defines the notion of a "machine"—a device for encoding the structures in discrete processes. It has been argued that, due to the inherent limitations of scientific instruments, all an observer can know of a process in nature is a discrete-time, discrete-space series of measurements. Fortunately, this is precisely the kind of thing—strings of discrete symbols, a "formal" language—that computation theory analyzes for structure.

How does this apply to nature? Given a discrete series of measurements from a process, a machine can be constructed that is the best description or predictor of this discrete time series. The structure of this machine can be said to be the best approximation to the original process's information-processing structure, using the model size and apparent error minimization method discussed above. Once we have reconstructed the machine, we can say that we understand the structure of the process.

But what kind of structure is it? Has machine reconstruction discovered patterns in the data? Computation theory answers such questions in terms of the different classes of machines it distinguishes. There are machine classes with finite memory, those with infinite one-way stack memory, those with first-in first-out queue memory, and those with infinite random access memory, among others. When applied to the study of nature, these machine classes reveal important distinctions among natural processes. In particular, the computationally distinct classes correspond to different types of pattern or regularity.

Given this framework, one talks about the structure of the original process in terms of the complexity of the reconstructed machine. This is a more useful notion of complexity than measures of randomness, such as the Kolmogorov complexity, since it indicates the degree to which information is processed in the system, which accords more closely to our intuitions about what complexity should mean. Perhaps more importantly, the reconstructed machine describes *how* the information is processed. That is, the architecture of the machines themselves represents the organization of the information processing, that is, the intrinsic computation. The reconstructed machine is a model of the mechanisms by which the natural process manipulates information.

## 15.6 Computational Mechanics: Beyond Statistics, Toward Structure

Reference [1] reviews how a machine can be reconstructed from a series of discrete measurements of a process. Such a reconstruction is a way that an observer can model its environment. In the context of biological evolution, for example, it is clear that to survive agents must detect regularities in their environment. The degree to which an agent can model its environment in this way depends on its own computational resources and on what machine class or language it implicitly is restricted to or explicitly chooses when making a model. Reference [1] also shows how an agent can jump out of its original assumptions about the model class and, by induction, can leap to a new model class which is a much better way of understanding its environment. This is a formalization of what is colloquially called "innovation." The inductive leap itself follows a hierarchical version of machine reconstruction.

The overall goal, then, concerns how to detect structures in the environment—how to form an "internal model"—and also how to come up with true innovations to that internal model. There are applications of this approach to time series analysis and other areas, but the main goal is not engineering but scientific: to understand how structure in nature can be detected and measured and, for that matter, discovered in the first place as wholly new innovations in one's assumed representation.

What is new in this approach? Computation theorists generally have not applied the existing structure metrics to natural processes. They have mostly limited their research to analyzing scaling properties of computational problems; in particular, to how difficulty scales in certain information processing tasks. A second aspect computation theory has dealt with little, if at all, is measuring structure in stochastic processes. Stochastic processes, though, are seen throughout nature and must be addressed at the most basic level of a theory of modeling nature. The domain of computation theory—pure discreteness, uncorrupted by noise—is thus only a partial solution. Indeed, the order-randomness dichotomy indicates that the interpretation of any experimental data has an intrinsic probabilistic component which is induced by the observer's choice of representation. As a consequence probabilistic computation must be included in any structural description of nature. A third aspect computation theory has considered very little is measuring structure in processes that are extended in space. A fourth aspect it has not dealt with traditionally is measuring structure in continuous-state processes. If computation theory is to form the foundation of a physics of structure, it must be extended in at least these three ways. These extensions have engaged a number of workers in dynamical systems recently, but there is much still to do.[26–32]

## 15.7 The Calculi of Emergence

Reference [1] focuses on temporal information processing and the first two extensions—probabilistic and spatial computation—assuming that the observer is looking at a series of measurements of a continuous-state system whose states an instrument has discretized. The phrase "calculi of emergence" in its title emphasizes the tools required to address the problems which intrinsic emergence raises. The tools are (i) dynamical systems theory with its emphasis on the role of time and on the geometric structures underlying the increase in complexity during a system's time evolution, (ii) the notions of mechanism and structure inherent in computation theory, and (iii) inductive inference as a statistical framework in which to detect and innovate new representations.

First, Reference [1] defines a complexity metric that is a measure of structure in the way discussed above. This is called "statistical complexity," and it measures the structure of the minimal machine reconstructed from observations of a given process in terms of the machine's size. Second, it describes an algorithm—ε-machine reconstruction—for reconstructing the machine, given an assumed model class. Third, it describes an algorithm for innovation—called "hierarchical machine reconstruction"—in which an agent can inductively jump to a new model class. Roughly speaking, hierarchical machine reconstruction detects regularities in a *series* of increasingly-accurate models. The inductive jump to a higher computational level occurs by taking those regularities as the new representation. The bulk of Reference [1] analyzes several examples in which these general ideas are put into practice to determine the intrinsic computation in continuous-state dynamical systems, recurrent hidden Markov models, and cellular automata. It concludes with a summary of the implications of this approach to detecting and understanding structure in nature.

The goal throughout is a more refined appreciation of what "emergence" is, both when new computational structure appears over time and when agents with improved computational and modeling ability evolve. The interplay between computation, dynamics, and induction emphasizes a trinity of conceptual tools required for studying the emergence of complexity; presumably this is a setting that has a good chance of providing empirical application.

## 15.8 Discovery versus Emergence

The arguments and development turn on distinguishing several different levels of interpretation: (i) a system behaves, (ii) that behavior is modeled, (iii) an observer detects regularities and builds a model based on prior knowledge, (iv) a collection of agents model each other and their environment, and (v) scientists create artificial universes and try to detect the change in computational capability by constructing their own models of the emergent structures. It is all too easy to conflate two or more of these levels, leading to confusion or, worse, subtle statements seeming vacuous.

It is helpful to draw a distinction between discovery and emergence. The level of pattern formation and the modeling framework of computational mechanics concern discovery. Above it was suggested that innovation based on hierarchical machine reconstruction is one type of discovery, in the sense that new regularities across increasingly-accurate models are detected and then taken as a new basis for representation. Discovery, though, is not the same thing as emergence, which at a minimum is dynamical: over time, or over generations in an evolutionary system, something new appears. Discovery, in this sense, is atemporal: the change in state and increased knowledge of the observer are not the focus of the analysis activity; the products of model fitting and statistical parameter estimation are.

In contrast, emergence concerns the *process* of discovery. Moreover, intrinsic emergence puts the subjective aspects of discovery *into* the system under study. In short, emergence pushes the semantic stack down one level. In this view analyzing emergence is more objective than analyzing pattern formation in that detecting emergence requires modeling the dynamics of discovery, not just implementing a discovery procedure.

The arguments to this point can be recapitulated by an operational definition of emergence. A process undergoes emergence if at some time the architecture of information processing has changed in such a way that a distinct and more powerful level of intrinsic computation has appeared that was not present in earlier conditions.

It seems, upon reflection, that our intuitive notion of emergence is not captured by the "intuitive definition" given in the first section. Nor is it captured by the somewhat refined notion of pattern formation. "Emergence" is meaningless unless it is defined within the context of processes themselves; the only well-defined notion of emergence would seem to be intrinsic emergence. Why? Simply because emergence defined without this closure leads to an infinite regress of observers detecting patterns of observers detecting patterns.... This is not a satisfactory definition, since it is not finite. The regress must be folded into the system, it must be immanent in the dynamics. When this happens complexity and structure are no longer referred outside, no longer relative and arbitrary; they take on internal meaning and functionality.

## 15.9 Evolutionary Mechanics

Where in science might a theory of intrinsic emergence find application? Are there scientific problems that at least would be clarified by the computational view of nature outlined here?

In several ways the contemporary debate on the dominant mechanisms operating in biological evolution seems ripe. Is anything ever new in the biological realm? The empirical evidence is interpreted as a resounding "yes." It is often heard that organisms today are more complex than in earlier epochs. But how did this emergence of complexity occur? Taking a long view, at present there appear to be three schools of thought on what the guiding mechanisms are in Darwinian evolution that produce the present diversity of biological structure and that are largely responsible for the alteration of those structures.

Modern evolutionary theory continues to be governed by Darwin's view of the natural selection of individuals that reproduce with variation. This view emphasizes the role of fitness selection in determining which biological organisms appear. But there are really two camps: the Selectionists, who are Darwin's faithful heirs now cognizant of genetics, and the Historicists, who espouse a more anarchistic view.

The Selectionists hold that structure in the biological world is due primarily to the fitness-based selection of individuals in populations whose diversity is maintained by genetic variation.[33] In a sense, genetic variation is a destabilizing mechanism that provides the raw diversity of structure. Natural selection then is a stabilizing dynamic that acts on the expression of that variation. It provides order by culling individuals based on their relative fitness. This view identifies a source of new structures and a mechanism for altering one form into another. The adaptiveness accumulated via selection is the dominant mechanism guiding the appearance of structure.

The second, anarchistic camp consists of the Historicists who hold fast to the Darwinian mechanisms of selection and variation, but emphasize the accidental determinants of biological form.[34,35] What distinguishes this position from the Selectionists is the claim that major changes in structure can be and have been nonadaptive. While these changes have had the largest effect on the forms of present day life, at the time they occurred they conferred no survival advantage. Furthermore, today's existing structures needn't be adaptive. They reflect instead an accidental history. One consequence is that a comparative study of parallel earths would reveal very different collections of life forms on each. Like the Selectionists, the Historicists have a theory of transformation. But it is one that is manifestly capricious or, at least, highly stochastic with few or no causal constraints. For this process of change to work the space of biological structures must be populated with a high fraction which are functional.

Lastly, there are the Structuralists whose goal is to elucidate the "principles of organization" that guide the appearance of biological structure. They contend that energetic, mechanical, biomolecular, and morphogenetic constraints limit the infinite range of possible biological form.[16,36–40] The constraints result in a relatively small set of structure archetypes. These are something like the Platonic solids in that they pre-exist, before any evolution takes place. Natural selection then plays the role of choosing between these "structural attractors" and possibly fine-tuning their adaptiveness. Darwinian evolution serves, at best, to fill the waiting attractors or not depending on historical happenstance. Structuralists offer up a seemingly testable claim about the ergodicity of evolutionary processes: given an ensemble of earths, life would have evolved to a similar collection of biological structures.

The Structuralist tenets are at least consistent with modern thermodynamics.[24,25] In large open systems energy enters at low entropy and is dissipated. Open systems organize largely due to the reduction in the number of active degrees of freedom caused by the dissipation. Not all behaviors or spatial configurations can be supported. The result is a limitation of the collective modes, cooperative behaviors, and coherent structures that an open system can express. The Structuralist view is a natural interpretation of the many basic constraints on behavior and pattern indicated by physics and chemistry. For example, the structures formed in open systems such as turbulent fluid flows, oscillating chemical reactions, and morphogenetic systems are the product of this type of macroscopic pattern formation. Thus, open systems offer up a limited palette of structures to selection. The more limited the palette, the larger the role for "principles of organization" in guiding the emergence of life as we know it.

What is one to think of these conflicting theories of the emergence of biological structure? In light of the preceding sections there are several impressions that the debate leaves an outsider with.

1. Natural selection's culling of genetic variation provides the Selectionists with a theory of transformation. But the approach does not provide a theory of structure. Taking the theory at face value, in principle one can estimate the time it takes a *given* organism to change. But what is the mean time under the evolutionary dynamic and under the appropriate environmental pressures for a hand to appear on a fish? To answer this one needs a measure of the structure concerned and of the functionality it does or does not confer.

2. The Historicists also have a theory of transformation, but they offer neither a theory of structure nor, apparently, a justification for a high fraction of functionality over the space of structures. Perhaps more disconcerting, though, in touting the dominance of historical accident, the Historicists advocate an antiscientific position. This is not to say that isolated incidents do not play a role; they certainly do. But it is important to keep in mind that the event of a meteor crashing into the earth is extra-evolutionary. The explanation of its occurrence is neither the domain of evolutionary theory nor is its occurrence likely ever to be explained by the principles of dynamics: it just happened, a consequence of particular initial conditions. Such accidents impose constraints; they are not an explanation of the biological response.

3. In complementary fashion, the Structuralists do not offer a theory of transformation. Neither do they, despite claims for the primacy of organization in evolutionary processes, provide a theory of structure itself. In particular, the structure archetypes are not analyzed in terms of their internal components nor in terms of system-referred

functionality. Considering these lacks, the Structuralist hope for "deep laws" underlying biological organization is highly reminiscent of Chomsky's decades-long search for "deep structures" as linguistic universals, without a theory of cognition. The ultimate failure of this search[41] suggests a reconsideration of fundamentals rather than optimistic forecasts of Structuralist progress.

The overwhelming impression this debate leaves, though, is that there is a crying need for a theory of biological structure and a qualitative dynamical theory of its emergence.[42] In short, the tensions between the positions are those (i) between the order induced by survival dynamics and the novelty of individual function and (ii) between the disorder of genetic variation and the order of developmental processes. Is it just an historical coincidence that the structuralist-selectionist dichotomy appears analogous to that between order and randomness in the realm of modeling? The main problem, at least to an outsider, does not reduce to showing that one or the other view is correct. Each employs compelling arguments and often empirical data as a starting point. Rather, the task facing us reduces to developing a synthetic theory that balances the tensions between the viewpoints. Ironically, evolutionary processes themselves seem to do just this sort of balancing, dynamically.

The computational mechanics of nonlinear processes can be construed as a theory of structure. Pattern and structure are articulated in terms of various types of machine classes. The overall mandate is to provide both a qualitative and a quantitative analysis of natural information processing architectures. If computational mechanics is a theory of structure, then innovation via hierarchical machine reconstruction is a computation-theoretic approach to the transformation of structure. It suggests one mechanism with which to study what drives and what constrains the appearance of novelty. The next step, of course, would be to fold hierarchical machine reconstruction into the system itself, resulting in a dynamics of innovation, the study of which might be called "evolutionary mechanics."

## 15.10 The Mechanics

By way of summarizing the main points, let's question the central assumption of this approach to emergence.

Why talk about "mechanics"? Aren't mechanical systems lifeless, merely the sum of their parts? One reason is simply that scientific explanations must be given in terms of mechanisms. Explanations and scientific theories without an explicit hypothesis of the underlying causes—the mechanisms—are neither explanations nor theories, since they cannot claim to entail falsifiable predictions.[43] Another, more constructive reason is that modern mathematics and physics have made great strides this century in extending the range of Newtonian mechanics to ever more complex processes. When computation is combined with this, one has in hand a greatly enriched notion of mechanism.

It might seem implausible that an abstract "evolutionary mechanics" would have anything to contribute to (say) biological evolution. A high-level view at least suggests a fundamental, if indirect, role. By making a careful accounting of where the observer and system-under-study are located in various theories of natural phenomena, a certain regularity appears which can be summarized by a hierarchy of mechanics. The following list is given in the order of increasing attention to the context of observation and modeling in a classical universe. The first two are already part of science proper; the second two indicate how computation and innovation build on them.

1. Deterministic mechanics (dynamical systems theory) The very notions of cause and mechanism are defined in terms of state space structures. This is Einstein's level: the observer is entirely outside the system-under-study.

2. Statistical mechanics (probability theory) Statistical mechanics is engendered by deterministic mechanics largely due to the emergence of irreducible uncertainty. This occurs for any number of reasons. First, deterministic mechanical systems can be very large, too large in fact to be usefully described in complete detail. Summarizing the coarse, macroscopic properties is the only manageable goal. The calculus for managing the discarded information is probability theory. Second, deterministic nonlinear systems can be chaotic, communicating unseen and uncontrollable microscopic information to affect observable behavior.[44] Both of these reasons lead to the necessity of using probabilistic summaries of deterministic behavior to collapse out the irrelevant and accentuate the useful.

3. Computational mechanics (theory of structure for statistical mechanics) As discussed at some length, it is not enough to say that a system is random or ordered. What is important is how these two elements, and others, interact to produce complex systems. The information processing mechanisms distinguished by computation theory give a (partial) basis for being more objective about detecting structure, quantifying complexity, and the modeling activity itself.

4. Evolutionary mechanics (dynamical theory of innovation) As noted above, evolutionary mechanics concerns how genuine novelty occurs. This is the first level at which emergence takes on its intrinsic aspect. Building on the previous levels, the goal is to delineate the constraints guiding and the forces driving the emergence of complexity.

A typical first question about this hierarchy is, Where is quantum mechanics? The list just given assumes a classical physical universe. Therefore, quantum mechanics is not listed despite its undeniable importance. It would appear, however, either as the most basic mechanics, preceding deterministic mechanics, or at the level of statistical mechanics, since that is the level at which probability first appears. In a literal sense quantum mechanics is a theory of the deterministic dynamics of complex "probabilities" that can interfere over spacetime. The interference leads to new phenomena, but the goals of and manipulations used in quantum mechanics are not so different from that found in stochastic processes and so statistical mechanics. My own prejudice in these issues will be resolved once a theory of measurement of nonlinear processes is complete. There are several difficulties that lie in the way. The effect of measurement distortion can be profound, for example, leading to irreducible indeterminacy in completely deterministic systems.[31]

So is anything ever new? I would answer "most definitely." With careful attention to the location of the observer and the system-under-study, with detailed accounting of intrinsic computation, with quantitative measures of complexity, we can analyze the patterns, structures, and novel information processing architectures that emerge in nonlinear processes. In this way, we demonstrate that something new has appeared.

## Acknowledgments

Many thanks are due to Melanie Mitchell for a critique of Reference [1] that led to the present essay. The author is also indebted to Dan McShea for his thoughts on the evolutionary trichotomy. Comments from Lisa Borland, Don Glaser, Jim Hanson, Blake LeBaron, and Dan Upper are gratefully acknowledged.

## Note

This chapter originally appeared in George A. Cowan, David Pines and David Meltzer (eds.) *Complexity: Metaphors, Models and Reality*, 515–533, Redwood City, Westview Press. 1994.

## Bibliography

1. J. P. Crutchfield. The calculi of emergence: Computation, dynamics, and induction. *Physica D*, 1994. In press; Santa Fe Institute Report SFI-93-11.

2. C. W. Reynolds. Flocks, herds, and schools: A distributed behavioral model. *Computer Graphics*, 21:25–34, 1987.

3. B. Holldobler and E. O. Wilson. *The Ants*. Belknap Press of Harvard University Press, Cambridge, Mass., 1990.

4. E. F. Fama. Efficient capital markets II. *J. Finance*, 46:1575–1617, 1991.

5. E. Land. The retinex. *Am. Scientist*, 52:247–264, 1964.

6. B. A. Wandell. Color appearance: The effects of illumination and spatial pattern. *Proc. Nat. Acad. Sci.*, 10:2458–2470, 1993.

7. I. Kovacs and B. Julesz. A closed curve is much more than an incomplete one—effect of closure in figure ground segmentation. *Proc. Nat. Acad. Sci.*, 90:7495–7497, 1993.

8. J. P. Crutchfield, N. H. Packard, J. D. Farmer, and R. S. Shaw. Chaos. *Sci. Am.*, 255:46, 1986.

9. H. L. Swinney and J. P. Gollub, editors. *Hydrodynamic Instabilities and the Transition to Turbulence*, Berlin, 1981. Springer Verlag.

10. A. M. Turing. The chemical basis of morphogenesis. *Trans. Roy. Soc., Series B*, 237:5, 1952.

11. A. T. Winfree. *The Geometry of Biological Time*. Springer-Verlag, Berlin, 1980.

12. Q. Ouyang and H. L. Swinney. Transition from a uniform state to hexagonal and striped turing patterns. *Nature*, 352:610–612, 1991.

13. H. E. Stanley. *Introduction to Phase Transitions and Critical Phenomena*. Oxford University Press, Oxford, 1971.

14. P. Bak and K. Chen. Self-organized criticality. *Physica A*, 163:403–409, 1990.

15. J. J. Binney, N. J. Dowrick, A. J. Fisher, and M. E. J. Newman. *The Theory of Critical Phenomena*. Oxford University Press, Oxford, 1992.

16. D. W. Thompson. *On Growth and Form*. Cambridge University Press, Cambridge, 1917.

17. H. Meinhardt. *Models of Biological Pattern Formation*. Academic Press, London, 1982.

18. S. Forrest, editor. *Emergent Computation: Self-organizing, Collective, and Cooperative Behavior in Natural and Artificial Computing Networks: Introduction to the Proceedings of the Ninth Annual CNLS Conference*, Amsterdam, 1990. North-Holland.

19. J. G. Kemeny. The use of simplicity in induction. *Phil. Rev.*, 62:391, 1953.

20. C. S. Wallace and D. M. Boulton. An information measure for classification. *Comput. J.*, 11:185, 1968.

21. J. Rissanen. Modeling by shortest data description. *Automatica*, 14:462, 1978.

22. J. P. Crutchfield and B. S. McNamara. Equations of motion from a data series. *Complex Systems*, 1:417, 1987.

23. J. Rissanen. *Stochastic Complexity in Statistical Inquiry*. World Scientific, Singapore, 1989.

24. G. Nicolis and I. Prigogine. *Self-Organization in Nonequilibrium Systems*. Wiley, New York, 1977.

25. H. Haken. *Synergetics, An Introduction*. Springer, Berlin, third edition, 1983.

26. J. P. Crutchfield and K. Young. Inferring statistical complexity. *Phys. Rev. Let.*, 63:105, 1989.

27. S. Wolfram. Computation theory of cellular automata. *Comm. Math. Phys.*, 96:15, 1984.

28. L. Blum, M. Shub, and S. Smale. On a theory of computation over the real numbers. *Bull. AMS*, 21:1, 1989.

29. M. G. Nordahl. Formal languages and finite cellular automata. *Complex Systems*, 3:63, 1989.

30. J. E. Hanson and J. P. Crutchfield. The attractor-basin portrait of a cellular automaton. *J. Stat. Phys.*, 66:1415, 1992.

31. J. P. Crutchfield. Unreconstructible at any radius. *Phys. Lett. A*, 171:52–60, 1992.

32. C. Moore. Real-valued, continuous-time computers: A model of analog computation, part I. Technical Report 93-04-018, Santa Fe Institute, 1993.

33. J. Maynard-Smith. *Evolutionary Genetics*. Oxford University Press, Oxford, 1989.

34. J. Monod. *Chance and Necessity: An Essay on the Natural Philosophy of Modern Biology*. Vintage Books, New York, 1971.

35. S. J. Gould. *Wonderful Life*. Norton, New York, 1989.

36. C. H. Waddington. *The Strategy of the Genes*. Allen and Unwin, London, 1857.

37. B. Goodwin. Evolution and the generative order. In B. Goodwin and P. Sanders, editors, *Theoretical Biology: Epigenetic and Evolutionary Order from Complex Systems*, pages 89–100, Baltimore, Maryland, 1992. Johns Hopkins University Press.

38. S. A. Kauffman. *Origins of Order: Self-Organization and Selection in Evolution*. Oxford University Press, New York, 1993.

39. W. Fontana and L. Buss. What would be conserved if the tape were played twice? *Proc. Nat. Acad. Sci.*, 91:757–761, 1994.

40. W. Fontana and L. Buss. "The Arrival of the Fittest": Toward a theory of biological organization. *Bull. Math. Bio.*, 56:1–64, 1994.

41. R. A. Harris. *The Linguistic Wars*. Oxford University Press, New York, 1993.

42. B. Goodwin and P. Sanders, editors. *Theoretical Biology: Epigenetic and Evolutionary Order from Complex Systems*, Baltimore, Maryland, 1992. Johns Hopkins University Press.

43. K. R. Popper. *The Logic of Scientific Inquiry*. Basic Books, New York, 1959.

44. R. Shaw. Strange attractors, chaotic behavior, and information flow. *Z. Naturforsh.*, 36a:80, 1981.

# 16 Design, Observation, Surprise! A Test of Emergence

**Edmund M. A. Ronald, Moshe Sipper, and Mathieu S. Capcarrère**

## 16.1 Introduction

When a bank's accounting program goes seemingly independent and does its own thing, the programmer scratches his head, sighs, and prepares for doing overtime with the debugger. But when a society of agents does something surprising, Alife researchers may solemnly document this "emergent behavior," and move on to other issues without always seeking to determine the cause of their observations. Indeed, overly facile use of the term emergence has made it controversial. Arkin recently observed that:

> Emergence is often invoked in an almost mystical sense regarding the capabilities of behavior-based systems. Emergent behavior implies a holistic capability where the sum is considerably greater than its parts. It is true that what occurs in a behavior-based system is often a surprise to the system's designer, but does the surprise come because of a shortcoming of the analysis of the constituent behavioral building blocks and their coordination, or because of something else? ([1], p. 105)

Altogether, it seems the emergence tag has become a great attention grabber, thanks to the striking behaviors demonstrated in artificial life experiments. We do not think, however, that emergence should be diagnosed ipso facto whenever the unexpected intrudes into the visual field of the experimenter; nor should the diagnosis of emergence immediately justify an economy of explanation. Such abuse and overuse of the term eventually will devalue its significance, and bring work centered on emergence into disrepute. Therefore, we contend that, in the absence of an acceptable definition, researchers in the field would be well served by adopting an emergence certification mark that would garner approval from the Alife community.

Motivated by this wish to standardize the tagging task, we propose an *emergence test*, namely, criteria by which one can justify conferring the emergence label [24]. Our criteria are motivated by an examination of published work in the field of Alife.

The emergence test is presented in the next section and followed in section 16.3 by a host of case studies demonstrating its applicability. Finally, in section 16.4, we discuss

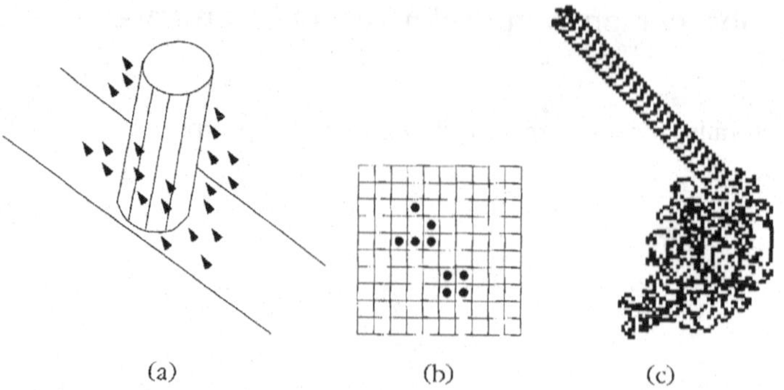


**Figure 16.1**
Examples of emergence. (a) A flock of simulated birds parts smoothly when faced with an obstacle and "flows" around it to then reunite again (after Reynolds [23]). (b) Two Game-of-Life patterns, known as a "glider" and a "block." (c) The trail created by the highway-constructing Langton ant.

our test in the light of previous work, and establish that it indeed articulates and clarifies existing views on the matter of emergence.

## 16.2 An Operant Definition of Emergence for Alife Researchers

### 16.2.1 Examples of Emergence in the Alife Literature

Before presenting the emergence test itself, we outline below a sampling of work in the field, which serves to delineate the scope of our investigation. The reader is urged to keep these in mind while perusing the emergence test, described in section 16.2.4.

- Emergence of flocking behavior in simulated birds, from a set of three simple steering behaviors (figure 16.1a) [23].
- Emergence of wall-following behavior in an autonomous, mobile robot, from the simultaneous operation of two simple behavior systems: obstacle avoidance and wall seeking [29].
- Emergence of cooperation in the iterated prisoner's dilemma, from the application of simple game strategies [3].
- Emergence of self-replicating structures from simple basic components [10, 19, 27].
- Emergence of a menagerie of patterns in the Game of Life (e.g., gliders, spaceships, puffer trains) from simple, local rules (figure 16.1b) [4].
- Emergence of team behavior (foraging, flocking, consuming, moving material, grazing) in autonomous, mobile robots, from simple rules [2].

- Emergence of social structures and group behaviors in the artificial society of "Sugarscape," from the interactions of individuals (agents) [14].
- Emergence of a "highway" created by the artificial Langton ant, from simple movement rules (figure 16.1c) [30].
- Emergence of complex behaviors in machines known as Braitenberg vehicles, from simple internal structures and mechanisms [7].
- Emergence of a nest structure in a simulated wasp colony, from the interactions taking place between individual wasps [31].
- Emergence of a solution to a character-recognition problem in artificial neural networks, from the interactions of the individual neurons.
- Emergence of a solution to the density problem in cellular automata, from simple, local interactions (figure 16.2) [8, 26, 28].
- Minsky's theory, according to which mind emerges from a society of myriad, mindless components [21].

### 16.2.2 Artificial Life as a Science of the Artificial

As seen from the examples in the previous subsection, we are restricting the scope of our study of emergence to instances of artificial life. Alife is a constructive endeavor: Some researchers aim at evolving patterns in a computer; some seek to elicit social behaviors in real-world robots; others wish to study life-related phenomena in a more controllable setting, while still others are interested in the synthesis of novel lifelike systems in chemical, electronic, mechanical, and other artificial media. Alife is an experimental discipline, fundamentally consisting of the observation of run-time behaviors, those complex interactions generated when populations of man-made, artificial creatures are immersed in real or simulated environments. Published work in the field usually relates the conception of a model, its instantiation into real-world or simulated objects, and the observed behavior of these objects in a collection of experiments.

The field of artificial life thus quintessentially exemplifies a science of the artificial, as it accords with the four indicia given by Herbert Simon in his influential work, *The Sciences of the Artificial* ([25], p. 8):

1. Artificial things are synthesized (though not always or usually with full forethought) by man.
2. Artificial things may imitate appearances in natural things while lacking, in one or many respects, the reality of the latter.
3. Artificial things can be characterized in terms of functions, goals, adaptation.
4. Artificial things are often discussed, particularly when they are being designed, in terms of imperatives as well as descriptives.

We agree with Simon in that the artificial can—and should—be treated differently from the natural. In this spirit, our emergence test below is aimed only at the artificial, and in particular at artificial life.

### 16.2.3 Preliminary Remarks on the Ontology of Emergence

Before formulating a definition of emergence, let us make some preliminary remarks:

- Emergence is not a specific thing; for example, like a stone in your pocket.
- Emergence is not one specific behavior, known to occur at some moment in time.
- Neither is emergence a well-defined category, like the stones on some particular beach.
- The category of all emergent behaviors is of interest; yet debate will occur on which behaviors should be included, and which should be excluded.
- Hence it is each observer who decides to include or exclude a given behavior in his own category of emergent behaviors.

The description of a phenomenon as emergent is contingent, then, on the existence of an observer; being *a visualization constructed in the mind of the observer*, emergence can be described as a *concept*, like beauty or intelligence. Such concepts are slippery.

### 16.2.4 Formulating the Emergence Test

The difficulties we face in adopting a definition of the concept of emergence are reminiscent of the complications faced by early Artificial Intelligence (AI) researchers in defining intelligence. Nonetheless, where the equally elusive concept of intelligence is concerned, Alan Turing found a way to cut the Gordian knot, by means of an *operant definition* that is useful *within the limited context of man-machine interaction* [32]. Debate concerning the concept of intelligence is unlikely to subside in the foreseeable future, and the same, we believe, holds for emergence. We deem, however, that viewing the world through Turing-colored glasses might improve our vision as regards the concept of emergence—at least where modern-day Alife practice is concerned.

The Turing test focuses on a human experimenter's incapacity at discerning human from machine when holding what we now would call an Internet chat session. Our emergence test centers on an observer's avowed incapacity (amazement) to reconcile his perception of an experiment in terms of a global world view with his awareness of the atomic nature of the elementary interactions.

Assume that the scientists attendant upon an Alife experiment are just two: a system designer and a system observer (both of whom in fact can be one and the same), and that the following three conditions hold:

1. Design The system has been constructed by the designer, by describing *local* elementary interactions between components (e.g., artificial creatures and elements of the environment) in a language $\mathscr{L}_1$.

2. Observation The observer is *fully aware* of the design, but describes *global* behaviors and properties of the running system, over a period of time, using a language $\mathscr{L}_2$.

3. Surprise The language of design $\mathscr{L}_1$ and the language of observation $\mathscr{L}_2$ are distinct, and the causal link between the elementary interactions programmed in $\mathscr{L}_1$ and the behaviors observed in $\mathscr{L}_2$ is *non-obvious* to the observer—who therefore experiences surprise. In other words, there is a cognitive dissonance between the observer's mental image of the system's design stated in $\mathscr{L}_1$ and his contemporaneous observation of the system's behavior stated in $\mathscr{L}_2$.

When assessing this clause of our test one should bear in mind that as human beings we are quite easily surprised (as any novice magician will attest). The question reposes rather on how *evanescent* the surprise effect is; that is, how easy (or strenuous) it is for the observer to bridge the $\mathscr{L}_1$–$\mathscr{L}_2$ gap, thus reconciling his global view of the system with his awareness of the underlying elementary interactions. One can draw an analogy with the concept of intelligence and the Turing test: the chatty terminal at first might appear to be carrying on like an intelligent interlocutor, only to lose its "intelligence certificate" once the tester has pondered upon the true nature of the ongoing conversation.

The above three clauses relating design, observation, and surprise describe our conditions for diagnosing emergence, that is, for accepting that a system is displaying emergent behavior. Some of the above points deserve further elaboration, or indeed invite debate. Before treating these issues in section 16.4, we wish to demonstrate the application of our test to several cases.

## 16.3 Administering the Emergence Test: Case Studies

In this section we administer the emergence test to eight examples (some of which are taken from section 16.2.1), thus demonstrating its application. Each example ends with a "test score," a diagnosis constituting our own assertion as observers of whether we are indeed surprised, that is, of whether emergent behavior is indeed displayed—or not.

### 16.3.1 Emergence of a Nest Structure in a Simulated Wasp Colony, from the Interactions Taking Place between Individual Wasps

- Design The design language $\mathscr{L}_1$ is that of local wasp interactions, including movement on a three-dimensional cubic lattice and placement of bricks. A wasp's decision is based upon a local configuration of bricks, which lie in its "visual" field. Actions to be taken are prewired under the form of a lookup table with as many entries as there are stimulating configurations.

▪ Observation The observation language $\mathscr{L}_2$ is that of large-scale geometry, as employed to describe nest architectures.

▪ Surprise While fully aware of the underlying wasp interaction rules, the observer nonetheless marvels at the sophistication of the constructions and at their striking similarity to naturally occurring nests.

▪ ? Diagnosis Emergent behavior is displayed by the nest-building wasps [31].

### 16.3.2 Emergence of a "Highway" Created by the Artificial Langton Ant, from Simple Movement Rules

▪ Design The design language $\mathscr{L}_1$ is that of single moves of a simple, myopic ant. The ant starts out on the central cell of a two-dimensional, rectangular lattice, heading in some selected direction. It moves one cell in that direction and looks at the color of the cell it lands on—black or white. If it lands on a black cell, it paints it white and turns 90 degrees to the left; if it lands on a white cell, it paints it black and turns 90 degrees to the right. These simple rules are iterated indefinitely.

▪ Observation The observation language $\mathscr{L}_2$ is that of global behavioral patterns, extended over time and space (i.e., tens of thousands of single ant moves, spanning thousands of cells). Specifically, the ant was observed to construct a "highway," that is, a repeating pattern of fixed width that extends indefinitely in a specific direction (figure 16.1c).

▪ Surprise While fully aware of the very simple ant rules, the observer is nonetheless surprised by the appearance of a highway.

▪ ? Diagnosis Emergent behavior is displayed by the highway-constructing ant [30].

### 16.3.3 Emergence of a Menagerie of Patterns in the Game of Life (e.g., Gliders, Spaceships, Puffer Trains), from Simple, Local Rules

▪ Design The Game of Life is played out on a two-dimensional, rectangular lattice, each cell of which can be colored either white or black (as with the Langton ant). The design language $\mathscr{L}_1$ is that of local color changes; this language is employed to delineate the rules according to which a cell changes its color in light of its immediate surrounding cells.

▪ Observation The observation language $\mathscr{L}_2$ is that of global behavioral patterns, including such observed structures as gliders, spaceships, and puffer trains (figure 16.1b).

▪ Surprise While fully aware of the simple color-transformation rules, the observer is nonetheless amazed by the appearance of this bestiary of critters.

▪ ? Diagnosis Emergent behavior is displayed by numerous instantiations of the Game of Life [4].

### 16.3.4 Emergence of Complex Behaviors in Machines Known as Braitenberg Vehicles, from Simple Internal Structures and Mechanisms

▪ Design The design language $\mathscr{L}_1$ is that of simple internal structures and mechanisms (sensors, actuators, and computational devices).

▪ Observation The observation language $\mathscr{L}_2$ is that of global behavioral patterns, to which Braitenberg playfully ascribed such anthropomorphic terms as "fear," "aggression," and "love."

▪ Surprise While fully aware of the vehicles' simple internal workings, the observer is nonetheless amazed by the appearance of lifelike behaviors. This is true for vehicles 3 through 14; on the contrary, the behaviors of vehicles 1 and 2 can be straightforwardly divined by the observer.

▪ ? Diagnosis Emergent behavior is displayed where vehicles 3 through 14 are concerned, while vehicles 1 and 2 are non-emergent [7].

### 16.3.5 Minsky's Theory, According to which Mind Emerges from a Society of Myriad, Mindless Components

▪ Design The design language $\mathscr{L}_1$ is that of simple (putative) processes, which Minsky calls agents.

▪ Observation The observation language $\mathscr{L}_2$ is the common language of discourse used to describe intelligent, human behavior (examples from Minsky's book are nonverbal reasoning, language learning, and humor [21]).

▪ ? Surprise and Diagnosis Mind is an emergent phenomenon, *par excellence*, since the observer always marvels at its appearance [21].

### 16.3.6 Emergence of Flocking Behavior in Simulated Birds, from a Set of Three Simple Steering Behaviors

▪ Design The design language $\mathscr{L}_1$ is that of local bird interactions, the three rules being: *separation*—steer to avoid crowding local flockmates; *alignment*—steer toward the average heading of local flockmates; *cohesion*—steer to move toward the average position of local flockmates. A bird's decision is based upon its nearby neighbors, that is, those that are in its "visual" field.

- Observation The observation language $\mathscr{L}_2$ is that of flocking behaviors, such as the flock's parting smoothly when faced with an obstacle, and "flowing" around it—to then reunite again (figure 16.1a).
- Surprise While fully aware of the underlying bird interaction rules, the observer nonetheless marvels at the lifelike flocking behaviors.
- ? Diagnosis The flocking behavior exhibited by the artificial birds was considered a clear case of emergence when it first appeared in 1987. However, one now could maintain that it no longer passes the emergence test, since widespread use of this technique in computer graphics has obviated the element of surprise. This example demonstrates that the diagnosis of emergence is contingent upon the sophistication of the observer [23].

### 16.3.7 Emergence of Wall-Following Behavior in an Autonomous, Mobile Robot, from the Simultaneous Operation of Two Simple Behavior Systems: Obstacle Avoidance and Wall Seeking

- Design The design language $\mathscr{L}_1$ is that of simple robot behaviors, including—in this case—obstacle avoidance and wall seeking.
- Observation The observation language $\mathscr{L}_2$ is that of more elaborate robot behaviors, consisting—in this case—of wall following.
- Surprise Steels wrote that "Wall following is emergent in this case because the category 'equidistance to the (left/right) wall' is not explicitly sensed by the robot or causally used in one of the controlling behavior systems" ([29], p. 92).
- ? Diagnosis Steels diagnosed emergence in this case as it accords with his own definition, namely, that a behavior is emergent if it necessitates the use of new descriptive categories that are not needed to describe the behavior of the constituent components [29]. While thus alluding to the language dichotomy rendered explicit by our definition (i.e., the existence of two distinct languages—that of design and that of observation), we maintain that the surprise element is missing: The wall-following behavior can be quite readily deduced by an observer aware of the two underlying simpler behaviors. We thus conclude that emergent behavior is *not* displayed by the wall-following robot [29].

### 16.3.8 Emergence of a Solution to the Density Problem in Cellular Automata, from Simple, Local Interactions

This final example is delineated in more detail than the previous ones so as to demonstrate a number of interesting points concerning our test [8, 26, 28]. The example is based on the model known as cellular automata (CA), originally conceived by Ulam and von Neumann in the 1940s to provide a formal framework for investigating the behavior of complex, extended systems [27, 33]. CAs have been widely used over the years in many fields of inquiry, including physics, biology, and computer science; in particular, they figure prominently in Alife research. We first describe briefly the workings of a CA, followed by the presentation of a specific problem, known as density classification, which has received much attention in the CA literature. We delineate three CA solutions to this problem, concluding that the first two pass the emergence test while the last one does not.

Cellular automata are discrete, dynamical systems that perform computations in a distributed fashion on a spatially extended grid. A cellular automaton consists of an array of cells, each of which can be in one of a finite number of possible states, updated synchronously in discrete time steps according to a local, identical interaction rule [26]. The *state* of a cell at the next time step is determined by the current states of a surrounding neighborhood of cells. This transition is usually specified in the form of a *rule table*, delineating the cell's next state for each possible neighborhood configuration. The cellular array (grid) is $n$-dimensional, where $n = 1, 2, 3$ is used in practice.

CAs are one of the prime models used to study emergent behavior and computation. One oft-cited problem involving (putative) emergent computation is for a CA to determine the global density of bits in an initial state configuration. This problem, known as density classification, has been studied intensively over the past few years [26]. In a recent paper, Sipper, Capcarrère, and Ronald [28] described two previous versions of the problem along with their CA solutions, and then went on to show that there exists yet a third version—which admits a simple solution. Below, we summarize their results, after which we will administer the emergence test to all three versions.

**16.3.8.1 Version I** In the original statement of the problem [22], a one-dimensional, two-state CA (meaning that each cell can be in one of two states, 0 or 1) is presented with an arbitrary initial configuration of states (the input). The CA then should converge in time to a state of all 1s if the initial configuration contains a density of 1s $> 0.5$ (i.e., a majority of 1s), and converge to all 0s if this density $< 0.5$ (i.e., a majority of 0s); for an initial density of 0.5, the CA's behavior is undefined (figure 16.2(I)). The final configuration is considered as the output of the computation. Spatially periodic boundary conditions are used, resulting in a circular grid. It has been proven that no perfect CA solution exists for this problem version, though high-performance CAs have been designed by hand as well as found by means of artificial evolution [26] (these CAs do not perform perfect classification, i.e., they misclassify some of the initial configurations; the CA solution demonstrated in figure 16.2(I), known as the GKL rule, in fact does not classify correctly all initial configurations).

**16.3.8.2 Version II** Capcarrère, Sipper, and Tomassini [8] showed that a perfect, one-dimensional, two-state CA density classifier does exist, upon defining a different output specification (again, periodic boundary conditions are assumed). This CA is

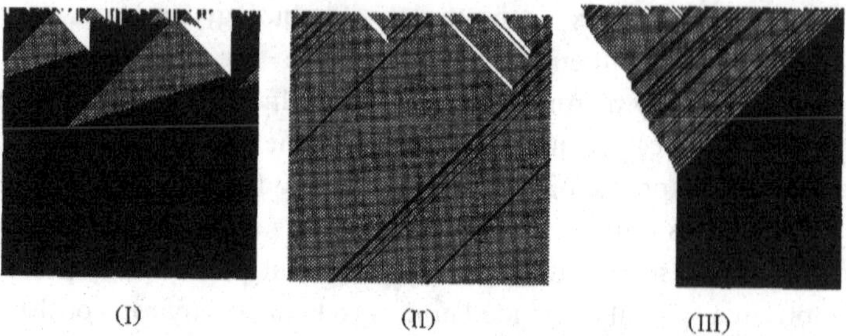


**Figure 16.2**
CA solutions to three versions of the density classification problem. The one-dimensional grid is of size $N = 149$ cells. White squares represent cells in state 0; black squares represent cells in state I. The two-dimensional images shown above are space-time diagrams, a common method of depicting the behavior of one-dimensional CAs over time. In the images, the horizontal axis depicts the configuration of states at a certain time $t$, and the vertical axis depicts successive time steps (time thus increases down the figure). The initial density in all three examples is $>0.5$.

demonstrated in figure 16.2(II): Upon presentation of an arbitrary initial configuration, the $N$-cell grid relaxes to a limit-cycle, within $\lceil N/2 \rceil$ time steps, that provides a classification of the initial configuration's density of 1s. If this density $> 0.5$ (respectively, $<0.5$), then the final configuration consists of one or more blocks of at least two consecutive 1s (0s), interspersed by an alternation of 0s and 1s; for an initial density of exactly 0.5, the final configuration consists of an alternation of 0s and 1s. The computation's output is given by the state of the consecutive block (or blocks) of same-state cells: If the same-state cells are in state 1 (respectively, 0), then this signifies a majority of 1s (0s) in the input; if there is but an alternation of 0s and 1s, then this signifies that the input contains an equal number of 0s and 1s.

**16.3.8.3 Version III** More recently, Sipper, Capcarrère, and Ronald [28] described yet another modification of the original problem (version I), with a different output specification, as well as fixed boundary conditions, rather than the periodic ones previously assumed. These two modifications give rise to a simple density classifier, demonstrated in figure 16.2(III): The CA of size $N$ (boundary cells excluded) converges in at most $N - 1$ time steps to a configuration $0^{\alpha}1^{\beta}$, where $\alpha$, $\beta$ denote the number of 0s and 1s at time step 0, respectively; $\alpha, \beta \in \{0, \ldots, N\}$, $\alpha + \beta = N$. For $N$ odd, the density classification of the input is attained by considering the middle cell's final state: 0 signifies a majority of 0s in the input, 1 signifies a majority of 1s; for $N$ even, we consider the two middle cells: 00 signifies a majority of 0s in the input, 11 signifies a majority of 1s, 01 signifies equality, and 10 is impossible. This example in fact initiated our current study of emergence, ultimately culminating in the proposed emergence test.

**16.3.8.4 Emergent or Not?** As CA researchers, we observed the three versions depicted in figure 16.2, experiencing unease with respect to version III—as though we were "cheating." And yet, why do the first two solutions seem complex and emergent, whereas the third one seems simple and non-emergent?[1] Density is a global property of a configuration (the 1s can be distributed throughout the grid), whereas a CA relies solely on local interactions; this property holds true for all three versions of the problem. Moreover, the three CA solutions presented are all legal, in that they violate none of the CA principles of operation. In short, we are faced here with three perfectly valid solutions (albeit an approximate one where version I is concerned).

The crux of the matter lies in the surprise phase of the emergence test: we maintain that there is no (lasting) surprise where version III is concerned, whereas we are surprised by versions I and II. The language of design $\mathscr{L}_1$ and the language of observation $\mathscr{L}_2$ are obviously distinct in all three cases, and yet the causal link between the elementary interactions programmed in $\mathscr{L}_1$ and the behaviors observed in $\mathscr{L}_2$ is rather straightforward for version III, whereas it is non-obvious for versions I and II. An observer even but slightly versed in CA dynamics will divine quickly the workings of the version-III CA (whose rule is fully specified in [28]): basically, one can picture it as a horizontally-oriented, transparent tube full of tennis balls, where each ball represents a state of 1 (the absence of a ball represents a state of 0). If one places the tube in the vertical position, the balls will roll down, with classification of the initial density then given by simply observing the absence or presence of balls in the central section of the tube. The CA rule in question implements this metaphor (the fixed boundary conditions are crucial in that they stop the balls dead at the tube's end). It is the quickly fading surprise effect experienced with the version-III solution that evoked in us the uneasy feeling that something was not "right." Armed with the emergence test, we can now pinpoint the source of our uneasiness: Version III is non-emergent (no lasting surprise) whereas versions I and II are (surprise!). (We emphasize again that version III—while non-emergent—is "right" in the sense that it is a bona fide solution, which perfectly accords with the CA principles.) This example illustrates the subtleties involved in our test. Our conclusions regarding this CA example are recapitulated below within the emergence-test framework:

- **Design** The design language $\mathscr{L}_1$ is that of local CA rules.
- **Observation** The observation language $\mathscr{L}_2$ is that of global behavioral patterns. Specifically, the patterns of interest involve those that represent a solution to the density classification problem.
- **Surprise** While fully aware of the underlying CA rules, the observer is nonetheless puzzled by the intricate behaviors of versions I and II, whereas version III straightforwardly yields its "secret."

▪ ? Diagnosis Emergent behavior is displayed where versions I and II are concerned, while version III is non-emergent.

## 16.4 Discussion

Our conception of the emergence test builds on a number of ideas that have been addressed over the years in the literature; the relationship between our test and these works is described in this section. While a number of researchers have remarked upon some or other ingredient of our definition, to the best of our knowledge nobody has yet put together all the constituent elements into an emergence test such as ours. Moreover, those elements that have been discussed in the literature are highly intermingled—which has rendered it impossible for us to separate the discussion below into totally orthogonal clauses; indeed, untangling this vicious circle of reasoning is, in our opinion, one of the major contributions of our test.[2]

### 16.4.1 The Operant Nature of the Test

In this we have drawn our inspiration from Turing, who—concerning intelligence—opted for an operant, informal, "social" definition, deliberately eschewing rigor. Turing's definition has served the AI community well, and it is still considered one of the seminal works in the field, almost half a century after its publication [32].

### 16.4.2 Emergence as a Property of Artificial Systems

In his book, *Emergence: From Chaos to Order*, Holland wrote that "Emergence occurs in systems that are generated" ([17], p. 225). Reviewing Holland's book, Mallot also opined that "In this context [the construction of artificial systems], the problem of emergence may actually be a genuine one" [20]. As noted in section 16.2.2, we have chosen to limit the scope of our test to the sciences of the artificial, and in particular to artificial life; this restriction is embodied in clause (1) of the test.

### 16.4.3 The Existence of an Observer

Artificial systems are constructed to be beheld—usually one does not build one's system, then walk away nonchalantly without ever looking back. Hence, there exists an observer ipso facto (who need not necessarily be the constructor himself), a fundamental aspect that has not escaped researchers in the field. In a paper discussing emergence and artificial life, Cariani wrote that "The interesting emergent events that involve artificial life simulations reside not in the simulations themselves, but in the ways that they change the way we think and interact with the world" ([9], p. 790). He goes on to say that "computer simulations are catalysts for emergent processes in our own minds..." ([9], p. 790).

Another author, Emmeche, in an introductory monograph on artificial life, examines the case for emergence "in the eye of the beholder" ([13], p. 145). Also, Crutchfield, in an article devoted to the subject of emergence, asks: "But for whom has the emergence occurred? More particularly, to whom are the emergent features 'new'? ... The newness in both cases is in the eye of an observer ..." ([12], p. 517).

Bonabeau, Dessalles, and Grumbach, in an article presenting a conceptual framework for characterizing emergent phenomena, noted the difference between "actors: interacting agents with *local perception* and the ability to *act locally*" and "spectators: one or several entities sensitive to the emergent phenomenon, and possessing *global perception*" ([6], p. 348). They wrote that "the emergent aspect of a phenomenon is related to the point of view of an observer of this phenomenon: it is not intrinsic to the phenomenon, but related to the global system (phenomenon + observer)" ([6], pp. 348–349).

Holland brings up the issue of the observer circuitously, when writing that "The whole is more than the sum of the parts in these generated systems.... Said another way, there are regularities in system behavior that are not revealed by direct inspection of the laws satisfied by the components" ([17], p. 225). One may ask *direct inspection by whom?* Why, by the observer of course![3] Clearly, the existence of an observer is a sine qua non for the issue of emergence to arise at all.

We wish to point out that one can make a case for the analogy between the concepts of emergence and complexity, as regards the presence of a baffled observer. For example, Kolen and Pollack, considering the highly formal notion of computational complexity, wrote: "Computational complexity, often used to separate cognitive behaviors from other types of animal behavior, will be shown to be dependent upon the observation mechanism as well as the process under examination" ([18], p. 254).

### 16.4.4 The Language Dichotomy

A number of authors have alluded to the existence of a language of observation as distinct from the language of design. In his paper on behavior-oriented artificial intelligence, Steels put forward a definition of emergence, writing that "A behavior is emergent if new categories are needed to describe this underlying regularity that are not needed to describe the behaviors (i.e., the regularities) generated by the underlying behavior systems on their own" ([29], p. 89).

In a book, entitled *Frontiers of Complexity*, Coveney and Highfield, upon discussing the behavior of collective systems of simple, interacting units, wrote that "Their interactions lead to coherent collective phenomena, so-called emergent properties that can be described only at higher levels than those of the individual units" ([11], p. 7).

Holland emphasized the distinct nature of these two languages, noting that one can "converse" in the language of observation without resorting to the language of design:

"When a macrolaw can be formulated, the behavior of the whole pattern can be described without recourse to the microlaws (generators and constraints) that determine the behavior of its components" ([17], p. 227).

### 16.4.5 The Observer's Reasoning Abilities

Writing on intelligence as an emergent behavior, Hillis contends that "The emergent behaviors exhibited by these systems are a consequence of the simple underlying rules defined by the program. Although the systems succeed in producing the desired results, their detailed behaviors are beyond our ability to analyze and predict" ([16], pp. 188–189). Hillis thus can be seen to allude to the observer's reasoning abilities.

As regards our test, the extent of the observer's reasoning abilities does indeed influence the diagnosis of emergence. To render a diagnosis established by a single judge more credible, we could replace one judge with a collective; moreover, membership of such an emergence jury could be restricted to the suitably qualified. On the jury issue, Turing also noted that "A number of interrogators could be used, and statistics compiled to show how often the right identification was given" [32]. Judgments often will come with a statute of limitations—a phenomenon might be reclassified from emergent to non-emergent with the progress of science (as with the artificial-flock example in section 16.3). Here again, the analogy with intelligence: Tasks that were once considered intelligent—such as doing sums—nowadays are considered to be but a job for dullards.

Minsky, in his book on the emergence of mind—*The Society of Mind* [21]—nicely illustrates the role of the observer's reasoning abilities. He provides two sets of examples, which he denotes subjective and objective, to which we might refer within our framework as emergent and non-emergent, respectively. Minsky's first set of examples (subjective or emergent) includes such questions as "What makes a drawing more than just its separate lines?" while the second set of examples (objective or non-emergent) includes such questions as "What makes a tower more than separate blocks?" Minsky goes on to explain that the development of the observer's reasoning abilities nullifies the emergence quality where questions of the second type are concerned: "To explain how walls and towers work, we just point out how every block is held in place by its neighbors and by gravity.... These explanations seem almost self-evident to adults. However, they did not seem so simple when we were children.... We regard such knowledge as 'obvious' only because we cannot remember how hard it was to learn" ([21], p. 27).

### 16.4.6 Surprise

The surprise element also has received attention in a number of works. We noted in section 16.1 Arkin's view: "[W]hat occurs in a behavior-based system is often a surprise to the system's designer..." ([1], p. 105). Minsky wrote that:

> We're often told that certain wholes are "more than the sum of their parts." We hear this expressed with reverent words like "holistic" and "gestalt," whose academic tones suggest that they refer to clear and definite ideas. But I suspect the actual function of such terms is to anesthetize a sense of ignorance. We say "gestalt" when things combine to act in ways we can't explain, "holistic" when we're caught off guard by unexpected happenings and realize we understand less than we thought we did. ([21], p. 27)

By bringing the observer's *emotion* of surprise into play, our emergence test widens the focal beam of discussion, now shining both on the *system's behavior* as well as on the experimenter and her *internalized expectations*. This relates to Cariani's nutshell description of emergence relative to a model as "the deviation of the behavior of a physical system from an observer's model of it" ([9], p. 779). An author subscribing to said deviation-from-model view would wish to document her a priori expectations before diagnosing emergence and abandoning attempts at explanation. Our emergence test then might be reformulated as *Design (Expectations), Observation, Surprise.*

### 16.4.7 Non-Obviousness

A key element of our test is the non-obviousness experienced in the surprise phase by the observer. The study of complex systems is revealing common causes of non-obviousness. Known categories to date include:

1. computational undecidability (e.g., in the Game of Life and cellular automata)
2. self-organizing phenomena
3. sensitivity to initial conditions, known as chaos (e.g., as in weather patterns, and in predator-prey oscillations)

## 16.5 Summary

To summarize, the three clauses of our emergent test are grounded in previous work: The design clause expresses our wish to restrict the test to artificially constructed systems, the observation clause reflects the necessity of there being an observer for emergence to arise at all, and the surprise clause embodies both the deliberation and the emotion implied by human judgments of value.

### Acknowledgments

We thank Bernard Besnier, Dario Floreano, Daniel Mange, Eduardo Sanchez, Marco Tomassini, and the anonymous reviewers for their helpful remarks and suggestions.

From *The Adventure of the Dancing Men*, by Arthur Conan Doyle:

He wheeled round upon his stool, with a steaming test-tube in his hand and a gleam of amusement in his deep-set eyes.
"Now, Watson, confess yourself utterly taken aback," said he.
"I am."
"I ought to make you sign a paper to that effect."
"Why?"
"Because in five minutes you will say that it is all so absurdly simple."
"I am sure that I shall say nothing of the kind."
"You see, my dear Watson"—he propped his test-tube in the rack and began to lecture with the air of a professor addressing his class . . .
. . .
"How absurdly simple!" I cried.

## Notes

This chapter originally appeared in *Artificial Life* 5 (1999), 225–239.

While not disowning this work, Dr. Capcarrère has explicitly distanced himself from the content of this chapter.

1. We note in passing that this "feeling" is also supported by the complexity of the proofs concerning the behaviors of CA versions I and II, versus the simplicity of the proof of CA III's behavior [8, 28].

2. This section is by no means intended to serve as a review on the subject of emergence. Rather, we have cited what we believe to be major works in this area that tie in with our emergence test. For a good critical review, the reader is referred to Bonabeau, Dessalles, and Grumbach [5].

3. Holland also cites a passage from Gell-Mann's book, *The Quark and the Jaguar* [15], which brings up indirectly the role of the observer: "In an astonishing variety of contexts, apparently complex structures or behaviors emerge from systems characterized by simple rules." ([17], p. 238). Gell-Mann's use of the qualifier "apparently" suggests that the quality in question necessitates a judgment call—that is, an observer.

## References

1. Arkin, R. C. (1998). *Behavior-Based Robotics*. Cambridge, MA: MIT Press.

2. Arkin, R. C. (1998). *Behavior-Based Robotics* (pp. 359–420). Cambridge, MA: MIT Press.

3. Axelrod, R. (1984). *The Evolution of Cooperation*. New York, NY: Basic Books.

4. Berlekamp, E. R., Conway, J. H., and Guy, R. K. (1982). *Winning Ways for Your Mathematical Plays, Volume 2*, (pp. 817–850). New York, NY: Academic Press.

5. Bonabeau, E., Dessalles, J. L., and Grumbach, A. (1995). Characterizing emergent phenomena (1): A critical review. *Revue Internationale de Systémique*, *9*, 327–346.

6. Bonabeau, E., Dessalles, J. L., and Grumbach, A. (1995). Characterizing emergent phenomena (2): A conceptual framework. *Revue Internationale de Systémique*, *9*, 347–371.

7. Braitenberg, V. (1984). *Vehicles: Experiments in Synthetic Psychology*. Cambridge, MA: MIT Press.

8. Capcarrère, M. S., Sipper, M., and Tomassini, M. (1996). Two-state, r = 1 cellular automaton that classifies density. *Physical Review Letters, 77*, 4969–4971.

9. Cariani, P. (1992). Emergence and artificial life. In C. G. Langton, C. Taylor, J. D. Farmer, and S. Rasmussen (Eds.), *Artificial Life II, Volume X, SFI Studies in the Sciences of Complexity* (pp. 775–797). Redwood City, CA: Addison-Wesley.

10. Chou, H. H., and Reggia, J. A. (1997). Emergence of self-replicating structures in a cellular automata space. *Physica D, 110*, 252–276.

11. Coveney, P., and Highfield, R. (1995). *Frontiers of Complexity: The Search for Order in a Chaotic World*. London: Faber and Faber.

12. Crutchfield, J. P. (1994). Is anything ever new? considering emergence. In G. Cowan, D. Pines, and D. Melzner (Eds.), *Complexity: Metaphors, Models, and Reality* (pp. 515–537). Reading, MA: Addison-Wesley.

13. Emmeche, C. (1994). *The Garden in the Machine: The Emerging Science of Artificial Life*. Princeton, NJ: Princeton University Press.

14. Epstein, J. M., and Axtell, R. (1996). *Growing Artificial Societies: Social Science from the Bottom Up*. Washington, DC: Brookings Institution Press.

15. Gell-Mann, M. (1994). *The Quark and the Jaguar: Adventures in the Simple and the Complex*. New York, NY: Freeman.

16. Hillis, W. D. (1988). Intelligence as an emergent behavior; or, the songs of eden. *Daedalus, Journal of the American Academy of Arts and Sciences, 117*, 175–189.

17. Holland, J. H. (1998). *Emergence: From Chaos to Order*. Reading, MA: Addison-Wesley.

18. Kolen, J. F., and Pollack, J. B. (1995). The observers' paradox: Apparent computational complexity in physical systems. *Journal of Experimental and Theoretical Artificial Intelligence, 7*, 253–277.

19. Koza, J. R. (1994). Artificial life: Spontaneous emergence of self-replicating and evolutionary self-improving computer programs. In C. G. Langton (Ed.), *Artificial Life III*, volume XVII of *SFI Studies in the Sciences of Complexity* (pp. 225–262). Reading, MA: Addison-Wesley.

20. Mallot, H. (1998). Life is like a game of chess: Review of *Emergence: From Chaos to Order* by John Holland. *Nature, 395*, 342.

21. Minsky, M. (1986). *The Society of Mind*. New York: Simon and Schuster.

22. Packard, N. H. (1988). Adaptation toward the edge of chaos. In J. A. S. Kelso, A. J. Mandell, and M. F. Shlesinger (Eds.), *Dynamic Patterns in Complex Systems* (pp. 293–301). Singapore: World Scientific.

23. Reynolds, C. W. (1987). Flocks, herds, and schools: A distributed behavioral model. *Computer Graphics, 21*, 25–34.

24. Ronald, E. M. A., Sipper, M., and Capcarrère, M. S. (1999). Testing for emergence in artificial life. In D. Floreano, J.-D. Nicond, and F. Mondada (Eds.), *Proceedings of the Fifth European Conference on Artificial Life (ECAL'99)* (pp. 13–20). Heidelberg: Springer-Verlag.

25. Simon, H. A. (1981). *The Sciences of the Artificial* (2nd ed). Cambridge, MA: MIT Press.

26. Sipper, M. (1997). *Evolution of Parallel Cellular Machines: The Cellular Programming Approach.* Heidelberg: Springer-Verlag.

27. Sipper, M. (1998). Fifty years of research on self-replication: An overview. *Artificial Life, 4,* 237–257.

28. Sipper, M., Capcarrère, M. S., and Ronald, E. (1998). A simple cellular automaton that solves the density and ordering problems. *International Journal of Modern Physics C, 9,* 899–902.

29. Steels, L. (1994). The artificial life roots of artificial intelligence. *Artificial Life, 1,* 75–110.

30. Stewart, I. (1994). The ultimate in anty-particles. *Scientific American, 271,* 104–107.

31. Theraulaz, G., and Bonabeau, E. (1995). Coordination in distributed building. *Science, 269,* 686–688.

32. Turing, A. M. (1950). Computing machinery and intelligence. *Mind, 59,* 433–460.

33. von Neumann, J. (1966). *Theory of Self-Reproducing Automata,* ed. A. W. Burks. Champaign, IL: University of Illinois Press.

# 17 *Ansatz* for Dynamical Hierarchies

Steen Rasmussen, Nils A. Baas, Bernd Mayer, and Martin Nillson

## 17.1 Introduction

### 17.1.1 Dynamical Hierarchies in Biological Systems

In biological systems hierarchies with multiple functionalities at different scales can be found everywhere. Clearly there is a coarse-grained hierarchy (which has many refinements and substructures) that may be expressed as ecosystems (including human sociotechnical systems), organisms, organs, tissues, cells, organelles, molecules, atoms. Obviously, the properties associated with each level are generated by the collective dynamics of the elements in these dynamical hierarchies. One problem here is to create a formal framework for consistently describing such hierarchical systems, with some coarse-graining procedure for moving between levels.

A second issue is how these complex and robust functionalities are generated in biological (and proto-biological) systems. We know that novel functionalities in molecular systems can arise in at least two ways: by assembly and by evolution. Molecular self-assembly processes are probably most central as a mechanism, for example, for bridging nonliving and living matter, for the transition from prokaryotic to eukaryotic organisms, and perhaps for other major biological transition processes. Self-assembly and self-organization allow limited inherited information to code for complex functionalities, from enzyme catalysts to the brain. It is probably necessary to embrace assembly and evolution as two complementary and perhaps equally important aspects of how to generate bio-complexity if one is to account for the generation and prevalence of dynamical hierarchies in living systems.

### 17.1.2 Dynamical Hierarchies in Molecular Systems

Evolution seems to have favored systems that have been built up by subunits at several levels. In that respect even the simplest cells are extremely complicated. To understand the basic principles leading to such structures we should start looking at the molecular level where we find many examples of supramolecular structures built up by subunits at several levels, which hence qualify for being intuitively called higher-order structures. Some guiding examples follow:

1. Polymers form from monomers, as for example, ethylene molecules forming polyethylene.

2. The monomeric protein actin spontaneously associates into linear, helical polymers in the presence of ATP. These polymers are called actin filaments. The filaments are cross-linked by other proteins—fodrin or filamin—to side-by-side aggregates like bundles or mesh-works. These certainly qualify as natural third-order structures. They greatly increase the viscosity of the medium in which the filaments are suspended [24].

3. Microtubules spontaneously form from monomeric subunits. The process starts with the tubulin $\alpha$- and $\beta$-subunits forming $\alpha$- and $\beta$-dimers that then form linear polymers associating side-by-side to form the hollow microtubules [23].

4. Viruses are also built up through aggregation of subunits. This applies to the simple tobacco mosaic virus and the much more complicated bacteriophage T4 [18].

5. Lipids dispersed in water form microscopic aggregates. Lipid molecules cluster together with their hydrophobic moieties in contact with each other and their hydrophilic groups interacting with surrounding water. These clusters may be micelles, in other cases bilayers, liposomes or even more complicated forms [26].

Many other and much more complicated examples of higher-order structures exist [48], but is seems reasonable to start with the simpler structures and try to model the generation of them first. In the following we study two formal and closely related examples where the first is chosen because of its conceptual clarity and the second because of its realism. Together they demonstrate fundamental properties of dynamical hierarchies. Our first system has hydrophobiclike (water "fearing") and hydrophiliclike (water "loving") monomers that can polymerize into amphiphilic polymers (one end of the polymer hydrophobic and the other end hydrophilic). These polymers in turn can aggregate to micellelike structures, thus forming a system with three distinctive levels: monomers, polymers, and aggregates. The second system is a more realistic water/amphiphilic system that also demonstrates micellation and exhibits three distinctive levels: monomers (and water), polymers, and micelles. To parallel laboratory procedures for creating micelles, we typically initialize the second system with water and amphiphilic polymers constructed out of monomers. However, it should be noted that both systems compromise simplicity to achieve greater physical and chemical realism, because our ultimate goal is to develop a simulation platform that promotes detailed comparison with experimental self-assembly systems. Understanding molecular self-assembly processes are crucial in the construction of proto-organisms in the laboratory.

Through simulation we have come a long way in understanding the nature of evolutionary processes. However, the origin of evolutionary processes is still not understood. How can self-replicating processes be generated through self-organization? In molecular systems, self-assembly processes are able to generate the necessary higher-order structures that in turn are able to self-replicate. To our knowledge there is no simple model either in vitro or in silico that integrates these two processes. Are there universal properties in origins, as we have discovered universal properties in evolution? And can evolution generate or explain all novelty after self-replication becomes possible or is self-assembly still necessary after the onset of evolution?

### 17.1.3 Related Work

Many groups are engaged in the modeling and simulation of molecular self-assembly processes mostly using classical molecular dynamics (mostly on an atomic scale resolution [13, 19, 49, 50, 53]) or traditional lattice gas (polymers as fundamental objects) methods (see, for example, [9, 10, 11, 17, 46, 54, 55]). However, the theoretical question we pose and address about the nature of dynamical hierarchies is not discussed in this work. The closest tradition within the artificial life community consists of artificial chemistries, both past and ongoing work, which are focused on the question of origins. This includes work using cellular automata, secondary folding algorithms, finite state automata (on the lattice or movable), random graphs, ordinary differential equations with meta dynamics, von Neumann machines, Turing machines, and the lambda calculus (see, for example, [6–8, 14–16, 18, 22, 23, 31, 32, 35, 38, 39, 43, 44, 47, 52]). For a nice comprehensive discussion of the artificial chemistry tradition, see the review by Dittrich et al. [13].

## 17.2 Higher-Order Dynamical Hierarchies

### 17.2.1 Physico-Chemical Examples

The dynamical hierarchies we are going to discuss have the following properties. We start by identifying a collection of objects (perhaps heterogeneous) and a set of dynamical rules for how those objects interact as a first "level" of phenomena. In reality, the first-order objects may well have internal structure and their interactions might well be explainable in terms of more fundamental interactions among the components of these objects. But we have to start our analysis somewhere, so when we identify a dynamical hierarchy we *treat* the first-order objects and their interactions *as if* axiomatically given. Now, the dynamics of the first-order objects creates various kinds of new, second-order objects, typically formed as certain kinds of aggregates of first-order objects. These second-order objects have new kinds of properties, not present in the first-order objects, including new kinds of interactions with other second-order objects and/or first-order objects. Intuitively speaking, second-order objects have emergent second-order properties, which could be described by a second-order dynamics. Similar to the emergence of second-order phenomena from first-order phenomena, the dynamics of second-order objects creates various kinds of new, third-order objects, typically formed as certain kinds of aggregates of second-order objects and possibly also first-order objects. As before, these third-order objects have emergent third-order properties, including third-order interactions, all of which could be described by a third-order dynamics.

Aggregation is certainly not the only way higher-order structures may arise. For example, in networks there may be functionally higher-order properties that are topologically distibuted.

A system consisting of water and monomers can be organized into a simple physical example of a third-order dynamical hierarchy. The first-order objects are individual water and monomer molecules. The second-order objects are polymers, that is, linear chains of monomers. The third-order objects are micelles and vesicles, that is, objects composed of organized amphiphilic polymers with single- or double-layer membranes, respectively.

Monomers have properties such as being hydrophilic or hydrophobic, and their interactions include various degrees of intermolecular attraction. The interactions between individual water and monomer molecules give rise to a number of emergent phenomena caused by the so-called hydrophobic effect. One simple example is the phase separation between water molecules and monomers that can occur if the monomers are highly hydrophobic. The principal reason for this separation is that water likes (attracts) water much more that it likes the hydrophobic monomers and the water "pushes" the monomers out of the way of the strong water–water interactions. Note that this separation cannot be observed at the level of the individual molecules. It is a collective property generated by the dynamics of all the individual molecules.

The interactions between water and monomers also give rise to second-order objects, the polymers, and in the process these polymers come to have properties that cannot be possessed by individual water molecules or individual monomers. A trivial example is the phase separation that can occur between water and hydrocarbon polymers, analogous to the phase separation between water and hydrophobic monomers.

A third example of a novel second-order property is elasticity. Even if we assume that the monomer–monomer bindings are nonelastic, the whole polymer will be elastic because the polymer is flexible and it is floating in a (water) "heat bath." The polymer obviously has many more folded than stretched configurations. The random motion between the water molecules and the polymer generates an elastic force that can be measured as a tendency for both ends of the polymer to move away from highly stretched positions.

More interesting phenomena are possible when the polymers have an orientation, for example, they have different properties in different ends. Some polymers are amphiphilic, that is, have one end that binds strongly with water and another end that binds weakly with water. The interactions among such polymers give rise to third-order objects, as micelles (and, for example, bilayer membranes and vesicles), and once again new properties are created. As before, we can observe a phase separation between water and micelles. But a more interesting novel feature of micelles is the property of having an inside and an outside, which then creates the property of permeability. Both permeability and having an inside and outside are properties that no monomer or polymer can possess.

Thus, in this system the first-order structures (water and monomers) generate second-order structures (polymers) that in turn generate third-order structures (micelles and vesicles). Note that the dynamics at each order of objects again generates emergent properties. This illustrates a third-order dynamical hierarchy with two successive orders of emergence.

### 17.2.2 Definition of Higher-Order Emergence

This section is a brief introduction to a more precise conceptual framework for describing higher-order emergent structures and how they are generated. Our concepts will enable us experimentally to study emergent dynamical hierarchies in both formal and natural systems. In general, higher-order phenomena occur when objects and properties of one order or kind give rise (by aggregation, for example) to new objects and properties. These new objects and properties themselves may give rise to further new objects and properties at a still higher order, and so on. The key question here is what it means for an object or property to be *new*, and the intuitive idea behind our definitions is simple and natural: A property that applies at a given level is emergent if it does not apply at any lower level. That is, the property applies to a composite entity but not to any of its components. We shall use the term *order* when referring to a particular observable property or functionality and *level* when we are referring to a particular length scale (level of description).

To treat this issue formally, we start by assuming the existence of certain objects and properties. In particular, we assume the existence of a *first-order* family of $n$ objects or structures $S_r^1$

$$S_r^1 = S_r^1(s_r, f_{rs}, \tau_r), \qquad r, s = 1, 2, \ldots, n \tag{17.1}$$

where $s_r$ is the state of the object (e.g., position, orientation, molecular type), $f_{rs}$ defines the object–object interactions, and $\tau_r$ is the local time for object $r$, and the indices $r$, $s$ refer to the different objects. As in the previous section, a monomer might be an example of a first-order structure. We also assume the existence of a set of observable properties of those structures. For example, being hydrophobic or hydrophilic are observable properties of monomers that can be observed using an observation function $O^1$. Finally, we assume that the dynamics of the system is given by the interaction rules $(f_{rs})$ and a functional $U$ that schedules the object updates (e.g., parallel, random, or event driven). Examples of interactions among molecules can be attraction and repulsion, and monomer–monomer bonds in the polymers.

The notion of an observable is a general one as explained in [3]. It is a mechanism to measure the properties of the system resulting from various interactions. They could be real-valued functions, algorithms, operators, functors, or simply an "investigator" (a person) that determines whether a certain condition about the dynamics is fulfilled or not. Each gives rise to mappings between the system state and some property. Thus, by properties we mean the resulting values of the observables applied to the system. A family $S_r^1$ of objects with specified interactions and properties represents a process, and the stable outcome or result of this process—in some situations the attractor for the system—will be represented by $R(S_r^1)$. To define $R$ formally will in most cases be very difficult. The purpose of the *conceptual* formalism presented here is to make explicit the basic ingredients required to produce emergence and higher-order structures in the various systems. This is of value when designing and analyzing these complex systems. However, much more formal rigor is necessary in each specific case to make these concepts mathematically precise.

The interactions in the system dynamics may now generate a family of new *second-order* objects or structures $S_v^2$

$$S_v^2 = R(S_v^1), \qquad r = 1, 2, \ldots, n, \; v = 1, 2, \ldots, m \, (n > m) \tag{17.2}$$

where $R$ is the process that generates $S_v^2$. This new family of objects itself has a set of observable properties, $O^2$, that may be equal, overlap, or be disjoint from $O^1$. A polymer formed from molecular bonds among monomers is an example of a second-order object, and an example of its second-order properties is its elasticity. We say that a property $P$ is *emergent* iff

$$P \in O^2(S_v^2) \qquad \text{and} \qquad P \notin O^2(S_r^1), \tag{17.3}$$

and we call objects or structures $S_v^2$ that are generated in the dynamics *emergent* if they have emergent properties.

Strictly speaking, we will refer to emergent properties and structures in Definition 17.3 as *second-order* emergence, since they apply to "wholes" composed of first-order "parts." This formalism can be iterated to define emergent structures and properties of third and higher orders, for example, order $N$:

$$S_{r_N}^N = R(S_{r_{N-1}}^{N-1}, S_{r_{N-2}}^{N-2}, \ldots). \tag{17.4}$$

We call such third- or higher-order structures *hyper-structures*.

It should be noted that a *trivial* example of an $N$th-order structure is one that has no new emergent properties over and above those possessed by $(N-1)$th-order structures. Such trivial $N$th-order structures should be avoided by treating them as equivalent to $(N-1)$th-order structures. It should also be noted that the specification of observable properties is somewhat arbitrary. This is quite analogous to the arbitrariness involved in the specification of the primitive objects and their interactions. So, a certain kind of phenomena can be viewed as $N$th-order emergent only relative to arbitrary choices about what counts as first-order objects and properties, what counts as second-order objects and properties, and so forth. For more details about this concept of emergence and its relation to dynamics, we refer to [1, 2, 41].

## 17.3 Dynamical Hierarchies in Molecular Dynamics Lattice Gases

### 17.3.1 Background

In the mathematical framework for our simulations—which we call molecular dynamics (MD) lattice gases—all molecular interactions are modeled by mediating particles [27, 45]. We model both matter and force fields as "information particles" that propagate locally along the edges of a lattice and interact with one another at nodes. The rules that generate the dynamics are the rules that propagate the information particles and define the local reactions together with an update functional $U$ that schedules the object updates. This general view of matter is similar to that described in [51]. Our modeling approach has the flavor of the classic lattice gas automata (LGA) [17, 55] except that we use more types of information particles than usual as we model detailed molecular properties similar to what is done in the classical MD simulations. For a discussion of some of the mathematical consequences of this general form of a dynamical system, see [41].

The force-communicating particles propagate locally, that is, between neighboring lattice sites. Transmitting force field particles between the molecules enables individual molecules to be updated independently using only local information. By choosing the mediating particles properly, a variety of different kinds of molecular interactions may be formulated. For instance, all molecules must obey an excluded volume constraint and the monomers in a polymer must obey a connectivity constraint.

Many lattice varieties have been tested, some with a combined matter and force particle lattice (two-dimensional square and triangular) and some with separate matter and field lattices (three-dimensional cubic). For simplicity the following discussion assumes a single two-dimensional triangular lattice where both matter and force fields live. It should be noted that the formalism easily generalizes.

### 17.3.2 The State Space

This lattice $\mathcal{L}$ is the "physical-space" in which the molecular configurations form, and it is represented directly by a set of coordinate pairs $(i, j)$ of integers, $i, j \in N$. The current state of each site in the triangular lattice is represented with several variables. If object $r$ is at lattice location $(i, j)$, the values of the variables at $(i, j)$ correspond directly to $s_r$ in Equation 17.1. So, the reidentification and movement of each primitive object is represented implicitly by the local propagation of state information about each site, just as a glider's reidentification and movement is represented in a cellular automaton like Conway's game of life.

The state of an object in the system can be described by the following set of internal states:

$$s_r = (i, j, x_1, \ldots, x_q) \in \mathcal{S}, \tag{17.5}$$

where $(i, j)$ indicates the lattice site location and $x_1, \ldots, x_q$ are values of the $q$-many aspects of the internal state of $s$ of the object, $x_l \in X_l$, $l = 1, \ldots, q$. The total state of a system containing $n$ primitive objects $z^{\text{objects}}$ (e.g., monomers or water molecules) is given by $n$ such sequences

$$z^{\text{objects}} \in \mathcal{Z}^{\text{objects}} = \prod_{\text{objects}} S. \tag{17.6}$$

Viewing the lattice, instead of the objects, as the fundamental structure in the system gives an alternative description. Each lattice site is associated with a data structure $\mathcal{D}_{\rangle|}$. The data structure $\mathcal{D}_{\rangle|}$ represents the internal states of the object at lattice point $(i, j)$, that is, $\mathcal{D}_{\rangle|} \in \mathcal{X}_\infty \times \cdots \times \mathcal{X}_{\text{II}} = \mathcal{D}$. The internal states have now been modified to enable represenation of vacant lattice sites. In this framework the total state of the system is described by

$$z^{\text{lattice}} \in \mathcal{Z}^{\text{lattice}} = \prod_{\text{lattice sites}} \mathcal{D}. \tag{17.7}$$

The two different state-space formulations given above are of course equivalent. However, when simulating the time evolution of the system, the two interpretations give rise to different simulation techniques. This difference is especially important in parallel computation. In traditional MD, for example, the objects (atoms or molecules) are viewed as the fundamental objects and the time steps in the update cycle sweep over these objects, that is, $\mathcal{Z}^{\text{objects}}$ is the natural state space. Formally we write

$$z^{\text{objects}}(t + \Delta t) = U^o(z^{\text{objects}}(t)) \tag{17.8}$$

where $U^o$ is the update operator $U^o : \mathcal{Z}^{\text{objects}} \to \mathcal{Z}^{\text{objects}}$.

In our lattice gas simulation the update function sweeps over all the lattice points. This means that $\mathcal{Z}^{\text{lattice}}$ is the natural state-space description and the update operator $U^l : \mathcal{Z}^{\text{lattice}} \to \mathcal{Z}^{\text{lattice}}$ is written

$$z^{\text{lattice}}(t + \Delta t) = U^l(z^{\text{lattice}}(t)). \tag{17.9}$$

### 17.3.3 Nature of the Dynamics

The general dynamical system question is this: If we start out with a random lattice configuration, how will the dynamics transform it—what is the transient and what is the asymptotic behavior? Our goal with this system is to look for the production of polymers (second-order structures) and aggregates of polymers (third-order structures) within one formal dynamical system. Therefore our observers are external mechanisms (algorithms or humans) looking for monomer configurations, polymer configurations, and aggregate polymer configurations. The interactions are all determined by the local object models but are at the same time using the data structures as a kind of internal observer that makes the interactions context dependent.

In the simulation we observe that polymers and polymer aggregates are created. We now want to put this into a dynamical system scheme. We suggest that the observed dynamics can be interpreted as follows: There exists a subspace $\mathscr{Y} \subset \mathscr{X}$ such that $U(\mathscr{Y}) \subset \mathscr{Y}$, where $U$ is the only implicitly given dynamic on the configurations and where

$$\mathscr{Y} = M_1 \cup M_2 \cup M_3 \cup P_1 \cup P_2 \cup A_1. \tag{17.10}$$

$M_1$, $M_2$, and $M_3$ are sets of monomer configurations of different complexity in a well-defined sense;[1] $M_1$ is of the lowest complexity, then $M_2$ and then $M_3$. Similarly $P_1$ and $P_2$ are sets of polymer configurations with $P_2$ more complex than $P_1$. $A_1$ is a set of polymer-aggregate configurations.

$\tilde{\mathscr{Y}} = A_1 \cup P_1 \cup M_1$ is invariant and the dynamics follows the scheme:

$$\begin{array}{ccccc} M_3 & \cup & M_2 & \cup & M_1 \\ \downarrow & & \downarrow & & \circlearrowleft \\ P_2 & \cup & P_1 & & \\ \downarrow & & \circlearrowleft & & \\ A_1 & & & & \\ \circlearrowleft & & & & \end{array} \tag{17.11}$$

which can easily be generalized to an arbitrary number of levels. This scheme should be interpreted as follows:

1. The low complexity monomers in $M_1$ simply remain within their class under iteration of the dynamics. They do not have enough structure (object complexity) to evolve.

2. The more complex monomers in the $M_2$ class form polymers of type $P_1$, but under further iteration they just remain in the class and do not evolve further.

3. The most complex monomers in $M_3$ transform into $P_2$ polymers, but only as an intermediate stage. The dynamics now give rise to new, context-interpreted interactions that under further iteration form polymer aggregates $A_1$. This class remains stable

under iteration. The evolution from monomers has stopped. The elements of $A_1$ represent the system's maximal ability to generate higher-order complexity.

The elements of $A_1$ have been formed by the process

$$M_3 \rightarrow P_2 \rightarrow A_1 \circlearrowleft \qquad (17.12)$$

by interactions and observations and form a genuine hyper-structure of order 3 in the sense of Baas [1].

It would be interesting to look into general dynamical systems with this kind of structure—especially the scheme given by Equation 17.11—and ask whether they would be natural generators of higher-order structures. In general dynamical systems one could say that this scheme would correspond to

$M_1$ An invariant set.

$P_1$ Attracting domain of $M_2$.

$P_2$ "Semi-attracting" domain of $M_3$—to be mathematically defined.

$A_1$ Attracting domain of $P_2$ and second attracting domain of $M_3$.

## 17.4 Two Examples of Physicochemical Dynamical Hierarchies

In this section we define and illustrate the behavior of two simulations that generate three-level dynamical hierarchies as discussed in sections 17.2 and 17.3. In each system, the primitive first-order structures are individual water and monomer molecules. The second-order structures are polymers—linear chains of monomers. The third-order structures are micellar aggregates—single-layer membrane objects composed of organized amphiphilic polymers.

The representation of the dynamical hierarchy in the first system we present is especially simple, to maximize the conceptual clarity of the underlying processes. The second system is a bit more complicated, to show that the same kind of processes also underlie dynamical hierarchies with much more realistic behavior.

### 17.4.1 Simple Two-Dimensional Example: Monomers, Polymers, Micelles

This first system starts out with isolated monomers in water, the first-order structures. These first-order monomers start to polymerize and generate the second-order structures, the polymers, because hydrophilic monomers act as nucleation centers to which hydrophobic monomers can attach or bind. Hydrophilic monomers cannot form chemical bonds with each other, but a hydrophobic monomer can bind to a hydrophilic monomer and a hydrophobic monomer can bind to a polymerized hydrophobic monomer. These bonding rules lead to the formation of second-order amphiphilic polymers, with one hydrophilic monomer as the head and a chain of hydrophobic monomers as a tail. The amphiphilic polymers can then assemble into third-order micelle-like aggregates when their hydrophobic tails try to hide from the water and their hydrophilic heads stick out into the water.

To simplify the first system as much as possible, we omit explicit water molecules and model the water only implicitly. Vacant lattice sites are interpreted as a mixture of vacuum and water in which the water–monomer interactions are modeled as mean field interactions.

Our formal system consists of a single lattice with matter and force particles modeled as propagating information particles that jump between the data structures at each lattice point (recall section 17.3). The dynamics is driven by a heat bath (modeling the random water dynamics) that randomly kicks each monomer in one of the possible lattice directions each time step. In addition, monomer collisions are also modeled as kicks in a random lattice direction away from each other due to their excluded volumes. This heat bath allows us to ignore the conservation of energy and momentum and worry only about how the monomers interact with each other.

Formally, the dynamical system is of the type defined in Equation 17.9. In this particular application the data structure we use has dimension $q = 7$ (+2 if we also count the lattice coordinates), and the object–object interaction algorithm, $f_{rs}$, consists of 10 substeps, 5 for each of two scheduling colors. The data structure space $\mathscr{S}$ on the lattice is defined as follows:

$x_1$ is the scheduling color. There are two colors that are alternating in a polymer. The alternating scheduling structure for polymers ensures that neighboring monomers in a polymer can be updated independently in each half of the parallel update cycle. The scheduling color is random for free monomers.

$x_2$ is the type of matter (if any). A value of 0 indicates vacuum and water in some fraction (a mean field approximation), 1 indicates a hydrophilic monomer, and 2 indicates a hydrophobic monomer.

$x_3$ is incoming excluded-volume particle ("repellons"), which are being propagated from every monomer.

$x_4$ is incoming force particles ("forceons"), which are being propagated from every monomer. Hydrophobic monomers propagate positive force particles and hydrophilic monomers propagate negative force particles.

$x_5$ is the current velocity, which is being influenced by the random kicks of the solvent (water).

$x_6$ is bond directions. A monomer can either be free or have one (at the end of a polymer) or two bonds (inside of a polymer).

$x_7$ is incoming binding-force particles ("bondons").

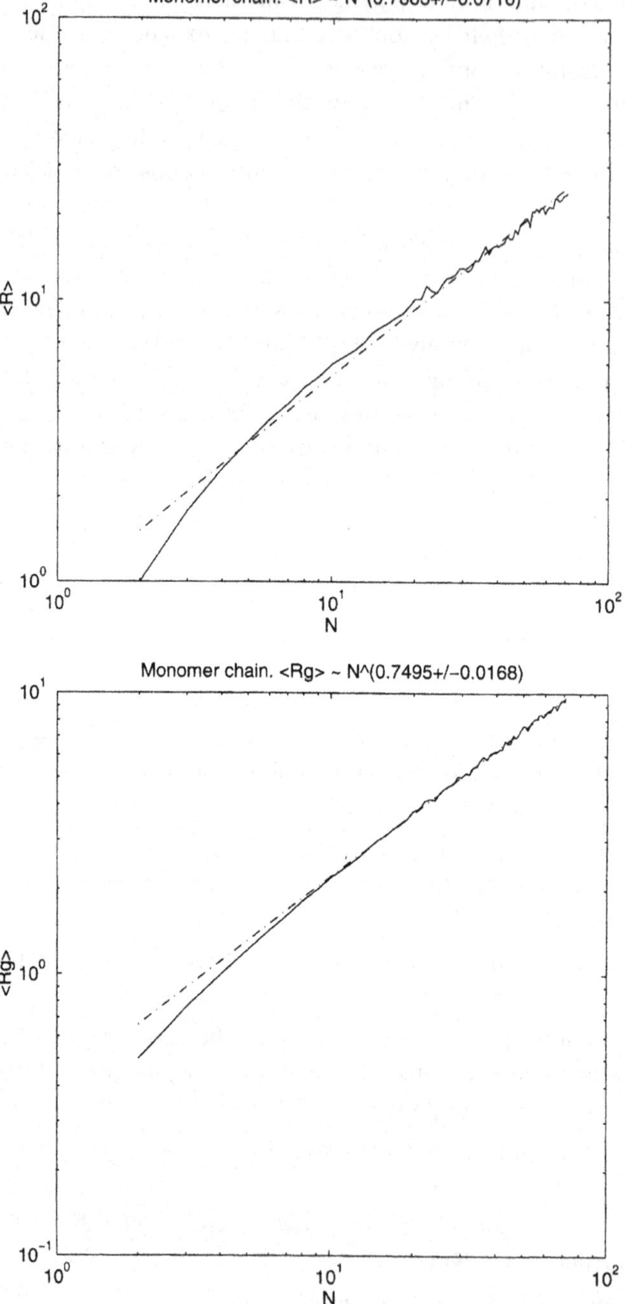
Monomer chain. <R> ~ N^(0.7883+/−0.0716)
<R>
N
Monomer chain. <Rg> ~ N^(0.7495+/−0.0168)
<Rg>
N

Each full update cycle consists of the following main steps for each scheduling color:

1. Propagate information particles (repellons, forceons, and bondons) and apply random kicks from monomers and water molecules.
2. Create new bonds (if any).
3. Compute the proper move direction (if any) for each monomer.
4. Move any molecules that need to move.
5. Clear the lattice of propagated information particles.
6. Repeat Steps 1–5 for the other scheduling color.

One iteration of the global update cycle is complete after each primitive object has been updated once.

As an example of an emergent property that is carried by a second-order structure in this two-dimensional system we can observe polymer elasticity. This elasticity is reflected in the way the average end-to-end length and the radius of gyration behave as a function of the polymer length. First, note that these formal polymers do not have any elasticity in their monomer–monomer bonds (which is close to correct in a hydrocarbon with covalent bonds between the carbon atoms). Second, note that many more folded polymer states exist than stretched states. Since the polymer is in a heat bath (random kicks) it always tends to "find" one of the many (and thus much) more likely folded states and it takes work to stretch it out (elasticity) into its very unlikely stretched state due to its interaction with the heat bath. See figure 17.1 where the expected 3/4 power law scaling of the end-to-end distance of the polymer as well as the radius of gyration is shown. The dynamics of the monomer polymerization followed by the polymer aggregation process is shown in figure 17.2, which demonstrates how a simple formal system can generate third-order emergent structures. For a more detailed discussion of similar two-dimensional systems, please see [27–30, 45]. A different simple cellular automaton formulation of polymer dynamics can be found in [36].

### 17.4.2 Realistic Three-Dimensional Example: Monomers, Polymers, Micelles

Again we model the three-level dynamical hierarchy consisting of water and monomers, polymers, and micellar aggregates. However, this model is significantly more realistic than the previous one as the water is simulated explicitly. The main computational

**Figure 17.1**

Solid lines: end-to-end distance (top) and radius of gyration (bottom) as a function of polymer length. Dotted lines: estimated scaling laws (parameters in the figures). Note the generated scaling law is very close to the known theoretical limit (power $\frac{3}{4}$). These plots show that the polymer is folded in the heat bath and an elastic force is needed to stretch it out. Elasticity is an example of a second-order functionality that cannot be observed at the level of the individual monomers. It is a result of the collective dynamics of the polymer (second-order structure) in water.

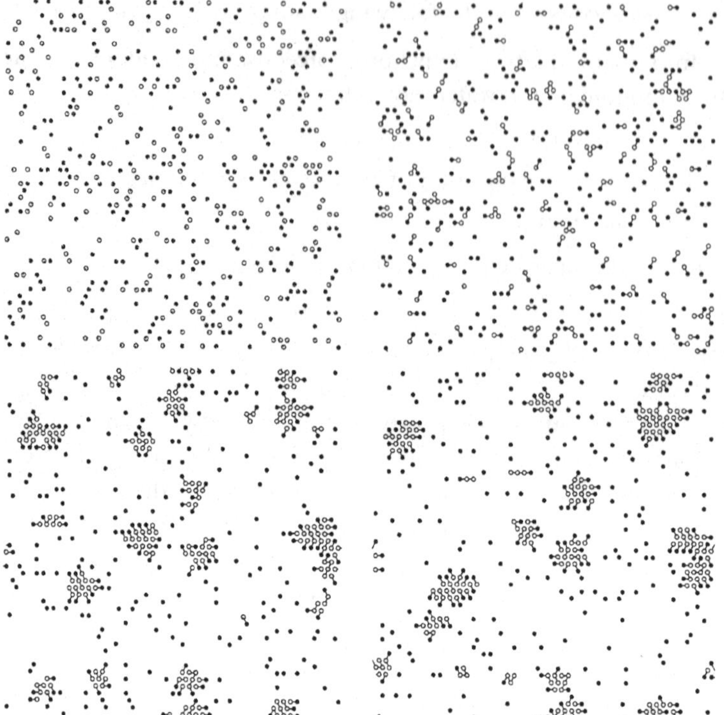

**Figure 17.2**
Generation of third-order structures on the two dimensional lattice $L_1$ (monomers at time $t = 0) \rightarrow L_2$ (polymers at time $t = 30) \rightarrow L_3$ (micellelike aggregates at time $t = 20{,}000$ and 60,000). Initially there are approximately the same number of hydrophobic (open) and hydrophilic (filled) monomers. Polymerization occurs when a hydrophilic monomer (which is the nucleation center) forms a bond to a hydrophobic monomer, which in turn polymerizes another hydrophobic monomer, etc. At time $t = 30$ there are still a few free hydrophilic monomers not yet polymerized. Note that the micellelike clusters that form are quite stable for these initial conditions and parameters, since they do not change much between $t = 20{,}000$ and 60,000.

effort actually goes into simulating the dynamics of the water molecules, for example, rotation to form energetically favorable hydrogen bonds. The dynamics is further complicated by the increased number of different intermolecular interactions described in more detail below.

The model is designed to fill the gap between traditional MD simulations on the one hand and LGA and lattice Boltzmann equations (LBEs) on the other. MD simulations generally address molecular processes that are studied on length scales of Ångstroms to nanometers and over time scales of pico- to nanoseconds. LGA and LBEs simulate complex fluids and fluid flow phenomena on length scales of submicrometers to meters and over time scales of microseconds to minutes. Our MD lattice gas simulations address molecular interactions on nanometer to micrometer length scales and over time scales up to milliseconds on workstations and seconds on supercomputers. This intermediate range suits them to modeling molecular self-organization and self-assembly processes involving ions, monomers, complex polymers, polymer aggregates (supermolecular structures), complex surfaces with charge, and chemical reactions.

The MD lattice gas is also defined in a discrete space. The dynamics involve information particles that move on two superimposed three-dimensional lattices: a molecular lattice and a field lattice. On the molecular lattice, particles carry explicit information about the structure of matter—for example, whether it consists of a water molecule, a hydrophobic or hydrophilic monomer, or a reactive radical. Particles (molecules) have excluded volumes: Only one can reside at a lattice site at any given time. They can also have an orientation; as neighboring molecules interact, they rotate until their potential energy reaches a local minimum. Water, for example, has the ability to form a maximum of four hydrogen bonds per molecule. Finally, molecules have an associated kinetic energy in each lattice direction that changes as they collide with one another.

On the field lattice, particles carry explicit information about the molecules' electromagnetic field structure, which determines how the molecules interact. The field structure and the potentials are defined by the truncated Schrödinger equation, just as in good MD simulations. For example, a water molecule has four well-defined hydrogen bonds that influence its movement and dictate how it interacts with other molecules, especially water. Our dynamics are driven by six intermolecular interactions: dipole–dipole, charge–charge, hydrogen bond, dipole–induced dipole, induced dipole–induced dipole, and cooperativity. Each molecular interaction is thus decomposed into a set of repelling and attracting particles of varying values depending on the type and position of the interaction. These interactions (potential energy terms) account for the physicochemical properties of our molecular species that are crucial to the species' self-assembly in a polar environment. See Equation (17.1.1) and discussion in section 17.5.4. The MD lattice gas interactions conserve mass, energy, and momentum and the resulting field at any lattice site influences the resident molecule and determines where it moves on the molecular lattice. If no molecule is present at a lattice site, no force fields will be propagated from that site. However, fields can reach and pass through unoccupied sites as well as be influenced (partially shielded) by the molecules at occupied sites, to model a changing dielectric constant.

Covalent bonds between monomers in polymers are also viewed as information particles that must be propagated so that the bonds do not break as the polymer moves around on the lattice. Only local interactions for molecules are used to move extended objects such as polymers and aggregates. That is, our model follows a bottom-up approach: Interactions are derived from the laws of physics, which have been modified only to accommodate the constraints of a three-dimensional lattice. A more detailed explanation of the MD lattice gas is given in [34].

The MD is calculated by iterating the following steps:

1. Propagate information to the field lattice.

2. Rotate the molecules to minimize their potential energy. Since this is done by sweeping the lattice once (i.e., $n$ time steps of a zero temperature simulated annealing), there will be a high degree of frustration (multiple exclusive maxima) in the system even after the rotations.

3. Account for kinetic energy exchanges in collisions.

4. Propagate the molecules according to their kinetic energy and the local field interactions. In this step we also ensure excluded volume and that no bonds break.

In short, the microscopic interactions we track are based on first principles and are both deterministic and reversible. The overall simulation corresponds to a microcanonical ensemble. We have used our MD lattice gas to model micelle formation, the hydrophobic effect, phase separations, hydrocarbons in water, amphiphilic fluids, complex fluids at mineral interfaces, self-reproducing micelles, membrane stability, and the dynamics of templating polymers such as RNA and PNA (ribonucleic and peptide nucleic acids).

In plate 17.1 we demonstrate a micellation process as it is generated by the three-dimensional MD lattice gas. Initially we have amphiphilic polymers (randomly) suspended in water that after some milliseconds of real time self-assemble to form micellar structures. The length scale of this single CPU simulation is about 9 nm and the simulation time is about 30 ms.

## 17.5 *Ansatz* for Generating Dynamical Hierarchies

### 17.5.1 Second- and Third-Order Interactions

All interactions between the first-order objects, the molecules, are by definition first-order interactions. The first-order interactions generate the second-order structures. For example the "glue" that makes up a polymer is generated by the monomer–monomer interactions that make up the bonds.

Second-order interactions are, for instance, the interactions we find between polymers. These interactions are generated from a *composition* of first-order interactions, since each pairwise interaction always occurs between the first-order objects. When a polymer communicates with another polymer the interaction is given by the monomer–monomer interactions: the interactions between the monomers from the two different polymers as well as the interaction between the monomers making up each of the polymers. The second-order interactions are therefore only *implicitly* given.

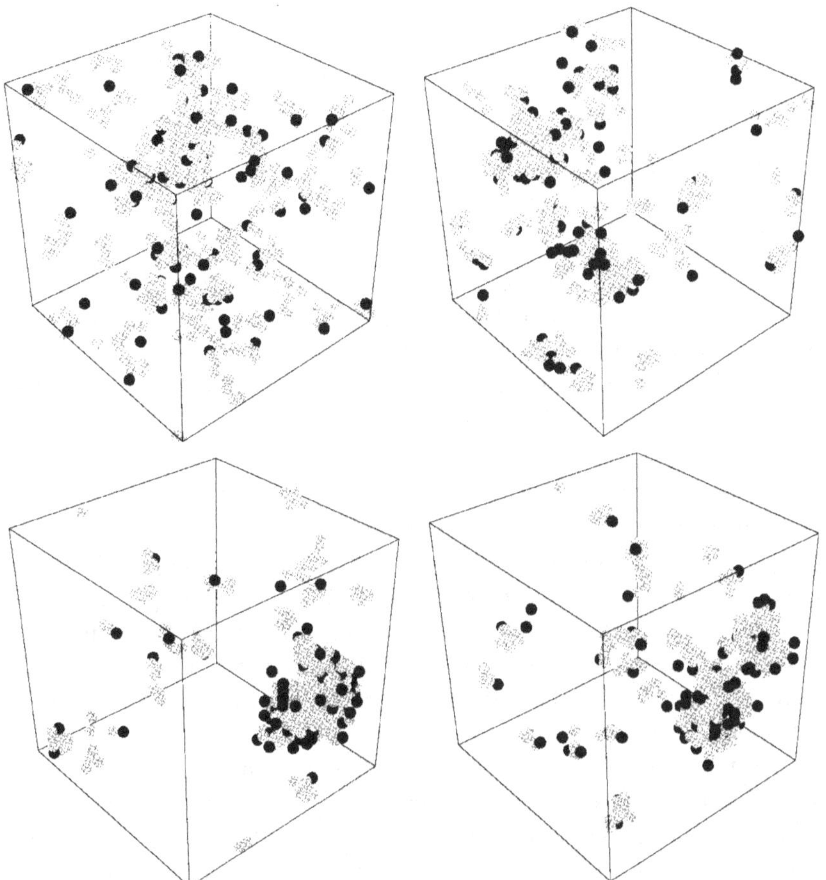

**Plate 17.1**

Generation of third-level structures on a $10^3$ three-dimensional lattice corresponding to a cube with 9-nm sides. Light areas represent $CH_2$ or $CH_3$ monomers in the hydrophobic hydrocarbon tails and dark areas represent hydrophilic COOH heads. Water is not shown although it is explicitly simulated. The simulation starts with a random initial distribution of amphiphilic pentamers at time 0. Snapshots are taken at times about 0, 120, 458, and 474 ms. Note how micelles are formed and a large micelle becomes unstable and divides between time 458 and 474 ms.

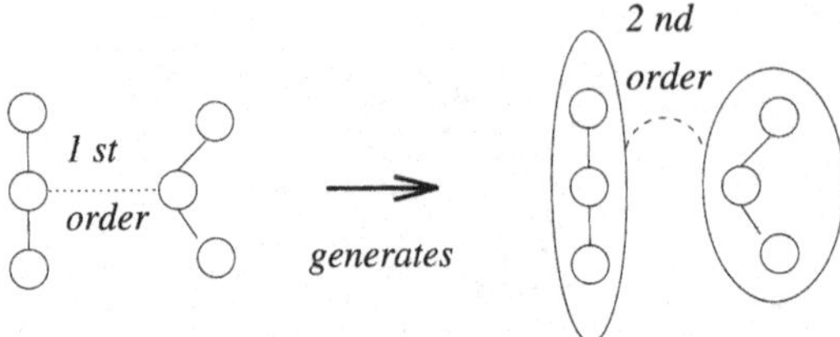


**Figure 17.3**
Data structures acting as internal observers (as opposed to external observers) generating second-order interactions. A first-order interaction gives rise to a communication channel between aggregates. The internal first-order interaction within the aggregates communicates this interaction through the aggregate. A second-order interaction between the aggregates is therefore induced by the first-order interactions—external as well as internal.

The local use—or *interpretation*—of the different kinds of communicated information defines the *operational semantics* of the information [45]. The solvent particles are, for instance, interpreted differently by the monomers depending on the context the monomers currently are in. If the monomer is free, the solvent particle (water) will induce a movement in its direction through a "kick" in the simple system and through a collision in the realistic example. However, if the monomer is polymerized it becomes a bit more complicated and the effect will depend on whether such a movement would violate bond restrictions.

Thus, the *meaning* of an interaction with a solvent molecule (a kick or a collision) depends on the context in which it occurs. We can interpret the data structure and the functions operating on it as an *internal* observer (figure 17.3). Data structures are internal observers in the sense that each data structure senses its neighborhood and acts accordingly by following an algorithm (e.g., the laws of physics) like an experimenter would have done from the outside. In particular, this can be stated as follows: The interpretation of the information depends on which hyper-structural level the communicating objects belong to. In this system we have three levels: $L_1$, the level of the free monomers and the solvent; $L_2$, the level of the polymers; and $L_3$, the level of the polymer aggregates.

In general, *new hyper-structural levels support new means of communication*, both within new levels and between old and new. This is why the *object complexity is bound to increase* as more hierarchical levels are to be generated, which we shall discuss further in the next section. The only way a unique, concurrent means of communication can be obtained is by a unique combination of the transmitted information. Each inter-level communication needs to have at least a partially independent information channel.

This seems to be true at least for a formal generation of these first hierarchical levels modeled in a physicochemical context. Whether more explicit internal object complexity always will be necessary to obtain yet higher-order emergent structures—and whether there is an object complexity limit above which any functional property can be produced, we do not know.

Note that different levels in general need not be part of a strict hierarchy. Interactions between first-order structures and second-order structures are of course also possible, but here the interactions are not symmetric. The first-order to second-order interaction is of second order whereas the second-order to first-order interaction is of first order because of the reasons given above. This interpretation of the information depending on the context (e.g., in the hyper-structure) of the interacting objects has profound consequences for how higher-order ($\geq 3$) emergent structures can self-assemble in formal self-programmable or constructive dynamical systems.

### 17.5.2 Third-Order Emergence

In figure 17.2 and plate 17.1 it is clear that genuine third-order structures are generated. As we know from the previous discussion each of these aggregates is generated through the interactions between first- and second-order objects as well as through the interaction between second-order objects.

In the simple example the system polymerizes the first-order structures into second-order structures that in turn assemble into third-level structures. In the realistic example the second-order structures assemble into third-order structures. Although we did not show it here, the water and the monomers from both models can produce emergent properties as phase separations where the water molecules and monomers occupy different volumes of the lattice [27]. Thus the dynamics at the first level can also generate emergent properties. As we saw through the calculation of the radius of gyration for the polymers in the simple example, these second-order structures indeed carry the property of elasticity as the end-to-end length scales with a power law of 3/4 as a function of polymer length. The micellar aggregates, the third-order structures, in both systems also clearly carry third-order emergent properties. The micelles have a characteristic size distribution that emerges out of the polymer and water dynamics; they have an inside and an outside, and if we placed a tracer molecule in the center of the hydrophobic part of the micelle we could observe a permeability for these structures. As we have shown elsewhere [28, 30], by adding a simple (micellar) surface-induced reaction these third-order micellar structures can self-reproduce as long as "fuel" molecules are provided, which has already been demonstrated experimentally by Luisi and coworkers [4, 5, 26]. Thus, self-reproduction is yet another emergent property these third-order structures can carry.

A conceptual clarification of the three-level dynamical hierarchy is given in figure 17.4. Note that the order of the generated structures corresponds to the level of description (typical length scale) in this particular system. This, however, will not be the case in general (e.g., see figure 17.7).

| Level of Description | Molecular Structure and Order | Observed (Emergent) Functionality |
|---|---|---|
| Level 3 | micelle (3rd order) | inside/outside, permeability, self-reproduction |
| Level 2 | polymer and water (2nd order) | elasticity, radius of gyration |
| Level 1 | water and monomers (1st order) | phase separation, pair distributions |

**Figure 17.4**
The three-level, physicochemical, dynamical hierarchy that is discussed conceptually and in simulation in this article.

Finally, it should be noted that a two-way causation exists in these dynamical hierarchies. On the one hand each of the elemental structures constitutes or composes the next higher level structures. On the other hand, these higher-level structures cause the lower-level structures to behave in a different manner. The dynamics of the monomers are more restricted once they form a polymer and polymers are more restricted once they form an aggregate. Thus, there is a clear *downward causation* in such systems as well.

### 17.5.3 Object Complexity

It is clear from the observations we have made of the dynamics of the MD lattice gas systems in two and three dimensions that their ability to produce emergent structures is highly dependent on the degree of detail of the object models. As more and more interactions—and more and more different molecules—are taken into account, more complex emergent structures in the lattice systems are produced.

In our MD lattice gas framework, allowing just a simple molecule–molecule interaction without an excluded volume enables us to define traditional lattice gases that can generate a variety of macroscopic fluid dynamics phenomena ($\mathcal{D}_1$ in figure 17.5). Defining an excluded volume for the monomers allows the production of a little more detailed monomer–monomer interaction dynamics, which is what the simplest two-dimensional MD lattice gas system has ($\mathcal{D}_2$ and $\mathcal{D}_3$ in figure 17.5). By the addition of binding and scheduling information to each of the data structures it becomes possible to generate polymer dynamics with inert monomers ($\mathcal{D}_4$ in figure 17.5). These are examples of second-order emergent phenomena, as we have discussed earlier. Without hydrophobic/hydrophilic interactions for the monomers, for example, third-order micellar structures cannot be generated.

The polymers, however, can in a direct way be generated by the dynamics if we allow an interaction that is specific for a polymerization process. Thus, it becomes possible to produce second-order structures from first-order structures ($\mathcal{D}_6$ in figure 17.5). If no

```
| * | | * | | * | ||||| ||||| ||||| ||||| scheduling color
||||| ||||| ||||| ||||| ||||| ||||| ||||| vacuum and molecules
| * | ||||| ||||| ||||| ||||| ||||| ||||| excluded volume particles
| * | | * | ||||| | * | ||||| | * | ||||| hydrophobic/philic interactions
| * | | * | | * | ||||| ||||| ||||| ||||| monomer-monomer bonds
| * | | * | | * | | * | | * | ||||| ||||| polymerization of monomers

 D1    D2    D3    D4    D5    D6    D7
```

**Figure 17.5**

Definition of object complexity as the number of active variables in the data structure. $\mathcal{D}_v$ labels classes of systems wherein the variables indicated by lines are activated and the variables marked by asterisks are empty. $\mathcal{D}_1$ produces fluid dynamics as defined through the traditional lattice gas automata. $\mathcal{D}_2$ produces monomer dynamics with excluded molecular volumes. $\mathcal{D}_3$ produces aggregates of hydrophobic monomers surrounded by water. $\mathcal{D}_4$ produces polymer dynamics. $\mathcal{D}_5$ produces micelles, vesicles, and membranes starting from polymer interactions. $\mathcal{D}_6$ produces polymers from monomers through a polymerization process that can assemble. $\mathcal{D}_7$ produces micelles, vesicles, and membranes from polymers that are polymerized from the initial monomers.

binding information is present in the data structures no polymers can form, but it is still possible to generate a phase separation or produce aggregates of hydrophobic monomers surrounded by water and hydrophilic monomers ($\mathcal{D}_3$ in figure 17.5). These aggregates are also examples of second-order structures.

If binding information is present and the initial configuration is a random configuration of polymers with hydrophilic heads and hydrophobic tails the formation of micellelike aggregates becomes possible ($\mathcal{D}_5$ in figure 17.5). In such a situation the system has produced third-order structures from the interactions of second-order structures (the polymers). If bond information together with polymerization interactions as well as hydrophobic/hydrophilic molecules all are defined, it becomes possible to generate polymer aggregates from an initial condition of random monomers ($\mathcal{D}_7$ in figure 17.5). Intermediate configurations of the dynamics will then be dominated by the newly polymerized hydrophobic/hydrophilic polymers. Thus, it is possible to produce third-order structures from first-order structures. The above data-structure discussion is of course meaningful only within a given framework, which in this situation is a MD lattice gas simulation. If a different framework were chosen, these details would also be different.

Another part of the local object complexity is given by the complexity of the $f_{rs}$ measured as the number of rules or functions together with their arguments that define the local transformation of the variables in the data structures. The number and the nature of the variables in $\mathcal{S}$ together with their transformation rules $f_{rs}$ (and $U$) characterize the computational complexity (in terms of time, space, and algorithmic complexity [20, 33]) of the dynamical system. So our observed relation between the object complexity and the generative power of the dynamics can also be stated in the following form: The higher the order of structures we want to generate, the more complex objects the system must have.

Another way of saying this is that a more detailed description of the physicochemical system is necessary to allow the formal system to produce higher-order structures. Weaker effects also have to be taken into consideration if more complex biomolecular structures have to be explained.

The above may at a first glance seem contradictory to the findings of, for example, very simple cellular automata rules capable of generating arbitrary complex behavior—behavior equivalent to a Universal Turing Machine [25]. However, this is only an apparent contradiction, since these findings only show that there *exist* simple cellular automata rules (objects with low complexity), capable of generating (any) complex structures. It does not say that *every* simple rule is capable of generating arbitrary complex structures. Obviously, the dynamics of any of the above rather complex (molecular) object descriptions given in figure 17.5 could be reduced to one of these simple cellular automata rules. This is true because any formal procedure can be simulated by a Universal Turing Machine formulated as a simple cellular automaton. But what would that give us? It would very likely not have any *explanatory* power in terms of what we know about the real world since the interpretation of these new, simple rules would (by definition, because they are simpler) be quite different from what they currently are. Also such a compression of the rules would probably have to be compensated for by increased time and space complexities.

Another approach to finding simple cellular automata (CA) rules that can generate dynamical hierarchies would be through evolution. One could add a selection pressure to an evolutionary process and ask the process to find such rules. Do such rules exist? We believe that they do exist but they are probably very rare, and we question the stability of the higher-order structures that are generated by such rules. However, such a CA evolutionary experiment would be a worthwhile exercise.

### 17.5.4 A New *Ansatz* for Dynamical Hierarchies

We have demonstrated how the MD lattice gases can be formulated so that they fit into a broader dynamical systems context. We have also seen how the dynamics of these mappings create stable, higher-order structures that can be generated in simulation. It follows that genuine third-order structures can be generated using MD lattice gases.

We have seen that increasing the object complexity of the primitives, the first-order objects, is crucial for enabling the MD lattice gas simulations to produce higher-order emergence. In other words, a more detailed description of the system is necessary to allow the system to produce higher-order structures. Also, weaker physical effects have to be taken into consideration if more complex structures have to be explained in the molecular self-assembly processes.

To make it a little more precise what we mean in a physical sense about including weaker effects, we can describe the total potential energy $V_{total}$ of the three-dimensional MD lattice gas system with $n$ molecules on a lattice with $q$ neighbors [34] as

$$V_{\text{total}} = \sum_{r=1}^{n}\sum_{s=1}^{q} V^{r,s}_{\text{dip.-dip., charge-charge, H-bond}} + \sum_{r=1}^{n}\sum_{s=1}^{q} V^{r,s}_{\text{dip.-ind.dip.}} + \sum_{r=1}^{n}\sum_{s=1}^{q} V^{r,s}_{\text{ind.dip.-ind.dip.}} + \sum_{r=1}^{n}\sum_{s=1}^{q} V^{r,s}_{\text{coop.}}, \tag{17.13}$$

where these terms stem from the first terms in the truncated Schrödinger equation. These potential energy terms are implemented to account for specific physicochemical properties of our molecular objects as, for example, dipoles, induced dipoles, hydrogen bond donor and acceptor sites, or polarizability volumes, all crucial parameters for micelle formation in a polar environment. This set of weak intermolecular interactions given in the above equation has generally proven to be suitable for a description of macromolecular systems [37] as well as being responsible for the emergence of bulk phase phenomena such as the structured hydrogen-bond network in water and the hydrophobic effect [27].

An *ansatz* is a hypothesis taken to be true but acknowledged to be unproven that is used to reach further conclusions. Our efforts to simulate dynamical hierarchies lead us to the following:

*Ansatz* Given an appropriate simulation framework, an appropriate increase of the object complexity of the primitives is necessary and sufficient for generation of successively higher-order emergent properties through aggregation.

Once a particular dynamical systems framework has been chosen, our *ansatz* states that additional object complexity is necessary for producing additional levels in a dynamical hierarchy. Obviously, a given higher-level phenomenon may be described in many ways. Changing the framework might enable us to describe the higher-level phenomena with different objects that are simple, so increasing object complexity might not be necessary if we change the framework. But appropriate additional object complexity is still sufficient in the original framework.

Complexity is here used in a general sense, and clearly it has to be made precise in particular cases. It is not obvious how to make complexity mathematically precise for our micellar systems. Also, the complexity depends on the simulation framework selected. The dogma described in figure 17.6 has been dominant from the very beginning of the study of complex systems and artificial life.

Can our "complex" and realistic hyper-structures be created starting only with very simple object states and interaction rules without any external interaction with the system?[2] We doubt it. We do not doubt, of course, that complex structures in formal

*Simple Rules and States* ⟶ *Complex Dynamical Structures*

**Figure 17.6**
Complex systems dogma.

systems *can* arise from very simple rule and state descriptions that are much simpler than the structures they generate. However, we question a stronger version of the dogma that essentially holds that a common minimal simplicity underlies all emergent structures.

This approach can of course be questioned. By introducing more object complexity do we not then script the outcome of the dynamics? Yes, it is scripted in the sense that we try only to include information relevant for the processes under investigation. But this is the case in any scientific inquiry. More importantly, our object models do not "cross" the levels and reference polymers or polymer aggregates. They all reference only water or monomer properties, the first-order structures.

### 17.5.5 Bridging Nonliving and Living Matter

Before we end this discussion we would like to demonstrate how these concepts can generalize and lead us across the bridge between nonliving and living matter. The following dynamical hierarchy defines a simple proto-organism that recently has been proposed by some of the authors as the best candidate to assemble in the laboratory (figure 17.7).

Note how this system has three levels of description, but five orders of structures with significant functionalities. For details we refer to [42]. It should be noted that this structure has been created neither in simulation nor in the laboratory at the time when this article was written. However, we firmly believe that it can be done.

## 17.6 Discussion

The presentation and the discussion of the *ansatz* have raised more questions than they have answered. In the following we seek to formulate and discuss a few of the most important open questions.

What is the *ansatz*'s range of validity? The molecular self-assembly examples apparently tell a clear story when we start from the basic molecular structures: water, and hydrophobic and hydrophilic monomers. This story easily generalizes to higher-order structures as cells, tissues, organs, organisms, and ecosystems. However, what happens if we choose our fundamental structures to be the atoms—or the elementary particles? If we consider the atomic level, we have a few atoms, hydrogen, oxygen, and carbon, that can all be fairly simply specified in terms of their electron orbitals, and which in turn generate the higher-order emergent structures of molecules needed: water and

| *Level of Description* | *Molecular Structure and Order* | *Observed (Emergent) Functionality* |
|---|---|---|
| Level 3 | vesicle, redox complexes and templating polymers (5th order) | information storage evolution |
| | vesicle and redox complexes (4th order) | auto-catalytic self-reproduction |
| | micelles or vesicles (polymers and water) (3rd order) | inside/outside permeability |
| Level 2 | Polymer or redox complex and water (2nd order) | elasticity, radius of gyration, electron transfer |
| Level 1 | water and monomers (1st order) | phase separation, pair distributions |

**Figure 17.7**
Dynamical hierarchy for a proto-organism. Note that this system has three natural levels of description. Nevertheless, it is a fifth-order structure because it is defined by functionalities that are observable only after two additional assembly processes defined after third-order lipid aggregates have occurred.

monomers. If, for instance, only hydrogen was present, no dynamical hierarchy of the kind discussed here would be possible. But we would only need a few additional atoms to be able to make the necessary molecules to assemble a proto-organism. Although the object complexity increases (more atoms) there seems to be a critical limit above which we can build anything. With about 100 atoms (and their parts) we can build everything in the known universe. Can we also find such a critical object complexity level at the next higher level, the molecular level we have been operating in this article? We don't know. The *ansatz* is still valid, but perhaps less meaningful as we push it further down. If we consider the level of elementary particles, the *ansatz* still holds since electrons, protons, and neutrons are needed to create all atoms.

Although we have not discussed possible application areas outside of physics and chemistry, we suspect that the conceptual framework as well as the *ansatz* itself could be useful in understanding sociotechnical systems and humanmade organizations as well as other systems that consist of a complex assembly of interacting parts. The fundamental quality of these systems is that they to a great degree constitute assemblies of people (with different skills), tools, natural and humanmade structures, as well as natural and humanmade rules (e.g., laws, manuals, accepted behaviors) for interaction.

Finally, the formal presentation in this article is more descriptive and heuristic than definitional. The amount of work needed to develop precise mathematical definitions of all the concepts in our examples would have prevented us from presenting the main results of this work for a long time. However, all the concepts *could* be made mathematically precise. It is perhaps less of a problem to see how to do this with the observables we use: the notions of a polymer and a micelle, the notions of elasticity, an inside and an outside, and permeability. We believe it becomes more problematic with the notion of object complexity. Here complexity is used in a very general sense that has to be made precise in each example. Depending on the system it could range from a simple count of variables or variable states over information complexity to geometric or topological complexity as measured, for example, by cohomology theories. There is another open question relating to object complexity: Given a fundamental level of description, a certain object complexity, and a set of observables, how do we know whether a particular order of emergence defines the upper limit in the dynamical hierarchy that can be generated from these primitives? We do not believe that general answers can be given to this question. However, we do believe that systems can be defined such that the existence of a limit can be proven.

## 17.7 Conclusion

We have demonstrated how stable third-order emergence can be generated in formal dynamical systems through aggregation and we have developed an *ansatz* for generating dynamical hierarchies in dynamical systems. The *ansatz* states that, given a particular framework, there is a tight correspondence between the complexity of the simple objects used and the system's ability to generate dynamical hierarchies.

There is a tension (but no contradiction) between our *ansatz* and the current complex systems (and artificial life) dogma that all complex structures and dynamics should be reducible to simple rules or equations of motion (figure 17.6). The complex systems dogma encourages those studying dynamical hierarchies always to seek models with the simplest possible elements. Our *ansatz*, by contrast, encourages us to add complexity to system elements to explore more levels of the hierarchy. When we operate within a given framework at a given level of description, the complex systems dogma is a powerful and successful hypothesis. However, if we stay within that framework and attempt to explain several additional levels of description, our *ansatz* states that we must make the fundamental rules more complex. Of course, we want to preserve the complex systems dogma to the extent that is possible; we want the simplest possible models of dynamical hierarchies. But we want to stress that the complex systems dogma should not block us from building simulations with enough object complexity to model multilevel dynamical hierarchies successfully.

How general is this *ansatz*? The simple answer is that we do not know. However, the general nature of the self-assembly problem we have addressed suggests that this *ansatz* may apply to other assembly problems. And molecular self-assembly is one of the two main processes that are known to generate novelty in biological systems, where evolution is the other. The *ansatz* is also supported by what we know about modeling physical systems. More physical details are needed to explain higher-other structures and functionalities. Finally, what we know about biology provides circumstantial evidence for our *ansatz*. Living systems indeed are composed of not just two but a large number of covalently bonded molecular objects. Had it been simpler or even practically possible to create life in a binary molecular system, we would probably have seen this reflected in contemporary biology.

## Acknowledgments

This paper has a long history. A different version was initially submitted to one of the traditional physics journals in 1996 (LA-UR 96-2921) and was after several rewrites and two resubmissions left to rest in peace. In the spring of 2000, during the Cerro Grande Wildfire in Los Alamos, Mark Bedau convinced us to rewrite the paper yet another time, include new results, and submit it to the *Artificial Life* journal. In this rewriting process we are indebted to Mark Bedau for helping us clarify and improve our story and to Norman Packard for the difficult questions asked in the final referee process. Chris Barrett, Kai Nagel, Maya Paczuski, Christian Reidys, Jay Riordan, and in particular Al Scott are thanked for constructive comments and discussions either during the work that led to earlier versions of the paper or in the earlier presentation of the results. We also thank the Norwegian Research Council, the Technical University of Denmark, the Austrian Research Council, Los Alamos National Laboratory, and the Santa Fe Institute for financial support.

### Notes

This chapter originally appeared in *Artificial Life* 7 (2001), 329–353.

1. See discussion of object complexity in section 5.3.

2. Note that we are considering fixed objects with no self-programming or "mutations" in the explicit rules and variables. Our objects do not learn, which seems reasonable, since they model simple invariant molecules. Recall the discussion about evolving simple CA rules in the previous section.

### References

1. Baas, N. A. (1994). Emergence, hierarchies and hyperstructures. C. G. Langton (Ed.), *Artificial Life III* (pp. 515–537). Redwood City, CA: Addison-Wesley.

2. Baas, N. A. (1994). Hyperstructures as a tool in nanotechnology. *Nanobiology, 3*, 49–60.

3. Baas, N. A., and Emmeche, C. (1997). On emergence and explanation, *Intellectica, 25*, 67–83.

4. Bachmann, P. A., Luisi, P. L., and Lang, J. (1992). Self-replicating reverse micelles and chemical autopoiesis. *Nature, 357*, 57.

5. Bachmann, P. A., Walde, P., Luisi, P. L., and Lang, J. (1990). Autocatalytic self-replicating micelles and models for prebiotic structures. *Journal of the American Chemical Society, 112*, 8200.

6. Bagly, R., and Farmer, J. D. (1992). Spontaneous emergence of a metabolism. In C. Langton, C. Taylor, J. D. Farmer, and S. Rasmussen (Eds.), *Artificial Life II* (pp. 93–140). Redwood City, CA: Addison-Wesley.

7. Bagley, R., Farmer, J. D., and Fontana, W. (1992). Evolution of a metabolism. In C. Langton, C. Taylor, J. D. Farmer, and S. Rasmussen (Eds.), *Artifical Life II* (pp. 142–158). Redwood City, CA: Addison-Wesley.

8. Banzhaf, W. (1994). Self-organization in a system of binary strings. In R. Brooks and P. Maes (Eds.), *Artificial Life IV* (pp. 109–119). Cambridge, MA: MIT Press.

9. Boghosian, B., Coveney, P., and Emerton, A. (1996). A lattice gas model of microemulsions. *Proceedings of the Royal Society of London A, 452*, 1221.

10. Coveney, P., Emerton, A., and Boghosian, B. (1996). *Simulation of self-reproducing micelles using a lattice gas automaton.* (Tech. Pap. 96-13S). Oxford: Oxford University.

11. Coveney, P., Maillet, J. B., Wilson, J., Fowler, P., Mushadani, O., and Boghosian, B. (1998). Lattice gas simulations of ternary amphiphilic fluids in porous media. *International Journal of Modern Physics, C, 9*, 1479–1490.

12. Dittrich, P., Ziegler, J., and Banzhaf, W. (2001). Artificial chemistries—a review. *Artificial Life, 7*, 225–275.

13. Evans, D. F., and Ninham, B. W. (1986). Molecular forces in the selforganization of amphiphiles. *Journal of Physical Chemistry, 90*, 226–234.

14. Farmer, J. D., Kauffman, S., and Packard, N. (1986). Autocatalytic replication of polymers. *Physica D, 22*, 50–67.

15. Fontana, W. (1992). Algorithmic chemistry. In C. Langton, C. Taylor, J. D. Farmer, and S. Rasmussen (Eds.), *Artificial Life II* (pp. 159–210). Redwood City, CA: Addison-Wesley.

16. Fontana, W., and Buss, L. W. (1994). The arrival of the fittest: Toward a theory of biological organization. *Bulletin of Mathematical Biology, 56*, 1–64.

17. Frisch, U., Hasslacher, B., and Pomeau, Y. (1986). Lattice gas automata for the Navier–Stokes equation. *Physical Review Letters, 56*, 1722–1728.

18. Goel, N. S., and Thompson, R. L. (1988). Movable finite automata (MFA): A new tool for computer modeling of living systems. In C. G. Langton (Ed.), *Artificial Life* (pp. 317–390). Santa Fe Institute Studies in the Sciences of Complexity Vol VI. Redwood City, CA: Addison-Wesley.

19. Goetz, R., Gompper, G., and Lipowsky, R. (1999). Mobility and elasticity of self-assembled membranes. *Physical Review Letters, 82*, 221–224.

20. Hopcroft, J., and Ullman, J. (1979). *Introduction to Automata Theory, Languages, and Computation.* Redwood City, CA: Addison-Wesley.

21. Hotani, H., Laholz-Beltra, R., Combs, B., Hameroff, S., and Rasmussen, S. (1992). Liposomes, microtubules, and artificial cells. *Nanobiology, 1*, 67.

22. Kauffman, S. (1986). Autocatalytic sets of proteins. *Journal of Theoretical Biology, 119*, 1–24.

23. Langton, C. (1984). Self-reproduction in cellular automata. *Physica D, 10*, 135–144.

24. Lehninger, A. L., Nelson, P. L., and Cox, M. M. (1993). *Principles of Biochemistry*. New York: Worth Publishers.

25. Lindgren, K., and Nordahl, M. G. (1990). Universal computation in simple one dimensional celluar automata. *Complex Systems, 4*, 299–318.

26. Luisi, P. L., Walde, P., and Oberholzer, T. (1994). Enzymatic RNA synthesis in self-replicating vesicles: An approach to the construction of a minimal cell. *Berichte der Bunsen-Gesellschaft-Physical Chemistry Chemical Physics, 98*, 1160–1165.

27. Mayer, B., Koehler, G., and Rasmussen, S. (1997). Simulation and dynamics of entropy driven molecular self-assembly processes. *Physical Review E, 55*, 1–11.

28. Mayer, B., and Rasmussen, S. (2000). Lattice molecular automata (LMA): A physico-chemical simulation system for constructive molecular dynamics. *International Journal of Modern Physics C, 9*, 157.

29. Mayer, B., and Rasmussen, S. (1998). Self-reproduction of dynamical hierarchies in chemical systems. In C. Adami, R. Belew, H. Kitano, and C. Taylor (Eds.), *Artificial Life VI* (pp. 123–129). Cambridge, MA: MIT Press.

30. Mayer, B., and Rasmussen, S. (2000). Dynamics and simulation of micellular self-reproduction. *International Journal of Modern Physics C, 11*, 809–826.

31. McCaskill, J. (1988). *Polymer Chemistry on Tape: A Computational Model for Emergent Genetics*. (Internal report). MPI for Biophysical Chemistry, Gottingen, Germany.

32. McMullin, B., and Varela, F. (1997). Rediscovering computational autopoiesis. In P. Husbands and I. Harvey (Eds.), *Proceedings of the 4th European Conference of Artificial Life* (pp. 38–47). Cambridge, MA: MIT Press.

33. Minsky, M. (1972). *Computation—Finite and Infinite Machines*. Upper Saddle River, NJ: Prentice-Hall.

34. Nilsson, M., Rasmussen, S., Mayer, B., and Whitten, D. (in press). Constructive molecular dynamics (MD) lattice gases: 3-D molecular self-assembly. In D. Griffeath and C. Moore (Eds.), *New constructions in cellular automata*. Oxford: Oxford University Press.

35. Ono, N., and Ikegami, T. (1999). Model of self-replicating cell capable of self-maintenance. In D. Floreano, J.-D. Nicoud, and F. Mondana (Eds.), *Proceedings of the 4th European Conference on Artificial Life* (pp. 399–406). Berlin: Springer.

36. Ostrovsky, O., Smith, M. A., and Bar-Yam, Y. (1995). Applications of parallel computing to biological problems. *Annual Review of Biophysics and Biomolecular Structure, 24*, 239–267.

37. Privalov, P. L., and Gill, S. J. (1988). Stability of protein structure and hydrophobic interaction. *Advances in Protein Chemistry, 39*, 1991.

38. Rasmussen, S. (1985). *Aspects of Instabilities and Self-Organizing Phenomena*. Unpublished doctoral dissertation, Technical University of Denmark, Lyngby Copenhagen.

39. Rasmussen, S. (1989). Elements of a quantitative theory of the origin of life. In C. Langton (Ed.), *Artificial Life* (pp. 79–104). Redwood City, CA: Addison-Wesley.

40. Rasmussen, S., Baas, N. A., Barrett, C. L., and Olesen, M. W. (1997). A note on simulation and dynamical hierarchies. In F. Schweitzer (Ed.), *Self-Organization of Complex Structures: from Individual to Collective Dynamics* (pp. 83–89). Gordon and Breach.

41. Rasmussen, S., and Barrett, C. (1995). Elements of a theory of simulation. In F. Moran, A. Moreno, J. J. Morelo, and P. Chacon (Eds.), *Advances in Artificial Life* (pp. 515–529). Berlin: Springer.

42. Rasmussen, S., Benz, W., Burde, S., Chen, L., Colgate, S., Copley, S., Deamer, D., Fulghum, J., Gell-Mann, M., Hersman, L., Lackner, K., Maurice, P., Mayer, B., Newman, W., Nielsen, P., Pohorille, A., Stadler, P., and Uwins, P. Bridging nonliving and living matter. Proposed NASA Astrobiology Institute at Los Alamos National Laboratory, http://www.lanl.gov/home/steen/astrobiology2.

43. Rasmussen, S., Knudsen, C., and Feldberg, R. (1992). Dynamics of programmable matter. In C. G. Langton, C. Tayler, J. D. Farmer, and S. Rasmussen (Eds.), *Artificial Life II* (pp. 211–254). Redwood City, CA: Addison-Wesley.

44. Rasmussen, S., Knudsen, C., Feldberg, R., and Hindsholm, M. (1990). The coreworld: Emergence and evolution of cooperative structures in a computational chemistry. *Physica D, 42,* 111–134.

45. Rasmussen, S., and Smith, J. R. (1994). Lattice polymer automata. *Berichte der Bunsen Gesellschaft-Physical Chemistry, 98,* 1185–1193.

46. Rothman, D., and Keller, J. (1988). Immiscible cellular automata fluids. *Physical Review Letters, 56,* 886.

47. Sayama, H. (2000). A new structural dissolvable self-reproducing loop evolving in a single cellular automata. *Artificial Life, 5,* 343–365.

48. Scott, A. C. (1995). *Stairway to the mind.* Berlin: Springer.

49. Smit, B., Hilberts, P. A. J., Esselink, K., Rupert, L. A., and van Os, N. M. (1991). Simulation of oil/water/surfactant systems. *Journal of Physical Chemistry, 95,* 6361–6368.

50. Tieleman, T., van der Spoel, D., and Brendsen, H. J. C. (2000). Molecular dynamics simulators of dodecyclphosphocholine micelles at three different aggregate sizes: Micellar structure and chain relaxation. *Journal of Physical Chemistry B, 104,* 6380–6388.

51. Toffoli, T., and Margolus, N. (1991). Programmable matter: Concepts and realization. *Physica D, 47,* 263–272.

52. Varela, F., Maturana, H., and Uribe, R. (1974). Autopoiesis: The organization of living systems. *BioSystems, 5,* 187–196.

53. Watanabe, K., Ferrario, M., and Klein, M. L. (1998). Molecular dynamics study of a sodium octanoate micelle in aqueous solution. *Journal of Physical Chemistry, 92,* 819–821.

54. Wisdom, B. (1986). Lattice models of microemulsions. *Journal of Chemical Physics, 84,* 6943–6954.

55. Wolfram, S. (1986). Cellular automata fluids I: Basic theory. *Journal of Statistical Physics, 45,* 471–526.

# III Background and Polemics

# Introduction to Background and Polemics

Many chapters in parts I and II take for granted certain philosophical and scientific assumptions and techniques, such as reduction, supervenience, multiple realization, downwards causation, complexity theory, and chaos. The main purpose of part III is to provide some classic resources on these topics, many of which also contain creative contributions to the topics of reduction and emergence. They also illustrate the polemics that often surround discussions of emergence.

## Physicalism and Reductionism

Within the natural sciences, the relations between the subject matters of physics, chemistry, and biology have long been disputed. If chemistry were reducible to physics, at least in principle, then although chemistry might have its own methods that must be used for practical reasons, in reality the truths about the chemical world just would be complex truths about physical objects and properties. In contrast, if some chemical phenomena were irreducible in principle to purely physical phenomena, then even knowing everything it was possible to know about physics would leave further facts about chemistry beyond those physical facts. And so a common requirement for a phenomenon to be emergent is the impossibility of reducing that phenomenon to other, more basic, phenomena. Although irreducibility is not, of course, sufficient for emergence—facts about Frenchmen cannot be reduced to facts about flounders, for example—the inability to reduce the subject matter of one science to that of another often is taken as a sign that the sciences in question have the kind of autonomous domains required for emergence.

Even if reduction cannot be carried out in a comprehensive fashion, a frequently expressed belief is that in some sense the physical facts are all the facts that exist. This belief lies at the core of the contemporary philosophical position called physicalism. One version of physicalism holds that fundamental physics captures the basic truths about the world and determines everything else in physics as well as in chemistry and in biology. More liberal versions of physicalism require only that the basic truths lie within the domain of the physical sciences, though not necessarily at the most fundamental level. This more liberal form of physicalism leaves some scope within the domain of physics itself for a form of emergence based on irreducibility. Thus, although much of the philosophical literature on emergence may leave the impression that the issue is primarily about esoteric phenomena such as human consciousness, debates about reduction and emergence occur in a wide variety of more basic areas of science. When considering whether emergent phenomena exist, it is important to note that some of the best-understood candidates for emergence occur within physics, chemistry, biology, various social sciences, and complexity theory. Examples include first-order phase transitions, such as the transition from a liquid to a solid crystal; second-order phase transitions, such as the appearance of ferromagnetism below the critical temperature; the rigidity of solids; coherent radiation in lasers; the Belousov-Zhabotinsky reaction; the thermoregulation of beehives; the ability of self-organizing slime molds to form a spore-bearing structure; the oscillations of synchronized firefly flashes; flock formation in birds; and optimal pricing in markets. The first four examples are from physics, the next from chemistry, the following four from biology, and the last from economics. With the development of explicit theories in physical chemistry, biochemistry, and molecular biology—as well as the introduction of formal interdisciplinary models in complexity theory that can be applied to physical, chemical, biological, and economic systems, amongst others—the plausibility of these phenomena as candidates for emergence can be addressed in detail.

The possibilities for reduction are also of particular interest to psychology, to sociology, and to anthropology. Emile Durkheim, one of the founders of sociology, argued for the existence of autonomous sociological facts, such as different rates of suicide in different countries, that could be explained only in sociological terms. In a rather different way, the quasi-reductionist position known as *methodological individualism* at one time was influential within the discipline of history as well as in the various social sciences. The core idea of methodological individualism was that the collective terms that appeared in the social sciences, such as "nation" or "guild," referred to groups of individuals or patterns of individual human behavior, thus avoiding the need to appeal to irreducible sociological or economic structures or to autonomous historical laws. The position was an influential component of the antiemergentist attitudes of the mid-twentieth century. In contrast, the methods of agent-based modeling, while sympathetic to the goals of methodological individualism, frequently are of use in generating emergent phenomena. It is instructive to read chapter 12 in part II with this issue in mind.

What are the requirements for reduction, and what is at stake in the debates over reduction? Chapter 18 by Stephen Weinberg illustrates how slippery can be the concept of reduction while simultaneously demonstrating how high the stakes can be. Weinberg's chapter originally appeared during the struggle in the late 1980s and early

1990s to fund the superconducting supercollider, an accelerator considered crucial for advances in elementary-particle physics. One of the proposed tasks for that accelerator was to explore the phenomenon known as spontaneous symmetry breaking, a phenomenon which, ironically, frequently is cited as giving rise to emergent phenomena including ferromagnetism. Two issues discussed in Weinberg's chapter are of particular interest. Some of the disputes within the scientific community over the money needed to fund the collider concerned whether the subject matter of elementary-particle physics is in some sense ontologically privileged, as the kind of narrow physicalism discussed above would maintain. The other noteworthy aspect of Weinberg's chapter is the emphasis on the need to introduce new theoretical terms in order to understand phenomena occurring at higher levels of organization or complexity. As Weinberg says: "As one goes to higher and higher levels of organization, new concepts emerge that are needed to understand the behavior at that level." This is a particularly clear statement of the idea that conceptual novelty is central to emergence.

## Nagel-Style Reduction and Multiple Realization

One possible source of confusion when considering the possibilities for reduction is that several different ideas about reduction are in play. The standard approach for many years was a view most closely associated with the philosopher Ernest Nagel, illustrated here in chapter 19. Nagel formulated his view of reduction in an era in which linguistic approaches dominated philosophy, and one of his major concerns was how to deal with what he called *inhomogeneous reduction*. This is the situation where the theory to be reduced contains terms that do not appear in the reducing theory and so some connection must be made between the vocabularies of the two theories. For example, the chemical term *acid* does not appear in physics, and several different definitions of *acid* have been developed. In order to reduce these novel terms, Nagel required the use of bridge laws that provided nomologically necessary and sufficient conditions for the application of the higher-level concept in terms of predicates occurring in the reducing theory. The difficulty of finding appropriate bridge laws eventually caused interest in Nagel-style reduction to collapse.

Many philosophers also were persuaded to abandon Nagel-style reduction by the multiple realizability argument, a classic source of which is chapter 22 by Jerry Fodor. The argument hinges on two points: first, that a predicate that occurs in economics, such as *money*, can be realized in many different ways, such as by banknotes, electronic credits, cattle, and other means, and, second, that these realizers have so little in common besides their function as money that no simple predicate covers the heterogeneous realizers. The absence of any such simple predicate suggests that no natural kinds within a lower-level science, such as physics or biology, correspond to psychological or economic kinds. Thus, if bridge laws must relate natural kinds (as Fodor supposed) there are no bridge laws of the kind required by Nagel-style reduction, and the laws within sciences like economics or psychology are autonomous from those of more basic sciences like physics and biology. Another important aspect of Fodor's chapter is its appeal to functionalism. In the example above, *money* is a functionally characterized predicate. That is, something can serve as money if and only if it can perform the function of money in an economy, and what is required for it to carry out that function is for it to serve as the appropriate kind of a causal intermediary between the transaction input and the transaction output. So, coins can serve that function, but it would be difficult in human economies to use Julius Caesar's dying breath or black holes as money. Because of the prevalence of functionally characterized predicates in the social and psychological sciences, if Fodor's multiple realizability argument is sound, it establishes the conceptual and nomological autonomy of at least those sciences. It also makes the Nagel-style reduction of macroeconomics to microeconomics impossible and the goals of methodological individualism quixotic.

## Supervenience and Downward Causation

After the discovery that Nagel-style reduction and bridge laws apparently are rare, philosophers shifted their attention to different kinds of relations between levels. One such commonly used relation is that of supervenience, which is described clearly in chapter 23, an excerpt from David Chalmer's book on consciousness. The basic idea behind supervenience is that, if a property B supervenes upon a set of properties A, then B is fixed once A is fixed; no two possible situations are identical with respect to A properties but differ with respect to B. Supervenience is consistent with multiple realizability, and many physicalists have used this fact to advocate for the combination of supervenience and various forms of nonreductive physicalism. Nonreductive physicalism based on supervenience is a physicalist view in the sense that once the physical facts are in place, everything else is determined, including chemical, biological, and psychological facts. It is nonreductive (rather than antireductive) in the sense that it leaves open for further exploration whether the supervening features can be reduced to physical features. A key question for nonreductive physicalists is then whether some interesting form of emergence is compatible with physicalism.

The connection between supervenience and emergence has two core issues. The first is whether using supervenience relations can capture emergent phenomena. Chapter 4 in part I is one attempt to explain emergence in terms of supervenience. In chapter 24 in this section, Kim argues against nonreductive physicalism in general, and against supervenience-based accounts of emergence in particular. This argument, which has come to be known as the *downward causation argument,* concludes that nonreductive physicalism and emergentism are committed to things at the higher levels causing things at the physical level, and this is incompatible with the widely held physicalist commitment to the causal closure of the physical realm. This argument has had a considerable influence, and the positions developed in chapters 6, 7, and 8 in part I are framed explicitly so as to avoid the downward causation argument. The very terminology of *downward causation* assumes a hierarchy of levels between which reduction can take place, or between which emergence occurs. This assumption is common in the literature, but finding explicit criteria for distinguishing the levels is rare (see the discussion of dynamical hierarchies in the introduction to part II). Any given argument must be clear about whether the talk of levels is to be taken literally or metaphorically. An alternative approach that carries fewer metaphysical dangers is to abandon talk of levels and to frame the issues in terms of one domain of phenomena emerging from another domain.

The second issue about the connection between supervenience and emergence concerns exactly what it is that supervenience relations relate. As a way of moving away from the linguistic orientation of Nagel-style reduction and toward a more directly realist account of the relation between levels of reality, many philosophers take supervenience to concern the relationship between properties rather than linguistic predicates. An alternative approach, suggested in the Chalmers's chapter, takes the relation to hold between concepts. Whether sympathetic to supervenience-based accounts of emergence or opposed to them, it is crucial to decide which of these is the right approach. Because the downward causation argument relies on what Kim has dubbed *Alexander's Dictum*—the claim that only things which have causal effects are real—the status of the supervening properties as properties with new causal powers or as merely reconceptualized or redescribed subvenient properties must be decided. The status of Alexander's Dictum itself has been much discussed, and readers should decide for themselves whether or not it is generally true. One question that is useful to consider in this context is what status the dictum attributes to abstract entities such as mathematical objects and such things as patterns in cellular automata.

## Computational Irreducibility and Sensitive Dependence on Initial Conditions

Perhaps because the early debates about reduction and emergence took place in the context of relations between theories, the philosophical literature has had a tendency to emphasize synchronic accounts of emergence, situations in which the emergent features exist simultaneously with that from which they emerge. In contrast, especially in the areas of complexity theory, artificial life, and related disciplines, the scientific literature has explored extensively examples of diachronic emergence, situations in which the emergent phenomenon appears over time (see the introduction to part II). The cluster of issues addressed in chapters 20 and 21 are relevant to specifically diachronic forms of emergence and serve as a background to many of the chapters in part II, as well as to chapters 8 and 9 in part I.

Of particular interest here is the idea that the physical world can be conceived of as containing, transmitting, and processing information; and the related idea that the world itself is a computational device. A core concept involved is that of computational irreducibility, an idea which can be captured in various ways. One approach uses the idea that a computationally irreducible sequence of outcomes is one for which any computer program that produces that sequence is of approximately the same length as the sequence itself. Another approach, which plays a prominent role in chapter 20, rests on the idea that the number of computational steps needed to predict a future state of the system is an increasing function of how far into the future that state lies. In this second approach, if the world is a computer, no predictive short cuts exist to producing the sequence (nor to any of its states) other than letting the world itself generate its own states.

It is instructive to consider computational irreducibility in the context of modern approaches to emergence that involve unpredictability. The idea behind these approaches is analogous to those of British Emergentists like C. D. Broad, but such approaches hold the promise of being grounded in objective criteria for unpredictability based on measures of computational complexity. This raises a question of fundamental importance for these views of emergence: How common are computationally irreducible processes?[1] Wolfram in chapter 21 argues that computationally irreducible processes are ubiquitous in nature, and he concludes that for many physical systems "their own evolution is effectively the most efficient procedure for determining their future." It is a fundamental mathematical fact that the behavior of universal (i.e., fully programmable) computers is generally undecidable; the only sure way to tell how they will respond to an arbitrary input is to observe their behavior. Wolfram's argument traces the unpredictability of physical systems back to this fact. He illustrates his arguments with cellular automata, extremely simple models that can generate extremely complicated behavior. Cellular automata have since acquired an iconic status as complex systems, and they figure centrally in chapters 8 and 9 in part I and chapters 13 and 16 in part II.

Dynamical systems theory, and chaotic dynamics in particular, serves as an inspiration for some contemporary views on emergence. The application of dynamical systems to explaining chaotic behavior, described in chapter 20, demonstrates one important reason why many complex systems are so unpredictable. A system exhibiting deterministic chaos displays extreme sensitivity to initial conditions. That is, a very small difference in the initial states of two otherwise identical systems quickly can lead to great differences in their future behavior. Because it is never possible to measure initial states exactly, it is impossible to predict precisely how chaotic systems will develop over time. As a result, exact quantitative prediction is supplanted by descriptive topological and geometrical characteristics, such as "strange attractors." This illustrates one reason for the rejection of in-principle reduction found in many of the chapters in part II, and it suggests that the linguistic preoccupation of many approaches to reduction should be replaced with a broader set of mathematical tools. This chapter also emphasizes that nonlinear systems cannot be understood in terms of the traditional analytic decomposition methods of science. Nonlinearity is a hallmark of complex systems that generate emergent behavior, and it represents one more way in which the appeals of the British Emergentists to vague notions of nonadditivity now can be replaced with much more precise analyses.

### Note

1. A further question is whether the era of relatively easy, computationally compressible, predictions in science is coming to an end. If so, this suggests the need to rethink the idealized "limit science" orientation often used to discuss positions in the philosophy of science.

# 18 Newtonianism, Reductionism and the Art of Congressional Testimony

**Stephen Weinberg**

My talk this afternoon will be about the philosophy of science, rather than about science itself. This is somewhat uncharacteristic for me, and, I suppose, for working scientists in general. I've heard the remark (although I forget the source) that the philosophy of science is just about as useful to scientists as ornithology is to birds.

However, at just this time a question has arisen in the United States that will affect the direction of physics research until well into the twenty-first century, and that I think hinges very largely on a philosophical issue. On 30 January of this year the present administration in Washington announced that it had decided to go ahead with the construction of a large new accelerator for elementary particle physics, the Superconducting Supercollider, or SSC for short. "Large" in this case means that its circumference would be about 53 miles. The circumference is determined by the necessity of accelerating protons to energies of 20 TeV ($2 \times 10^{13}$ electron volt). Within this ring there would travel two counter-rotating beams of protons, that would slam into each other at a number of intersection regions. The intensity of the beams is designed to be such that one would have a collision rate of about one per second for typical processes (with a cross-section of a nanobarn). All of these design parameters lead to a bottom line parameter: the cost in 1986 dollars is estimated to be 4,400 million dollars.

The chief reason for wanting to go ahead with this accelerator is that it would open up a new realm of high energy which we have not yet been able to study. Just as when astronomers start to study the sky at a new wavelength, or when solid state physicists go down another factor of ten in temperature, every time particle accelerators go up a factor of ten in energy we discover exciting new physics. This has generally been the rationale for new accelerators. Occasionally, one can also point to specific discoveries that can be anticipated from a particular new accelerator. One example is provided by the accelerator built in Berkeley over 30 years ago, the Bevatron, which for the first time was capable of producing particles with masses of 1 GeV. (In those days American physicists talked about BeV instead of GeV.) The Bevatron was designed to be able to produce antiprotons, and indeed it did so shortly after it went on the air. That was not the only exciting thing done at that accelerator. Quite unexpected was the discovery of a vast forest of new mesonic and baryonic states, that led to a change in our conception of what we mean by an elementary particle. But in planning the Bevatron, it was nice to know in advance that at least one important discovery could be counted on.

The same is true now of the SSC. The SSC is so designed so that it will discover the particle known as the Higgs boson, provided that the Higgs boson is not too heavy. If the Higgs boson is too heavy, then the SSC will discover something else equally interesting.

Let me explain these remarks further. As many people may have heard, there has been a certain measure of unification among the forces of nature. This unification entails the idea that the symmetry among the forces, specifically the weak nuclear force and the electromagnetic force, is spontaneously broken. It can't be spontaneously broken by the forces we know about, that is, the ordinary strong and weak nuclear forces and the electromagnetic force; therefore there must be a new force in nature which is responsible for the symmetry breaking, like the phonon exchange force in a superconductor. We don't know exactly what that force is. The simplest picture is that it has to do with the existence of a new kind of elementary scalar particle. The members of the multiplet of elementary scalar particles that would be observable as physical particles are called Higgs bosons.

Now, we are not sure that that is actually the correct picture of the mechanism for electroweak symmetry breaking, and we certainly do not know the mass of the Higgs boson. The SSC would be able to discover the Higgs boson if its mass is not greater than about 850 GeV, and, of course, if it exists. However, the SSC (to borrow a phrase from M. Chanowitz[1]) is a no-lose proposition, because if the Higgs boson does not exist, or is heavier than 850 GeV, there would have to be strong interactions among longitudinally polarized W particles, which the SSC could also discover. These strong interactions would reveal the nature of the spontaneous symmetry breaking between the weak and the electromagnetic interactions.

Now it remains for Congress to decide whether or not to authorize construction of the accelerator and to appropriate the money. Two committees of the two houses of the Congress, the Committee on Space, Science, and Technology of the House of Representatives and the Subcommittee on Energy Research and Development of the Senate Committee on Energy and Natural Resources, announced hearings on the SSC, both to begin on 7 April of this year. In March, about a month before these hearings, I was asked to testify at them. I may admit that I found this more frightening than inviting. I had been active for some time in working for the building of the SSC, and all this time it had been a nightmare of mine that I would be called up before some tribunal, and asked in a stern voice why it is worth 4.4 billion dollars to find the Higgs boson. Also, I had testified in Congress only once before, and I did not consider myself a master of the art of congressional testimony.

The particle physicists of the United States are in fact quite united behind the idea that this is the right accelerator to build next. (As I said, its purpose is not limited to finding the Higgs boson, which is just one target, but, rather, it is to open up a new range of energies.) But there has been substantial opposition to the SSC from other physicists in the United States. I have read that this is perhaps the most divisive issue that has ever faced American physicists.[2] I believe that in Britain there is a similar debate—not about building an SSC but about whether Britain should remain in CERN, an issue on which I gather not all British scientists agree.

## Heavyweights

I knew at the hearings in Washington there would be two heavyweights who would be testifying vigorously against going ahead with the SSC. One would be Philip Anderson, known to everyone as among the leading condensed matter physicists in the world. Anderson has over many years opposed the large sums that are spent on high energy physics. Another to testify would be James Krumhansl, also a distinguished solid state physicist. He, as it happens, taught me physics when I was a freshman at Cornell, but in addition, and this I suspect counts for more, he is slated the year after next to be the president of the American Physical Society.

Both Anderson and Krumhansl I knew would oppose the SSC, and they would be making arguments with which I really couldn't disagree. In particular, I expected that they would argue that money spent on elementary particle physics, high energy physics, whatever you want to call it, is not as sure to yield immediate technological advances as the same money spent on condensed matter physics, and some other fields. I would have to agree with that (though I would put more emphasis on the benefits of unpredictable discoveries and spin-offs). I expected that they would also argue that elementary particle physics is not more intellectually profound than other areas of physics like, say, condensed matter physics. I would also agree with that. In fact, we've seen in the last few decades a continual trading back and forth of ideas between elementary particle physics and condensed matter physics. We learned about broken symmetry from them, they learned about the renormalization group from us. And now we're all talking about conformal quantum field theories in two dimensions (I don't know who learned that from whom). But it is clear that there's no lack of mathematical profundity in condensed matter physics as compared with elementary particle physics.

The case for spending large sums of money on elementary particle physics has to be made in a different way. It has to be at least in part based on the idea that particle physics (and here, parenthetically, I should say that under "particle physics" I include quantum field theory, general relativity, and related areas of astrophysics and cosmology) is in *some* sense more fundamental than other areas of physics. This was denied more or less explicitly by Anderson and Krumhansl in their testimony and also by most of the opponents of the SSC. I didn't see how I could avoid this issue in making a case for the SSC. But it's a dangerous argument. It tends to irritate one's friends in other areas of science. Let me give an example, and here I will quote from myself because then I want to quote some comments on my own remarks.

In 1974, shortly after the standard model was put into its final form with the success of quantum chromodynamics, I wrote an article[3] for *Scientific American* called "Unified Theories of Elementary Particle Interactions." Just to get the article started I began it with some platitudes, as follows: "One of man's enduring hopes has been to find a few simple general laws that would explain why nature with all its seeming complexity and variety is the way it is. At the present moment the closest we can come to a unified view of nature is a description in terms of elementary particles and their mutual interactions." I really didn't intend to make any important point by this; it was just the sort of thing one says (as, for instance, Einstein: "The supreme test of the physicist is to arrive at those universal elementary laws from which the cosmos can be built up by pure deduction"). Then a decade later I was asked by the MIT Press to review a proposed book, a collection of articles by various scientists. In the manuscript I found an article[4] by a friend of mine at Harvard, Ernst Mayr, who is one of the most eminent evolutionary biologists of our times. I found that Mayr cited the remarks in the *Scientific American* article as "a horrible example of the way physicists think." He called me "an uncompromising reductionist."

## Agreement

Now, I strongly suspect that there is no real disagreement between Ernst Mayr and myself, and that in fact we are simply talking past each other, and we should try to understand how we agree rather than fight over this. I don't consider myself an uncompromising reductionist. I consider myself a compromising reductionist. I would like to try to formulate in what way elementary particle physics is more fundamental than other areas of physics, trying to narrow this down in such a way that we can all agree on it.

Let me first take up some of the things I don't mean. And here it is useful to look back at some more of Ernst Mayr's writing, because he is in fact the leading opponent of the reductionist tendency within biology, as well as in science in general. He wrote a book[5] in 1982, *The Growth of Biological Thought*, that contains a well-known attack on reductionism, and so I looked at it to see what Mayr thought reductionism was, and whether or not I consider myself, in his terms, a reductionist.

The first kind of reductionism that Mayr opposes is called by him "theory reductionism." As far as I can understand it, it's the notion that the other sciences will eventually lose their autonomy and all be absorbed into elementary particle physics; they will all be seen as just branches of elementary particle physics.

Now I certainly don't believe that. Even within physics itself, leaving aside biology, we certainly don't look forward to the extinction of thermodynamics and hydrodynamics as separate sciences; we don't even imagine that they are going to be reduced to molecular physics, much less to elementary particle physics. After all, even if you knew everything about water molecules and you had a computer good enough to follow how every molecule in a glass of water moved in space, all you would have would be a mountain of computer tape. How in that mountain of computer tape would you ever recognize the properties that interest you about the water, properties like vorticity, turbulence, entropy and temperature?

There is in the philosophical literature a term, emergence, that is used to describe how, as one goes to higher and higher levels of organization, new concepts emerge that are needed to understand the behaviour at that level. Anderson summarized this neatly in the title of an interesting article[6] in *Science* in 1972: "More is Different."

Another kind of reductionism is called by Mayr "explanatory reductionism." As I understand it, it is the idea that progress at the smallest level, say the level of elementary particle physics, is needed to make progress in other sciences, like hydrodynamics, condensed matter physics and so on.

I don't believe that either. I think we probably know all we need to know about elementary particle physics for the purposes of the solid state physicist, for instance, and the biologist. Mayr in his book makes a point that surprised me (but I suppose it's true; he knows a lot more about this than I do), that even the discovery of DNA was not really of much value in the science of transmission genetics. Mayr writes, "To be sure the chemical nature of a number of black boxes in the classical genetic theory were filled in by the discovery of DNA, RNA, and others, but this did not affect in any way the nature of transmission genetics."

I don't disagree with any of this, but it seems to me that in their attacks on reductionism, Mayr and also physicists like Anderson, Krumhansl and others, are missing the point. In fact, we all do have a sense that there are different levels of fundamentalness. For instance, even Anderson[7] calls DNA the "secret of life." We do have a feeling that DNA is fundamental to biology. It's not that it's needed to explain transmission genetics, and it's certainly not needed to explain human behaviour, but DNA is fundamental nonetheless. What is it then about the discovery of DNA that was fundamental to biology? And what is it about particle physics that is fundamental to everything?

Having spoken at length about what I don't mean, now I want to say what I do mean. But I'm not trying here to say anything new, that you don't all already know. What I'm trying to do is precisely the opposite: to identify what we can all agree on.

In all branches of science we try to discover generalizations about nature, and having discovered them we always ask why are they true. I don't mean why we believe they are true, but why they *are* true. Why is nature that way? When we answer this question the answer is always found partly in contingencies, that is, partly in just the nature of

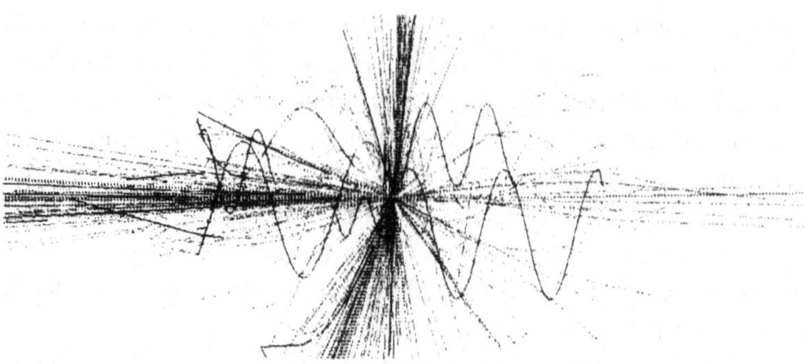

**Figure 18.1**
Hunting the Higgs—computer simulation of a proton-proton collision in the SSC. The picture is taken from "To the Heart of Matter," issued by the Universities Research Association.

the problem that we pose, but partly in other generalizations. And so there is a sense of direction in science, that some generalizations are "explained" by others.

To take an example relative to the tercentenary celebration of the *Principia*: Kepler made generalizations about planetary motion, Newton made generalizations about the force of gravity and the laws of mechanics. There is no doubt that historically Kepler came first and that Newton, and also Halley and Wren and others, derived the inverse square law of gravity from Kepler's laws. In formal logic, since Kepler's laws and Newton's laws are both true, either one can be said to imply the other. (After all, in formal logic the statement "A implies B" just means that it never happens that A is true and B isn't, but if A and B are both true then you can say that A implies B and B implies A.)

### Intuition

Nevertheless, quite apart from formal logic, and quite apart from history, we intuitively understand that Newton's laws of motion and law of gravity are more fundamental than Kepler's laws of planetary motion. I don't know exactly what I mean by that; presumably it has something to do with the greater generality of Newton's laws, but about this also it's hard to be precise. But we all know what we mean when we say that Newton's laws "explain" Kepler's. We probably could use help from professional philosophers in formulating exactly what that statement means, but I do want to be clear that it is a statement about the way the Universe is, not about the way physicists behave. In the same way, even though new concepts "emerge" when we deal with fluids or many-body systems, we understand perfectly well that hydrodynamics and thermodynamics are what they are because of the principles of microscopic physics. No one thinks that the phenomena of phase transitions and chaos (to take two examples quoted by Krumhansl) could have been understood on the basis of atomic physics without creative new scientific ideas, but does anyone doubt that real materials exhibit these phenomena because of the properties of the particles of which the materials are composed?

Another complication in trying to pin down the elusive concept of "explanation" is that very often the "explanations" are only in principle. If you know Newton's laws of motion and the inverse square law of gravity you can deduce Kepler's laws—that's not so hard. On the other hand, we also would say that chemical behaviour, the way molecules behave chemically, is explained by quantum mechanics and Coulomb's law, but we don't really deduce chemical behaviour for very complex molecules that way. We can for simple molecules; we can explain the way two hydrogen atoms interact to form a hydrogen molecule by solving Schrödinger's equation, and these methods can be extended to fairly large molecules, but we can't work out the chemical behaviour of DNA by solving Schrödinger's equation. In this case we can at least fall back on the remark that although we don't in fact calculate the chemical behaviour of such complicated molecules from quantum mechanics and Coulomb's law, we could if we wanted to. We have an algorithm, the variational principle, which is capable of allowing us to calculate anything in chemistry as long as we had a big enough computer and were willing to wait long enough.

The meaning of "explanation" is even less clear in the case of nuclear behaviour. No one knows how to calculate the spectrum of the iron nucleus, or the way the uranium nucleus behaves when fissioning, from quantum chromodynamics. We don't even have an algorithm; even with the biggest computer imaginable and all the computer time you wanted, we would not today know how to do such calculations. Nevertheless, most of us are convinced that quantum chromodynamics does explain the way nuclei behave. We say it explains it "in principle," but I am not really sure of what we mean by that.

Still, relying on this intuitive idea that different scientific generalizations explain others, we have a sense of direction in science. There are arrows of scientific explanation, that thread through the space of all scientific generalizations. Having discovered many of these arrows, we can now look at the pattern that has emerged, and we notice a remarkable thing: perhaps the greatest scientific discovery of all. These arrows seem to converge to a common source! Start anywhere in science and, like an unpleasant child, keep asking "Why?" You will eventually get down to the level of the very small.

By the mid-1920s, the arrows of explanation had been traced down to the level of the quantum mechanics of electrons, photons, atomic nuclei and, standing somewhat off in the corner, the classical theory of gravity. By the 1970s we had reached a deeper level—a quantum field theory of quarks, leptons and gauge bosons known as the standard model, and with gravity still somewhat isolated, described by a not very satisfactory quantum field theory of gravitons. The next step, many of us think, is the theory of superstrings, still under development. I myself, although a late-comer to this field, confess my enthusiasm for it. I think it provides our best hope of making the next step beyond the standard model.

## Objective Reductionism

Now reductionism, as I've described it in terms of the convergence of arrows of explanation, is not a fact about scientific programmes, but is a fact about nature. I suppose if I had to give a name for it, I could call it objective reductionism. It is very far from a truism. In particular, these arrows of explanation might have led to many different sources. I think it's important to emphasize that, until very recently, most scientists thought that that was the case; this discovery, that the arrows of explanation point down to a common source, is quite new. (In a comment on an earlier version of this talk, Ernst Mayr informs me that what I call "objective reductionism" is what he means by "theory reductionism." Maybe so, but I prefer to keep the separate terms, because I wish to emphasize that what I am talking about here is not the future organization of the human scientific enterprise, but an order inherent in nature itself.)

To underscore this point, I'd like to mention a few examples of the contrary view surviving until well into the twentieth century. The first is biological vitalism, the idea that the usual rules of physics and chemistry need to be modified when applied to living organisms. One might have thought that this idea would have been killed off by the rise of organic chemistry and evolutionary biology in the nineteenth century. However, Max Perutz in his talk at the Schrödinger centenary in London in April reminded us that both Niels Bohr and Erwin Schrödinger believed that the laws of physics as then understood in the 1920s and 1930s were inadequate for understanding life.[8] Perutz explains that the problem of the orderliness of life that bothered Schrödinger was cleared up by advances in the understanding of enzymatic catalysis. Ernst Mayr was careful in his book to disavow any lingering attachment to vitalism, as follows: "Every biologist is fully aware of the fact that molecular biology has demonstrated decisively that all processes in living organisms can be explained in terms of physics and chemistry." (Mayr, by the way, is using the word "explained" in exactly the same sense as I am here.)

A second example, Lord Kelvin, in a speech to the British Association for the Advancement of Science, around 1900, said,[9] "There is nothing new to be discovered in physics now. All that remains is more and more precise measurement." There is a similar remark of Michelson's that is often quoted.[10] These remarks of Kelvin's and Michelson's are usually cited as examples of scientific arrogance and blindness, but I think this is based on a wrong interpretation of what Kelvin and Michelson meant. The reason that Kelvin and Michelson made these remarkable statements is, I would guess, that they had a very narrow idea of what physics was. According to their idea, the subject matter of physics is motion, electricity, magnetism, light and heat, but not much else. They felt that that kind of physics was coming to an end, and in a sense it really was. Kelvin could not possibly have thought in 1900 that physics had already explained chemical behaviour. He didn't think so, but he also didn't think that was a task for physics. He thought that physics and chemistry were sciences on the same level of fundamentalness. We don't think that way today, but it isn't long ago that physicists did think that way.

I said that these arrows of explanation could have led down to a number of separate sciences. They also could have gone around in a circle. This is still a possibility. There is an idea that's not quite dead among physicists and cosmologists, the "anthropic principle," according to which there are constants of nature whose value is inexplicable except through the observation that if the constants had values other than what they have the Universe would be so different that scientists would not be there to ask their questions. If the anthropic principle were true, there would be a kind of circularity built into nature, and one would then I suppose have to say that there is no one fundamental level—that the arrows of explanation go round in circles. I think most physicists would regard the anthropic principle as a disappointing last resort to fall back on only if we persistently fail to explain the constants of nature and the other properties of nature in a purely microscopic way. We'll just have to see.

Now although what I have called objective reductionism became part of the general scientific understanding only relatively recently (after the development of quantum mechanics in the 1920s), its roots can be traced back to Newton (who else?). Newton was the first to show the possibility of an understanding of nature that was both comprehensive and quantitative. Others before him, from Thales to Descartes, had tried to make comprehensive statements about nature, but none of them took up the challenge of explaining actual observations quantitatively in a comprehensive physical theory.

I don't know of any place where Newton lays out this reductionist programme explicitly. The closest I can come to it is a remark in the Preface to the first edition of the *Principia*, written in May 1686. Newton says, "I wish we could derive the rest of the phenomena of nature by the same kind of reasoning from mechanical principles [I suppose he means as in the *Principia*] for I am induced by many reasons to suspect that they may all depend on certain forces." I suppose that the most dramatic example of the opening up by Newton of the possibility of a comprehensive quantitative understanding of nature is in the third book of the *Principia* where Newton reasons that the moon is 60 times further away from the centre of the Earth than Cambridge is (either Cambridge) and therefore the acceleration of the Moon towards the Earth should be less than the acceleration of an apple in Cambridge by a factor of $60^2$. With this argument Newton unites celestial mechanics and observations of falling fruits in a way that

I think captures for the first time the enormous power of mathematical reasoning to explain not only idealized systems like planets moving in their orbits, but ultimately everything.

A digression. Since I have been talking about Newton, and also talking about the SSC, a prime example of "big science," I can't resist remarking that Newton himself was involved in big science.[11] In 1710, as President of the Royal Society, Newton by royal command was given control of observations at the largest national laboratory for science then in existence in England, the Greenwich Observatory. He was also given the responsibility of overseeing the repair of scientific instruments by the Master of Ordnance, an interesting connection with the military. (This arrangement, incidentally, infuriated the then Astronomer Royal, Flamsteed.)

## Gaps

There are many gaps, of course, and perhaps there always will be many gaps in what I have called the chains of explanation. The great moments in the history of science are when these gaps are filled in, as for example when Darwin and Wallace explained how living things, with all their adaptations to their environment, could develop without any continuing external intervention. But there are still gaps.

Also, sometimes it isn't so clear which way the arrows of explanation point. Here's one example, a small one, but one that has bothered me for many years. We know mathematically that as a consequence of Einstein's general theory of relativity gravitational waves should be waves of spin two, and therefore when quantized, the theory of gravity should have in it particles of mass zero and spin two. On the other hand, we also know that any particles of mass zero and spin two must behave as described by Einstein's general theory of relativity. The question is, which is the explanation of which? Which is more fundamental, general relativity or the existence of particles of mass zero and spin two? I've oscillated in my thinking about this for many years. At the present moment in string theory the fact that the graviton has mass zero and spin two appears as an immediate consequence of the symmetries of the string theory, and the fact that gravity is described by the formalism of Riemannian geometry and general relativity is a somewhat secondary fact, which arises in a way that is still rather mysterious. But I don't know if that is the final answer. I mention this example just to show that although we don't always know which truths are more fundamental, it's still a worthwhile question to ask, because it is a question about the logical order of nature.

I believe that objective reductionism, reductionism as a statement about the convergence of arrows of explanation in nature, is by now ingrained among scientists, not only among physicists but also among biologists like Ernst Mayr. Let me give an example. Here's a quote from the presidential address of Richard Owen to the British Association in 1858.[12] Owen was an anatomist, generally regarded as the foremost of his time, and a great adversary of Darwin. In his address, Owen says, "Perhaps the most important and significant result of palaeontological research has been the establishment of the axiom of the continuous operation of the ordained becoming of living things." I'm not too clear what precisely Owen means by this axiom. But my point is that today no biologist would make such a statement, even if he or she knew what the axiom meant, because no biologist today would be content with an axiom about biological behaviour that could not be imagined to have an explanation at a more fundamental level. That more fundamental level would have to be the level of physics and chemistry, and the contingency that the Earth is billions of years old. In this sense, we are all reductionists today.

Now, these reflections don't in themselves settle the question of whether the SSC is worth 4.4 billion dollars. In fact, this might be a difficult problem, if we were simply presented with a choice between 4.4 billion dollars spent on the SSC and 4.4 billion dollars spent on other areas of scientific research. However I don't think that that's likely to be the choice with which we are presented. There is evidence that spending on "big science" tends to increase spending on other science, rather than the reverse. We don't really know with what the SSC will compete for funds. In any case, I haven't tried here to settle the question of whether or not the SSC should be built for 4.4 billion dollars—it is a complicated question, with many side arguments. All I have intended to argue here is that when the various scientists present their credentials for public support, credentials like practical values, spinoff etc., there is one special credential of elementary particle physics that should be taken into account and treated with respect, and that is that it deals with nature on a level closer to the source of the arrows of explanation than other areas of physics. But how much do you weigh this? That's a matter of taste and judgement, and I'm not paid to make that final decision. However I would like to throw into the balance one more point in favour of the SSC.

I have remarked that the arrows of explanation seem to converge to a common source, and in our work on elementary particle physics we think we're approaching that source. There is one clue in today's elementary particle physics that we are not only at the deepest level we can get right now, but we are at a level which is in fact in absolute terms quite deep, perhaps close to the final source. And here again I would like to quote from myself, from my own testimony in Congress, because afterwards I am going to quote some comments on these remarks, and I want you to know what it is that the comments were about:

> There is reason to believe that in elementary particle physics we are learning something about the logical structure of the Universe at a very very deep level. The reason I say this is because as we have been going to higher and higher energies and as we have been studying structures that are smaller and smaller we have found that the laws, the physical principles, that describe what we learn become simpler and simpler. I am not saying that the mathematics gets easier, Lord knows

it doesn't. I am not saying that we always find fewer particles in our list of elementary particles. What I am saying is that the rules that we have discovered become increasingly coherent and universal. We are beginning to suspect that this isn't an accident, that it isn't just an accident of the particular problems that we have chosen to study at this moment in the history of physics but there is simplicity, a beauty, that we are finding in the rules that govern matter that mirrors something that is built into the logical structure of the Universe at a very deep level. I think that this kind of discovery is something that is going on in our present civilization at which future men and women and not just physicists will look back with respect.

After I made these remarks there were remarks by other witnesses, and then there were questions from members of the Committee on Space, Science, and Technology. I am going to quote from the remarks of two of them. The first is Harris W. Fawell, Republican congressman from Illinois. Fawell throughout his questioning had been generally favourable to the SSC. The second is representative Don Ritter, of Pennsylvania, also a Republican, who had been the congressman most opposed to the SSC throughout the morning. (I suppose you could regard this as a modern dialogue between Sagredo and Simplicio.) I quote here from the unedited transcript of the hearings.

Mr Fawell: Thank you very much. I appreciate the testimony of all of you. I think it was excellent. If ever I would want to explain to one and all the reasons why the SSC is needed I am sure I can go to your testimony. It would be very helpful. I wish sometimes we have some one word that could say it all and that is kind of impossible. I guess perhaps Dr. Weinberg you came a little close to it and I'm not sure but I took this down. You said you suspect that it isn't all an accident that there are rules which govern matter and I jotted down, will this make us find God? I'm sure you didn't make that claim, but it certainly will enable us to understand so much more about the universe?

Mr Ritter: Will the gentleman yield on that? [That's something congressmen say to each other.] If the gentleman would yield for a moment I would say...

Mr Fawell: I'm not sure I want to.

Mr Ritter: If this machine does that I am going to come round and support it.

Now while this dialogue was going on I thought of a number of marvellous observations that I could make to score points for the SSC. However, by the time Mr Ritter reached his final remark I had decided to keep my mouth shut. And that, my friends, is what I learned about the art of congressional testimony.

## Acknowledgments

I was greatly aided in preparing the talk at Cambridge and thus this article by conversations with G. Holton, H. Mark, E. Mayr and E. Mendelsohn. I am also grateful for helpful comments on an earlier written version by J. Krumhansl, E. L. Goldwasser, E. Mayr, M. Perutz and S. Wojicki.

## Notes

This chapter originally appeared in *Nature* 330 (1987), 433–437.

1. Chanowitz, M. S. presented at the 23rd International Conference on High Energy Physics, Berkeley, California, July 16–23, 1986 (Lawrence Berkeley Laboratory Publication No. 21973).

2. Dixon, B. *The Scientist* June 15, p. 13 (1987).

3. Weinberg, S. *Scientific American* 231, 50 (1974).

4. Mayr, E. in *Evolution at a Crossroads* (ed. Depew, D. J. and Weber, B. H.) (MIT Press, Cambridge, 1985).

5. Mayr, E. *The Growth of Biological Thought* 58–66 (Harvard University Press, Cambridge, 1982).

6. Anderson, P. *Science* 177, 393 (1972).

7. Anderson, P. Lettter to *New York Times* June 8 (1987).

8. Perutz, M. in *Schrödinger, Centenary Celebration of a Polymath* (ed. Kilmister, C. W.) 234 (Cambridge University Press, Cambridge, 1987).

9. Salam, A. in an address to the symposium The Challenge of Higher Energies, Oxford (1982).

10. Michelson, A. A. *Light Waves and Their Uses* (1903).

11. Mark, H. *Navigation* 26, 25 (1979).

12. *Edinburgh Review* 11, 487–532 (1960); see also Hull, D. L. in *Darwin and his Critics* (Harvard University Press, Cambridge, 1973).

# 19 Issues in the Logic of Reductive Explanations

Ernest Nagel

A recurrent theme in the long history of philosophical reflection on science is the contrast—voiced in many ways by poets and scientists as well as philosophers—between the characteristics commonly attributed to things on the basis of everyday enounters with them, and the accounts of those things given by scientific theories that formulate some ostensibly pervasive executive order of nature. This was voiced as early as Democritus, when he declared that while things are customarily said to be sweet or bitter, warm or cold, of one color rather than another, in truth there are only the atoms and the void. The same contrast was implicit in Galileo's distinction, widely accepted by subsequent thinkers, between the primary and secondary qualities of bodies. It was dramatically stated by Sir Arthur Eddington in terms of currently held ideas in physics, when he asked which of the two tables at which he was seated was "really there"—the solid, substantial table of familiar experience, or the insubstantial scientific table which is composed of speeding electric charges and is therefore mostly "emptiness."

Formulations of the contrast vary and have different overtones. In some cases, as in the examples I have cited, the contrast is associated with a distinction between what is allegedly only "appearance" and what is "reality"; and there have been thinkers who have denied that so-called "common-sense" deals with ultimate reality, just as there have been thinkers who have denied that the statements of theoretical science do so. However, a wholesale distinction between appearance and reality has never been clearly drawn, especially since these terms have been so frequently used to single out matters that happen to be regarded as important or valuable; nor have the historical controversies over what is to count as real and what as appearance thrown much light on how scientific theories are related to the familiar materials that are usually the points of departure for scientific inquiry. In any case, the contrast between the more familiar and manifest traits of things and those which scientific theory attributes to them need not be, and often is not, associated with the distinction between the real and the apparent; and in point of fact, most current philosophies of science, which in one way or another occupy themselves with this contrast, make little if any use of that distinction in their analyses.

But despite important differences in the ways in which the contrast has been formulated, I believe they share a common feature and can be construed as being addressed to a common problem. They express the recognition that certain relations of dependence between one set of distinctive traits of a given subject matter are allegedly explained by, and in some sense "reduced" to, assumptions concerning more inclusive relations of dependence between traits or processes not distinctive of (or unique to) that subject matter. They implicitly raise the question of what, in fact, is the logical structure of such reductive explanations—whether they differ from other sorts of scientific explanation, what is achieved by reductions, and under what conditions they are feasible. These questions are important for the understanding of modern science, for its development is marked by strong reductive tendencies, some of whose outstanding achievements are often counted as examples of reduction. For example, as a consequence of this reductive process, the theory of heat is commonly said to be but a branch of Newtonian mechanics, physical optics a branch of electromagnetic theory, and chemical laws a branch of quantum mechanics. Moreover, many biological processes have been given physicochemical explanations, and there is a continuing debate as to the possibility of giving such explanations for the entire domain of biological phenomena. There have been repeated though still unsuccessful attempts to exhibit various patterns of men's social behavior as examples of psychological laws.

It is with some of the issues that have emerged in proposed analyses of reductive explanations that this chapter is concerned. I will first set out in broad outlines what I believe is the general structure of such explanations; then examine some difficulties that have recently been raised against this account.

## 19.1

Although the term *reduction* has come to be widely used in philosophical discussions of science, it has no standard definition. It is therefore not surprising that the term encompasses several sorts of things which need to be distinguished. But before I do this, a brief terminological excursion is desirable. Scientists and philosophers often talk of deducing or inferring one phenomenon from another (e.g., of deducing a planet's orbital motion), of explaining events or their concatenations (e.g., of explaining the occurrence of rainbows), and of reducing certain processes, things, or their properties to others (e.g., of reducing the process of heat conduction to molecular motions). However, these locutions are elliptical, and sometimes lead to misconceptions and confusions. For strictly speaking, it is not phenomena which are deduced from other phenomena, but rather *statements* about phenomena from other statements. This is obvious if we remind ourselves that a given phenomenon can be subsumed under a variety of distinct descriptions, and that phenomena make no assertions or claims. Consequently, until the traits or relations of a phenomenon which are to be discussed are indicated, and predications about them are formulated, it is literally impossible to make any deductions from them. The same holds true for the locutions of explaining or reducing phenomena. I will therefore avoid these elliptic modes of speech hereafter, and talk instead of deducing, explaining, or reducing statements about some subject matter.

Whatever else may be said about reductions in science, it is safe to say that they are commonly taken to be explanations, and I will so regard them. In consequence, I will assume that, like scientific explanations in general, every reduction can be construed as a series of statements, one of which is the conclusion (or the statement which is being reduced), while the others are the premises or reducing statements. Accordingly, reductions can be conveniently classified into two major types: homogeneous reductions, in which all of the "descriptive" or specific subject matter terms in the conclusion are either present in the premises also or can be explicitly defined using only terms that are present; and inhomogeneous reductions, in which at least one descriptive term in the conclusion neither occurs in the premises nor is definable by those that do occur in them. I will now characterize in a general way what I believe to be the main components and the logical structure of these two types of reduction, but will also state and comment upon some of the issues that have been raised by this account of reduction.

A frequently cited example of homogeneous reduction is the explanation by Newtonian mechanics and gravitational theory of various special laws concerning the motions of bodies, including Galileo's law for freely falling bodies near the earth's surface and the Keplerian laws of planetary motion. The explanation is homogeneous, because on the face of it at any rate, the terms occurring in these laws (e.g., distance, time, and acceleration) are also found in the Newtonian theory. Moreover, the explanation is commonly felt to be a reduction of those laws, in part because these laws deal with the motions of bodies in restricted regions of space which had traditionally been regarded as essentially dissimilar from motions elsewhere (e.g., terrestrial as contrasted with celestial motions), while Newtonian theory ignores this traditional classification of spatial regions and incorporates the laws into a unified system. In any event, the reduced statements in this and other standard examples of homogeneous reduction are commonly held to be deduced logically from the reducing premises. In consequence, if the examples can be taken as typical, the formal structure of homogenous reductions is in general that of deductive explanations. Accordingly, if reductions of this type are indeed deductions from theories whose range of application is far more comprehensive and diversified than that of the conclusions derived from them, homogeneous reductions appear to be entirely unproblematic, and to be simply dramatic illustrations of the well understood procedure of deriving theorems from assumed axioms.

However, the assumption that homogeneous reductions are deductive explanations has been recently challenged by a number of thinkers, on the ground that even in the stock illustrations of such reductions the reduced statements do not in general follow from the explanatory premises. For example, while Galileo's law asserts that the acceleration of a freely falling body near the earth's surface is constant, Newtonian theory entails that the acceleration is not constant, but varies with the distance of the falling body from the earth's center of mass. Accordingly, even though the Newtonian conclusion may be "experimentally indistinguishable" from Galileo's law, the latter is in fact "inconsistent" with Newtonian theory. Since it is this theory rather than Galileo's law that was accepted as sound, Galileo's law was therefore *replaced* by a different law for freely falling bodies, namely the law derived from the Newtonian assumptions. A similar outcome holds for Kepler's third planetary law. The general thesis has therefore been advanced that homogeneous reductions do not consist in the deduction or explanation of laws, but in the total *replacement* of incorrect assumptions by radically new ones which are believed to be more correct and precise than those they replace. This thesis raises far-reaching issues, and I will examine some of them presently. But for the moment I will confine my comments on it to questions bearing directly on homogeneous reductions.

a. It is undoubtedly the case that the laws derivable from Newtonian theory do not coincide exactly with some of the previously entertained hypotheses about the motions of bodies, though in other cases there may be such coincidence. This is to be expected. For it is a widely recognized function of comprehensive theories (such as the Newtonian one) to specify the conditions under which antecedently established regularities hold, and to indicate, in the light of those conditions, the modifications that may have to be made in the initial hypotheses, especially if the range of application of the hypotheses is enlarged. Nevertheless, the initial hypotheses may be reasonably close approximations to the consequences entailed by the comprehensive theory, as is indeed the case with Galileo's law as well as with Kepler's third Law. (Incidentally, when Newtonian theory is applied to the motions of just two bodies, the first and second Keplerian laws agree fully with the Newtonian conclusions). But if this is so, it is correct to say that in homogeneous reductions the reduced laws are either derivable from the explanatory premises, or are good approximations to the laws derivable from the latter.

b. Moreover, it is pertinent to note that in actual scientific practice, the derivation of laws from theories usually involves simplifications and approximations of various kinds, so that even the laws which are allegedly entailed by a theory are in general only approximations to what is strictly entailed by it. For example, in deriving the law for the period of a simple pendulum, the following approximative assumptions are made: the weight of the pendulum is taken to be concentrated in the suspended bob; the gravitational force acting on the bob is assumed to be constant, despite variations in the distance of the bob from the earth's center during the pendulum's oscillation;

and since the angle through which the pendulum may oscillate is stipulated to be small, the magnitude of the angle is equated to the sine of the angle. The familiar law that the period of a pendulum is proportional to the square root of its length divided by the constant of acceleration is therefore derivable from Newtonian theory only if these various approximations are taken for granted.

More generally, though no statistical data are available to support the claim, there are relatively few deductions from the mathematically formulated theories of modern physics in which analogous approximations are not made, so that many if not all the laws commonly said by scientists to be deducible from some theory are not strictly entailed by it. It would nevertheless be an exaggeration to assert that in consequence scientists are fundamentally mistaken in claiming to have made such deductions. It is obviously important to note the assumptions, including those concerning approximations, under which the deduction of a law is made. But it does not follow that given those assumptions a purported law cannot count as a consequence of some theory. Nor does it follow that if in a proposed homogeneous reduction the allegedly reduced law is only an approximation to what is actually entailed by the reducing theory when *no* approximative assumptions are added to the latter, the law has not been reduced but is being replaced by a radically different one.

c. Something must also be said about those cases of homogeneous reduction in which the law actually derivable from the reducing theory makes use of concepts not employed in the law to be reduced. Thus, while according to Kepler's third (or harmonic) law, the squares of the periods of the planets are to each other as the cubes of their mean distances from the sun, the Newtonian conclusion is that this ratio is not constant for all the planets but varies with their *masses*. But the notion of mass was introduced into mechanics by Newton, and does not appear in the Keplerian law; and although the masses of the planets are small in comparison with the mass of the sun, and the Keplerian harmonic law is therefore a close approximation to the Newtonian one, the two cannot be equated. Nevertheless, while the two are not equivalent, neither are they radically disparate in content or meaning. On the contrary, the Newtonian law identifies a causal factor in the motions of the planets which was unknown to Kepler.

## 19.2

I must now turn to the second major type of reductive explanations. Inhomogeneous reductions, perhaps more frequently than homogenous ones, have occasioned vigorous controversy among scientists as well as philosophers concerning the cognitive status, interpretation, and function of scientific theories; the relations between the various theoretical entities postulated by these theories, and the familiar things of common experience; and the valid scope of different modes of scientific analysis. These issues are interconnected, and impinge in one way or another upon questions about the general structure of inhomogenous reductions. Since none of the proposed answers to these issues has gained universal assent, the nature of such reductions is still under continuing debate.

Although there are many examples of inhomogeneous reductions in the history of science, they vary in the degree of completeness with which the reduction has been effected. In some instances, all the assumed laws in one branch of inquiry are apparently explained in terms of a theory initially developed for a different class of phenomena; in others, the reduction has been only partial, though the hope of completely reducing the totality of laws in a given area of inquiry to some allegedly "basic" theory may continue to inspire research. Among the most frequently cited illustrations of such relatively complete inhomogenous reductions are the explanation of thermal laws by the kinetic theory of matter, the reduction of physical optics to electromagnetic theory, and the explanation (at least in principle) of chemical laws in terms of quantum theory. On the other hand, while some processes occurring in living organisms can now be understood in terms of physicochemical theory, the reducibility of all biological laws in a similar manner is still a much disputed question.

In any case, the logical structure of inhomogeneous reductive explanations is far less clear and is more difficult to analyze than is the case with homogeneous reductions. The difficulty stems largely from the circumstance that in the former there are (by definition) terms or concepts in the reduced laws (e.g., the notion of heat in thermodynamics, the term "light-wave" in optics, or the concept of valence in chemistry) which are absent from the reducing theories. Accordingly, if the overall structure of the explanation of laws is taken to be that of a deductive argument, it seems impossible to construe inhomogeneous reductions as involving essentially little more than the logical derivation of the reduced laws (even when qualifications about the approximative character of the latter are made) from their explanatory premises. If inhomogeneous reductions are to be subsumed under the general pattern of scientific explanations, it is clear that additional assumptions must be introduced as to how the concepts characteristically employed in the reduced laws, but not present in the reducing theory, are connected with the concepts that do occur in the latter.

Three broad types of proposals for the structure of inhomogeneous reductions can be found in the recent literature of the philosophy of science. The first, which for convenience will be called the "instrumentalist" analysis, is usually advocated by thinkers who deny a cognitive status to scientific laws or theories, regarding them as neither true nor false but as rules (or "inference tickets") for inferring so-called "observation statements" (statements about particular events or occurrences capable of being "observed" in some not precisely defined sense) from other such statements. According to this view, for example, the kinetic theory of gases is not construed as an account of the composition of gases. It is taken to be a complex set of rules for predicting, among other things, what the pressure of a given volume of gas will be if its temperature is kept constant but its volume is diminished. However, the scope of application of a given law or theory may be markedly more limited than the scope of another. The claim that a theory T (e.g., the corpus of rules known as thermodynamics) is reduced to another theory T′ (e.g., the kinetic theory of gases) would therefore be interpreted as saying that all the observation statements which can be derived from given data with the help of T can also be derived with the help of T′, but not conversely. Accordingly, the question to which this account of inhomogeneous reduction is addressed is not the ostensibly asserted content of the theories involved in reduction, but the comparative ranges of observable phenomena to which two theories are applicable.

Although this proposed analysis calls attention to an important function of theories and provides a rationale for the reduction of theories, its adequacy depends on the plausibility of uniformly interpreting general statements in science as rules of inference. Many scientists certainly do not subscribe to such an interpretation, for they frequently talk of laws as true and as providing at least an approximately correct account of various relations of dependence among things. In particular, this interpretation precludes the explanation of macro-states of objects in terms of unobservable micro-processes postulated by a theory. Moreover, the proposal is incomplete in a number of ways: it has nothing to say about how theoretical terms in laws (e.g., "electron" or even "atom") may be used in connection with matters of observation, or just how theories employing such notions operate as rules of inference; and it ignores the question of how, if at all, the concepts of a reduced theory are related to those of the reducing one, or in what way statements about a variety of observable things may fall within the scope of both theories. In consequence, even if the proposed analysis is adequate for a limited class of reductive explanations, it does not do justice to important features characterizing many others.

The second proposed analysis of inhomogeneous reductions (hereafter to be referred to—perhaps misleadingly—as the "correspondence" proposal) is also based on several assumptions. One of them is that the terms occurring in the conclusion but not in the premises of a reduction have "meanings" (i.e., uses and applications) which are determined by the procedures and definitions of the discipline to which reduced laws initially belong, and can be understood without reference to the ideas involved in the theories to which the laws have been reduced. For example, the term "entropy" as used in thermodynamics is defined independently of the notions characterizing statistical mechanics. Furthermore, the assumption is made that many subject matter terms common to both the reduced and reducing theories—in particular, the so-called observation terms employed by both of them to record the outcome of observation and experiment—are defined by procedures which can be specified independently of these theories and, in consequence, have "meanings" that are neutral with respect to the differences between the theories. For example, the terms "pressure" and "volume change" which occur in both thermodynamics and the kinetic theory of gases are used in the two theories in essentially the same sense. It is important to note, however, that this assumption is compatible with the view that even observation terms are "theory impregnated," so that such terms are not simply labels for "bare sense-data," but predicate characteristics that are not immediately manifest and are defined on the basis of various theoretical commitments. For example, if the expression "having a diameter of five inches" is counted as an observation predicate, its application to a given object implicitly involves commitment to some theory of spatial measurement as well as to some laws concerning the instrument used in making the measurement. Accordingly, the point of the assumption is not that there are subject-matter terms whose meanings or uses are independent of *all* theories, but rather that every such term has a meaning which is fixed by *some* theory but independent of others. A third assumption underlying the correspondence analysis of inhomogeneous reductions is that, like homogeneous reduction, and with similar qualifications referring to approximations, they embody the pattern of deductive explanations.

In view of these assumptions, it is clear that if a law (or theory) T is to be reduced to a theory T′ not containing terms occurring in T, T′ must be supplemented by what have been called "rules of correspondence" or "bridge laws," which establish *connections* between the distinctive terms of T and certain terms (or combinations of terms) in T′. For example, since the second law of thermodynamics talks of the transfer of heat, this law cannot be deduced from classical mechanics, which does not contain the term "heat," unless the term is connected in some way with some complex of terms in mechanics. The statement of such a connection is a correspondence rule. However, because of the first of the above three assumptions, a correspondence rule cannot be construed as an explicit definition of a term distinctive of T, which would permit the elimination of the term on *purely logical grounds* in favor of the terms in T′. Thus, the notion of entropy as defined in thermodynamics can be understood and used without any reference to notions employed in theories about the microstructure of matter; and no amount of logical analysis of the concept of entropy can show the concept to be constituted out of the ideas employed in, say, statistical mechanics. If this is indeed the case (as I believe it is), then the theory T is not derivable from (and hence not reducible to) the theory T′, although T may be derivable from T′ when the latter is conjoined with an appropriate set of bridge laws.

What then is the status of the correspondence rules required for inhomogeneous reduction? Different articulations of the theories involved in a reduction, as well as different stages in the development of inquiry into the subject matter of the theories, may require different answers; but I will ignore these complications. In general, however, correspondence rules formulate *empirical hypotheses*—hypotheses which state certain relations of dependence between things mentioned in the reduced and reducing theories. The hypotheses are, for the most part, not testable by confronting them which observed instances of the relations they postulate. They are nevertheless not arbitrary stipulations, and as with many other scientific laws their factual validity must be assessed by comparing various consequences entailed by the system of hypotheses to which they belong with the outcome of controlled observations. However, bridge laws have various forms; and while no exhaustive classification of their structure is available, two sorts of bridge laws must be briefly described.

a. A term in a reduced law may be a predicate which refers to some distinctive *attribute* or characteristic of things (such as the property of having a certain temperature or of being red) that is not denoted by any of the predicates of the reducing theory. In this case the bridge law may specify the conditions, formulated in terms of the ideas and assumptions of the reducing theory, under which the attribute occurs. For example, the kinetic theory of gases formulates its laws in terms of such notions as molecule, mass, and velocity, but does not employ the thermodynamical notion of temperature. However, a familiar bridge law states that a gas has a certain temperature when the mean kinetic energy of its molecules has a certain magnitude. In some cases, bridge laws of the sort being considered may specify conditions for the occurrence of an attribute which are necessary as well as sufficient; in other cases the conditions specified may be sufficient without being necessary; and in still other cases, the conditions stated may only be necessary. In the latter case, however, laws involving the attribute will, in general, not be deducible from the proposed reducing theory. (Thus, though some of the necessary conditions for objects having colors can be stated in terms of ideas belonging to physical optics in its current form, the physiological equipment of organisms which must also be present for the occurrence of colors cannot be described in terms of those ideas. Accordingly, if there are any laws about color relations, they are not reducible to physical optics.)

In any case, such bridge laws are empirical hypotheses concerning the *extensions* of the predicates mentioned in these correspondence rules—that is, concerning the classes of individual things or processes designated by those predicates. An attribute of things connoted by a predicate in a reduced law may indeed be quite different from the attribute connoted by the predicates of the reducing theory; but the class of things possessing the former attribute may nevertheless coincide with (or be included in) the class of things which possess the property specified by a complex predicate in the reducing theory. For example, the statement that a liquid is viscous is not equivalent in meaning to the statement that there are certain frictional forces between the layers of molecules making up the liquid. But if the bridge laws connecting the macroproperties and the microstructure of liquids is correct, the extension of the predicate "viscous" coincides with (or is included in) the class of individual systems with that microstructure.

b. Let me now say something about a second sort of correspondence rule. Although much scientific inquiry is directed toward discovering the determining conditions under which various traits of things occur, some of its important achievements consist in showing that things and processes initially assumed to be distinct are in fact the same. A familiar example of such an achievement is the discovery that the Morning Star and the Evening Star are not different celestial objects but are identical. Similarly, although the term "molecule" designates one class of particles and the term "atom" designates another class, molecules are structures of atoms, and in particular a water molecule is an organization of hydrogen and oxygen atoms denoted by the formula "$H_2O$"; and accordingly, the extension of the predicate "water molecule" is the same as the class of things designated by the formula. Correspondence rules of the second sort establish analogous identifications between classes of individuals or "entities" (such as spatiotemporal objects, processes, and forces) designated by different predicates. An oft cited example of such rules is a bridge law involved in the reduction of physical optics to electromagnetic theory. Thus, prior to Maxwell, physicists postulated the existence of certain physical propagations designated as "light waves," while electromagnetic theory was developed on the assumption that there are electromagnetic waves. An essential step in the reduction of optics to electrodynamics was the introduction by Maxwell of the hypothesis (or bridge law) that these are not two *different* processes but a *single* one, even though electromagnetic waves are not always manifested as visible light. Analogous bridge laws are assumed when a flash of lightning is said to be a surge of electrically charged particles, or when the evaporation of a liquid is asserted to be the escape of molecules from its surface; and while the full details for formulating a similar bridge law are not yet available, the hope of discovering them underlies the claim that a biological cell is a complex organization of physicochemical particles.

Correspondence rules of the second kind thus differ from rules of the first, in that unlike the latter (which state conditions, often in terms of the ideas of a micro-theory, for the occurrence of traits characterizing various things, often macroscopic ones), they assert that certain logically nonequivalent expressions describe identical entities. Although both sorts of rules have a common function in reduction and both are in general empirical assumptions, failure to distinguish between them is perhaps one reason for the persistence of the mistaken belief that reductive explanations establish the "unreality" of those distinctive traits of things mentioned in reduced laws.

## 19.3

This account of inhomogeneous reduction has been challenged by a number of recent writers who have advanced an alternate theory which rejects the main assumptions of both the instrumentalist and the correspondence analyses, and which I will call the "replacement" view. Since I believe the correspondence account to be essentially correct, I shall examine the fundamental contention of the replacement thesis, as presented by Professor Paul Feyerabend, one of its most vigorous proponents.

Feyerabend's views on reduction rest upon the central (and on the face of it, sound) assumption that "the meaning of every term we use depends upon the theoretical context in which it occurs."[1] This claim is made not only for "theoretical" terms like "neutrino" or "entropy" in explicitly formulated scientific theories, but also for expressions like "red" or "table" used to describe matters of common observation (i.e., for observation terms). Indeed, Feyerabend uses the word "theory" in a broad sense, to include such things as myths and political ideas.[2] He says explicitly that "even everyday languages, like languages of highly theoretical systems, have been introduced in order to give expression to some theory or point of view, and they therefore contain a well-developed and sometimes very abstract ontology."[3] "The description of every single fact," he declares, is "dependent on *some* theory."[4] He further maintains that "theories are meaningful independent of observations; observational statements are not meaningful unless they have been connected with theories."[5] There is, therefore, no "observation core," even in statements of perception, that is independent of theoretical interpretation,[6] so that strictly speaking each theory determines its own distinctive set of observation statements. And while he allows that two "low level" theories which fall within the conceptual framework of a comprehensive "background theory" may have a common interpretation for their observation statements, two "high level" theories concerning the nature of the basic elements of the universe "may not share a single observational statement."[7] It is therefore allegedly an error to suppose that the empirical adequacy of a theory can be tested by appeal to observation statements whose meanings are independent of the theory, and which are neutral as between that theory and some alternative competing theory. "The methodological unit to which we must refer when discussing questions of test and empirical context, is constituted by a *whole set of partly overlapping, factually adequate, but mutually inconsistent theories*."[8]

Moreover, a change in a theory is accompanied by a change in the meanings of all its terms, so that theories constructed on "mutually inconsistent principles" are in fact "incommensurable."[9] Thus, if T is classical celestial mechanics, and T′ is the general theory of relativity, "the meanings of all descriptive terms of the two theories, primitive as well as defined terms, will be different," the theories are incommensurable, and "not a single descriptive term of T can be incorporated into T′."[10] In consequence, Feyerabend believes the correspondence account of inhomogeneous reduction is basically mistaken in supposing that allegedly reduced laws or theories can be derived from the reducing theory with the help of appropriate bridge laws:

What happens . . . when transition is made from a theory T′ to a wider theory T (which, we shall assume, is capable of covering all the phenomena that have been covered by T′) is something

> much more radical than incorporation of the *unchanged* theory T′ (unchanged, that is, with respect to the meanings of its main descriptive terms as well as to the meanings of the terms of its observation language) into the context of T. What does happen is, rather, a *complete replacement* of the ontology (and perhaps even of the formalism) of T′ by the ontology (and the formalism) of T and a corresponding change of the meanings of the descriptive elements of the formalism of T′ (provided these elements and this formalism are still used). This replacement affects not only the theoretical terms of T′ but also at least some of the observational terms which occurred in its test statements.... In short: introducing a new theory involves changes of outlook both with respect to the observable and with respect to the unobservable features of the world, and corresponding changes in the meaning of even the most "fundamental" terms of the language employed.[11]

Accordingly, if these various claims are warranted, there is not and cannot be any such thing as the reduction of laws or theories; and the examples often cited as instances of reduction are in fact instances of something else: the exclusion of previously accepted hypotheses from the corpus of alleged scientific knowledge, and the substitution for them of incommensurably different ones.

But are these claims warranted? I do not believe they are. Feyerabend is patently sound in maintaining that no single statement or any of its constituent terms has a meaning in isolation, or independently of various rules or conventions governing its use. He is no less sound in noting that the meaning of a word may change when its range of application is altered. However, these familiar truisms do not support the major conclusion he draws from them. The presentation of his thesis suffers from a number of unclarities (such as what is to count as a change in a theory, or what are the criteria for changes in meaning), which cloud the precise import of some of his assertions. I shall, however, ignore these unclarities here[12] and will comment briefly only on two difficulties in Feyerabend's argument.

a. It is a major task of scientific inquiry to assess the adequacy of proposed laws to the "facts" of a subject matter as established by observation or experiment, and to ascertain whether the conclusions reached are consistent with one another. However, if two proposed theories for some given range of phenomena share no term with the same meaning in each of them, so that the theories have completely different meanings (as Feyerabend believes is commonly the case), it is not evident in what sense two such theories can be said to be either compatible or inconsistent with one another: for relations of logical opposition obtain only between statements whose terms have common meanings. Moreover, it is also difficult to understand how, if the content of observation statements is determined by the theory which is being tested (as Feyerabend maintains), those statements can serve as a basis for deciding between the theory and some alternative to it. For according to his analysis those observation statements will automatically corroborate the theory that happens to be used to interpret observational data, but will be simply irrelevant in assessing the empirical validity of an alternative theory. Theories thus appear to be self-certifying, and to be beyond the reach of criticism based on considerations that do not presuppose them. This outcome is reminiscent of Karl Mannheim's claim that truth in social matters is "historically relative": there are no universally valid analyses of social phenomena, since every such analysis is made within some distinctive social perspective which determines the meaning as well as the validity of what is said to be observed, so that those who do not share the same perspective can neither reach common conclusions about human affairs, nor significantly criticize each others' findings.

Feyerabend attempts to escape from such skeptical relativism by involving what he calls the "pragmatic theory of observation." In this theory, it is still the case that the meaning of an observation statement varies with the theory used to interpret observations. However, it is possible to describe the observational and predictive statements an investigator utters as *responses* to the situations which "prompt" the utterances, and to compare the order of these responses with the order of the physical situations that prompt them, so as to ascertain the agreements or disagreements between the two orders.[13] But if this account of the role of observation statements in testing theories is to outflank the relativism Feyerabend wants to avoid, the *secondary* statements (they are clearly observation statements) about the responses (or primary observation statements) of investigators cannot have meanings dependent on the theory being tested, and must be invariant to alternative theories. However, if secondary statements have this sort of neutrality, it is not evident why only such observation statements can have this privileged status.

b. Feyerabend has difficulties in providing a firm observational basis for objectively evaluating the empirical worth of proposed hypotheses. The difficulties stem from what I believe is his exaggerated view that the meaning of every term occurring in a theory or in its observation statements is wholly and uniquely determined by that theory, so that its meaning is radically changed when the theory is modified. For theories are not quite the monolithic structures he takes them to be—their component assumptions are, in general, logically independent of one another, and their terms have varying degrees of dependence on the theories into which they enter. Some terms may indeed be so deeply embedded in the totality of assumptions constituting a particular theory that they can be understood only within the framework of the theory: e.g., the meaning of "electron spin" appears to be inextricably intertwined with the characteristic ideas of quantum theory. On the other hand, there are also terms whose meanings seem to be invariant in a number of different theories: e.g., the term "electric charge" is used in currently accepted theories of atomic structure in the same sense as in the earlier theories of Rutherford and Bohr. Similar comments apply to observation terms, however these may be specified. Accordingly, although both "theoretical" and "observational" terms may be "theory laden," it does not follow that there can be no term in a theory which retains its meaning when it is transplanted into some other theory.

More generally, it is not clear how, on the replacement view of reduction, a theory T can be at the same time more inclusive than, and also have a meaning totally different from, the theory T′ it allegedly replaces—especially since according to Feyerabend the replacing theory will entail "that all the concepts of the preceding theory have extension zero, or ... it introduces rules which cannot be interpreted as attributing specific properties to objects within already existing classes, but which change the system of classes itself."[14] Admittedly, some of the laws and concepts of the "wider theory" often differ from their opposite numbers in the earlier theory. But even in this case, the contrasted items may not be "incommensurable." Thus, the periodic table classifies chemical elements on the basis of certain patterns of similarity between the properties of the elements. The description (or theoretical explanation) of those properties has undergone important changes since the periodic table was first introduced by Mendeleev. Nevertheless, though the descriptions differ, the classification of the elements has remained fairly stable, so that fluorine, chlorine, bromine, and iodine, for example, continue to be included in the same class. The new theories used in formulating the classification certainly do not entail that the concepts of the preceding ones have zero extension. But it would be difficult to understand why this is so if, because of differences between the descriptions, the descriptions were totally disparate.

Consider, for example, the argument that thermodynamics is not reducible to statistical mechanics, on the ground that (among other reasons) entropy is a statistical notion in the latter theory but not in the former one: since the meaning of the word "entropy" differs in the two theories, entropy laws in statistical mechanics are not derivable from entropy laws in thermodynamics (and in fact are said to be incompatible). Admittedly, the connotation of the word "entropy" in each of the two theories is not identical; and if the correspondence account of reduction were to claim that they are the same, it would be patently mistaken. But the fact remains that the two theories deal with many phenomena common to both their ranges; and the question is how is this possible? In brief, the answer seems to be as follows. The word "entropy" in thermodynamics is so defined that its legitimate application is limited to physical systems satisfying certain specified conditions, e.g., to systems such as gases, whose internal motions are not too "tumultuous" (the word is Planck's), a condition which is not satisfied in the case of Brownian motions. These conditions are relaxed in the definition of "entropy" in statistical mechanics, so that the extension of the Boltzmann notion of entropy includes the extension of the Clausius notion. In consequence, despite differences in the connotations of the two definitions, the theories within which they are formulated have a domain of application in common, even though the class of systems for which thermodynamical laws are approximately valid is more restricted than is the class for the laws of statistical mechanics. But it is surely not the case that the latter theory implies that the Clausius definition of entropy has a zero extension or that the laws of thermodynamics are valid for no physical systems whatsoever.

This difficulty of the replacement view in explaining how the "wider" theory, which allegedly replaces a "narrower" one, may nevertheless have a domain of common application, does not arise in the correspondence account of reduction. For the bridge laws upon which the latter sets great store are empirical hypotheses, not logically true statements in virtue of the connotations of the terms contained in them. Bridge laws state what relations presumably obtain between the *extensions* of their terms, so that in favorable cases laws of the "narrower" theory (with suitable qualifications about their approximate character) can be deduced from the "wider" theory, and thereby make intelligible why the two theories may have a common field of application. Accordingly, although I will not pretend that the correspondence account of reduction is free from difficulties or that I have resolved all of them, on the whole it is a more adequate analysis than any available alternative to it.

## Notes

This chapter originally appeared as chapter 6 in *Teleology Revisited and Other Essays in the Philosophy and History of Science*, pp. 95–113, 320–321, New York, Columbia University Press, 1979. Section 19.4, with its notes on the possibilities of reducing biology to physics and chemistry, has been omitted.

1. Paul Feyerabend, "Problems of Empiricism," in R. G. Colodny, ed., *Beyond the Edge of Certainty* (Englewood Cliffs: Prentice-Hall, Inc., 1965), p. 180.

2. Paul Feyerabend, "Reply to Criticism," *Boston Studies in the Philosophy of Science* 2 (1962), p. 252.

3. Paul Feyerabend, "Explanation, Reduction, and Empiricism," *Minnesota Studies in the Philosophy of Science* 3 (1962), p. 76.

4. Feyerabend, "Problems of Empiricism," p. 175.

5. Ibid., p. 213.

6. Ibid., p. 216.

7. Ibid.

8. Ibid., p. 175.

9. Ibid., p. 227.

10. *Boston Studies in the Philosophy of Science* p. 231; cf. also Feyerabend, "On the 'Meaning' of Scientific Terms," *Journal of Philosophy* 62 (1965), p. 271.

11. Feyerabend, "Explanation, Reduction and Empiricism," pp. 28–9, 59.

12. Many of them are noted by Dudley Shapere in his "Meaning and Scientific Change," in R. G. Colodny, ed., *Mind and Cosmos* (Pittsburgh: University of Pittsburgh Press, 1966).

13. Feyerabend, "Problems of Empiricism," p. 21; and Feyerabend, "Explanation, Reduction, and Empiricism," p. 24.

14. Feyerabend, "On the 'Meaning' of Scientific Terms," *Journal of Philosophy* 62 (1965), p. 268.

# 20 Chaos

James P. Crutchfield, J. Doyne Farmer, Norman H. Packard, and Robert S. Shaw

The great power of science lies in the ability to relate cause and effect. On the basis of the laws of gravitation, for example, eclipses can be predicted thousands of years in advance. There are other natural phenomena that are not as predictable. Although the movements of the atmosphere obey the laws of physics just as much as the movements of the planets do, weather forecasts are still stated in terms of probabilities. The weather, the flow of a mountain stream, the roll of the dice all have unpredictable aspects. Since there is no clear relation between cause and effect, such phenomena are said to have random elements. Yet until recently there was little reason to doubt that precise predictability could in principle be achieved. It was assumed that it was only necessary to gather and process a sufficient amount of information.

Such a viewpoint has been altered by a striking discovery: simple deterministic systems with only a few elements can generate random behavior. The randomness is fundamental; gathering more information does not make it go away. Randomness generated in this way has come to be called chaos.

A seeming paradox is that chaos is deterministic, generated by fixed rules that do not themselves involve any elements of chance. In principle the future is completely determined by the past, but in practice small uncertainties are amplified, so that even though the behavior is predictable in the short term, it is unpredictable in the long term. There is order in chaos: underlying chaotic behavior there are elegant geometric forms that create randomness in the same way as a card dealer shuffles a deck of cards or a blender mixes cake batter.

The discovery of chaos has created a new paradigm in scientific modeling. On one hand, it implies new fundamental limits on the ability to make predictions. On the other hand, the determinism inherent in chaos implies that many random phenomena are more predictable than had been thought. Random-looking information gathered in the past—and shelved because it was assumed to be too complicated—can now be explained in terms of simple laws. Chaos allows order to be found in such diverse systems as the atmosphere, dripping faucets and the heart. The result is a revolution that is affecting many different branches of science.

What are the origins of random behavior? Brownian motion provides a classic example of randomness. A speck of dust observed through a microscope is seen to move in a continuous and erratic jiggle. This is owing to the bombardment of the dust particle by the surrounding water molecules in thermal motion. Because the water molecules are unseen and exist in great number, the detailed motion of the dust particle is thoroughly unpredictable. Here the web of causal influences among the subunits can become so tangled that the resulting pattern of behavior becomes quite random.

The chaos to be discussed here requires no large number of subunits or unseen influences. The existence of random behavior in very simple systems motivates a reexamination of the sources of randomness even in large systems such as weather.

What makes the motion of the atmosphere so much harder to anticipate than the motion of the solar system? Both are made up of many parts, and both are governed by Newton's second law, $F = ma$, which can be viewed as a simple prescription for predicting the future. If the forces $F$ acting on a given mass $m$ are known, then so is the acceleration $a$. It then follows from the rules of calculus that if the position and velocity of an object can be measured at a given instant, they are determined forever. This is such a powerful idea that the 18th-century French mathematician Pierre Simon de Laplace once boasted that given the position and velocity of every particle in the universe, he could predict the future for the rest of time. Although there are several obvious practical difficulties to achieving Laplace's goal, for more than 100 years there seemed to be no reason for his not being right, at least in principle. The literal application of Laplace's dictum to human behavior led to the philosophical conclusion that human behavior was completely predetermined: free will did not exist.

Twentieth-century science has seen the downfall of Laplacian determinism, for two very different reasons. The first reason is quantum mechanics. A central dogma of that theory is the Heisenberg uncertainty principle, which states that there is a fundamental limitation to the accuracy with which the position and velocity of a particle can be measured. Such uncertainty gives a good explanation for some random phenomena, such as radioactive decay. A nucleus is so small that the uncertainty principle puts a fundamental limit on the knowledge of its motion, and so it is impossible to gather enough information to predict when it will disintegrate.

The source of unpredictability on a large scale must be sought elsewhere, however. Some large-scale phenomena are predictable and others are not. The distinction has nothing to do with quantum mechanics. The trajectory of a baseball, for example, is inherently predictable; a fielder intuitively makes use of the fact every time he or she catches the ball. The trajectory of a flying balloon with the air rushing out of it, in contrast, is not predictable; the balloon lurches and turns erratically at times and places that are impossible to predict. The balloon obeys Newton's laws just as much as the baseball does; then why is its behavior so much harder to predict than that of the ball?

The classic example of such a dichotomy is fluid motion. Under some circumstances the motion of a fluid is laminar—even, steady and regular—and easily predicted from equations. Under other circumstances fluid motion is turbulent—uneven, unsteady and irregular—and difficult to predict. The transition from laminar to turbulent behavior is familiar to anyone who has been in an airplane in calm weather and then suddenly encountered a thunderstorm. What causes the essential difference between laminar and turbulent motion?

To understand fully why that is such a riddle, imagine sitting by a mountain stream. The water swirls and splashes as though it had a mind of its own, moving first one way and then another. Nevertheless, the rocks in the stream bed are firmly fixed in place, and the tributaries enter at a nearly constant rate of flow. Where, then, does the random motion of the water come from?

The late Soviet physicist Lev D. Landau is credited with an explanation of random fluid motion that held sway for many years, namely that the motion of a turbulent fluid contains many different, independent oscillations. As the fluid is made to move faster, causing it to become more turbulent, the oscillations enter the motion one at a time. Although each separate oscillation may be simple, the complicated combined motion renders the flow impossible to predict.

Landau's theory has been disproved, however. Random behavior occurs even in very simple systems, without any need for complication or indeterminacy. The French mathematician Henri Poincaré realized this at the turn of the century when he noted that unpredictable, "fortuitous" phenomena may occur in systems where a small change in the present causes a much larger change in the future. The notion is clear if one thinks of a rock poised at the top of a hill. A tiny push one way or another is enough to send it tumbling down widely differing paths. Although the rock is sensitive to small influences only at the top of the hill, chaotic systems are sensitive at every point in their motion.

A simple example serves to illustrate just how sensitive some physical systems can be to external influences. Imagine a game of billiards, somewhat idealized so that the balls move across the table and collide with a negligible loss of energy. With a single shot the billiard player sends the collection of balls into a protracted sequence of collisions. The player naturally wants to know the effects of the shot. For how long could a player with perfect control over his or her stroke predict the cue ball's trajectory? If the player ignored an effect even as minuscule as the gravitational attraction of an electron at the edge of the galaxy, the prediction would become wrong after one minute!

The large growth in uncertainty comes about because the balls are curved, and small differences at the point of impact are amplified with each collision. The amplification is exponential: it is compounded at every collision, like the successive reproduction of bacteria with unlimited space and food. Any effect, no matter how small, quickly reaches macroscopic proportions. That is one of the basic properties of chaos.

It is the exponential amplification of errors due to chaotic dynamics that provides the second reason for Laplace's undoing. Quantum mechanics implies that initial measurements are always uncertain, and chaos ensures that the uncertainties will quickly overwhelm the ability to make predictions. Without chaos Laplace might have hoped that errors would remain bounded, or at least grow slowly enough to allow him to make predictions over a long period. With chaos, predictions are rapidly doomed to gross inaccuracy.

The larger framework that chaos emerges from is the so-called theory of dynamical systems. A dynamical system consists of two parts: the notions of a state (the essential information about a system) and a dynamic (a rule that describes how the state evolves with time). The evolution can be visualized in a state space, an abstract construct whose coordinates are the components of the state. In general the coordinates of the state space vary with the context; for a mechanical system they might be position and velocity, but for an ecological model they might be the populations of different species.

A good example of a dynamical system is found in the simple pendulum. All that is needed to determine its motion are two variables: position and velocity. The state is thus a point in a plane, whose coordinates are position and velocity. Newton's laws provide a rule, expressed mathematically as a differential equation, that describes how the state evolves. As the pendulum swings back and forth the state moves along an "orbit," or path, in the plane. In the ideal case of a frictionless pendulum the orbit is a loop; failing that, the orbit spirals to a point as the pendulum comes to rest.

A dynamical system's temporal evolution may happen in either continuous time or in discrete time. The former is called a flow, the latter a mapping. A pendulum moves continuously from one state to another, and so it is described by a continuous-time flow. The number of insects born each year in a specific area and the time interval between drops from a dripping faucet are more naturally described by a discrete-time mapping.

To find how a system evolves from a given initial state one can employ the dynamic (equations of motion) to move incrementally along an orbit. This method of deducing the system's behavior requires computational effort proportional to the desired length of time to follow the orbit. For simple systems such as a frictionless pendulum the equations of motion may occasionally have a closed-form solution, which is a formula that expresses any future state in terms of the initial state. A closed-form solution provides a short cut, a simpler algorithm that needs only the initial state and the final time to predict the future without stepping through intermediate states. With such a solution the algorithmic effort required to follow the motion of the system is roughly independent of the time desired. Given the equations of planetary and lunar motion and the earth's and moon's positions and velocities, for instance, eclipses may be predicted years in advance.

Success in finding closed-form solutions for a variety of simple systems during the early development of physics led to the hope that such solutions exist for any mechanical system. Unfortunately, it is now known that this is not true in general. The unpredictable behavior of chaotic dynamical systems cannot be expressed in a closed-form solution. Consequently there are no possible short cuts to predicting their behavior.

The state space nonetheless provides a powerful tool for describing the behavior of chaotic systems. The usefulness of the state-space picture lies in the ability to represent behavior in geometric form. For example, a pendulum that moves with friction eventually comes to a halt, which in the state space means the orbit approaches a point. The point does not move—it is a fixed point—and since it attracts nearby orbits, it is known as an attractor. If the pendulum is given a small push, it returns to the same fixed-point attractor. Any system that comes to rest with the passage of time can be characterized by a fixed point in state space. This is an example of a very general phenomenon, where losses due to friction or viscosity, for example, cause orbits to be attracted to a smaller region of the state space with lower dimension. Any such region is called an attractor. Roughly speaking, an attractor is what the behavior of a system settles down to, or is attracted to.

Some systems do not come to rest in the long term but instead cycle periodically through a sequence of states. An example is the pendulum clock, in which energy lost to friction is replaced by a mainspring or weights. The pendulum repeats the same motion over and over again. In the state space such a motion corresponds to a cycle, or periodic orbit. No matter how the pendulum is set swinging, the cycle approached in the long-term limit is the same. Such attractors are therefore called limit cycles. Another familiar system with a limit-cycle attractor is the heart.

A system may have several attractors. If that is the case, different initial conditions may evolve to different attractors. The set of points that evolve to an attractor is called its basin of attraction. The pendulum clock has two such basins: small displacements of the pendulum from its rest position result in a return to rest; with large displacements, however, the clock begins to tick as the pendulum executes a stable oscillation.

The next most complicated form of attractor is a torus, which resembles the surface of a doughnut. This shape describes motion made up of two independent oscillations, sometimes called quasi-periodic motion. (Physical examples can be constructed from driven electrical oscillators.) The orbit winds around the torus in state space, one frequency determined by how fast the orbit circles the doughnut in the short direction, the other regulated by how fast the orbit circles the long way around. Attractors may also be higher-dimensional tori, since they represent the combination of more than two oscillations.

The important feature of quasi-periodic motion is that in spite of its complexity it is predictable. Even though the orbit may never exactly repeat itself, if the frequencies that make up the motion have no common divisor, the motion remains regular. Orbits that start on the torus near one another remain near one another, and long-term predictability is guaranteed.

Until fairly recently, fixed points, limit cycles and tori were the only known attractors. In 1963 Edward N. Lorenz of the Massachusetts Institute of Technology discovered a concrete example of a low-dimensional system that displayed complex behavior. Motivated by the desire to understand the unpredictability of the weather, he began with the equations of motion for fluid flow (the atmosphere can be considered a fluid), and by simplifying them he obtained a system that had just three degrees of freedom. Nevertheless, the system behaved in an apparently random fashion that could not be adequately characterized by any of the three attractors then known. The attractor he observed, which is now known as the Lorenz attractor, was the first example of a chaotic, or strange, attractor.

Employing a digital computer to simulate his simple model, Lorenz elucidated the basic mechanism responsible for the randomness he observed: microscopic perturbations are amplified to affect macroscopic behavior. Two orbits with nearby initial conditions diverge exponentially fast and so stay close together for only a short time. The situation is qualitatively different for nonchaotic attractors. For these, nearby orbits stay close to one another, small errors remain bounded and the behavior is predictable.

The key to understanding chaotic behavior lies in understanding a simple stretching and folding operation, which takes place in the state space. Exponential divergence is a local feature: because attractors have finite size, two orbits on a chaotic attractor cannot diverge exponentially forever. Consequently the attractor must fold over onto itself. Although orbits diverge and follow increasingly different paths, they eventually must pass close to one another again. The orbits on a chaotic attractor are shuffled by this process, much as a deck of cards is shuffled by a dealer. The randomness of the chaotic orbits is the result of the shuffling process. The process of stretching and folding happens repeatedly, creating folds within folds ad infinitum. A chaotic attractor is, in other words, a fractal: an object that reveals more detail as it is increasingly magnified.

Chaos mixes the orbits in state space in precisely the same way as a baker mixes bread dough by kneading it. One can imagine what happens to nearby trajectories on a chaotic attractor by placing a drop of blue food coloring in the dough. The kneading is a combination of two actions: rolling out the dough, in which the food coloring is spread out, and folding the dough over. At first the blob of food coloring simply gets longer, but eventually it is folded, and after considerable time the blob is stretched and refolded many times. On close inspection the dough consists of many layers of alternating blue and white. After only 20 steps the initial blob has been stretched to more than a million times its original length, and its thickness has shrunk to the molecular level. The blue dye is thoroughly mixed with the dough. Chaos works the same way, except that instead of mixing dough it mixes the state space. Inspired by this picture of mixing, Otto E. Rössler of the University of Tübingen created the simplest example of a chaotic attractor in a flow.

When observations are made on a physical system, it is impossible to specify the state of the system exactly owing to the inevitable errors in measurement. Instead the state of the system is located not at a single point but rather within a small region of state space. Although quantum uncertainty sets the ultimate size of the region, in practice different kinds of noise limit measurement precision by introducing substantially larger errors. The small region specified by a measurement is analogous to the blob of blue dye in the dough.

Locating the system in a small region of state space by carrying out a measurement yields a certain amount of information about the system. The more accurate the measurement is, the more knowledge an observer gains about the system's state. Conversely, the larger the region, the more uncertain the observer. Since nearby points in nonchaotic systems stay close as they evolve in time, a measurement provides a certain amount of information that is preserved with time. This is exactly the sense in which such systems are predictable: initial measurements contain information that can be used to predict future behavior. In other words, predictable dynamical systems are not particularly sensitive to measurement errors.

The stretching and folding operation of a chaotic attractor systematically removes the initial information and replaces it with new information: the stretch makes small-scale uncertainties larger, the fold brings widely separated trajectories together and erases large-scale information. Thus chaotic attractors act as a kind of pump bringing microscopic fluctuations up to a macroscopic expression. In this light it is clear that no exact solution, no short cut to tell the future, can exist. After a brief time interval the uncertainty specified by the initial measurement covers the entire attractor and all predictive power is lost: there is simply no causal connection between past and future.

Chaotic attractors function locally as noise amplifiers. A small fluctuation due perhaps to thermal noise will cause a large deflection in the orbit position soon afterward. But there is an important sense in which chaotic attractors differ from simple noise amplifiers. Because the stretching and folding operation is assumed to be repetitive and continuous, any tiny fluctuation will eventually dominate the motion, and the qualitative behavior is independent of noise level. Hence chaotic systems cannot directly be "quieted," by lowering the temperature, for example. Chaotic systems generate randomness on their own without the need for any external random inputs. Random behavior comes from more than just the amplification of errors and the loss of the ability to predict; it is due to the complex orbits generated by stretching and folding.

It should be noted that chaotic as well as nonchaotic behavior can occur in dissipationless, energy-conserving systems. Here orbits do not relax onto an attractor but instead are confined to an energy surface. Dissipation is, however, important in many if not most real-world systems, and one can expect the concept of attractor to be generally useful.

Low-dimensional chaotic attractors open a new realm of dynamical systems theory, but the question remains of whether they are relevant to randomness observed in physical systems. The first experimental evidence supporting the hypothesis that chaotic attractors underlie random motion in fluid flow was rather indirect. The experiment was done in 1974 by Jerry P. Gollub of Haverford College and Harry L. Swinney of the University of Texas at Austin. The evidence was indirect because the investigators focused not on the attractor itself but rather on statistical properties characterizing the attractor.

The system they examined was a Couette cell, which consists of two concentric cylinders. The space between the cylinders is filled with a fluid, and one or both cylinders are rotated with a fixed angular velocity. As the angular velocity increases, the fluid shows progressively more complex flow patterns, with a complicated time dependence.

Gollub and Swinney essentially measured the velocity of the fluid at a given spot. As they increased the rotation rate, they observed transitions from a velocity that is constant in time to a periodically varying velocity and finally to an aperiodically varying velocity. The transition to aperiodic motion was the focus of the experiment.

The experiment was designed to distinguish between two theoretical pictures that predicted different scenarios for the behavior of the fluid as the rotation rate of the fluid was varied. The Landau picture of random fluid motion predicted that an ever higher number of independent fluid oscillations should be excited as the rotation rate is increased. The associated attractor would be a high-dimensional torus. The Landau picture had been challenged by David Ruelle of the Institut des Hautes Études Scientifiques near Paris and Floris Takens of the University of Groningen in the Netherlands. They gave mathematical arguments suggesting that the attractor associated with the Landau picture would not be likely to occur in fluid motion. Instead their results suggested that any possible high-dimensional tori might give way to a chaotic attractor, as originally postulated by Lorenz.

Gollub and Swinney found that for low rates of rotation the flow of the fluid did not change in time: the underlying attractor was a fixed point. As the rotation was increased the water began to oscillate with one independent frequency, corresponding to a limit-cycle attractor (a periodic orbit), and as the rotation was increased still further the oscillation took on two independent frequencies, corresponding to a two-dimensional torus attractor. Landau's theory predicted that as the rotation rate was further increased the pattern would continue: more distinct frequencies would gradually appear. Instead, at a critical rotation rate a continuous range of frequencies suddenly appeared. Such an observation was consistent with Lorenz' "deterministic nonperiodic flow," lending credence to his idea that chaotic attractors underlie fluid turbulence.

Although the analysis of Gollub and Swinney bolstered the notion that chaotic attractors might underlie some random motion in fluid flow, their work was by no means conclusive. One would like to explicitly demonstrate the existence in experimental data of a simple chaotic attractor. Typically, however, an experiment does not record all facets of a system but only a few. Gollub and Swinney could not record, for example, the entire Couette flow but only the fluid velocity at a single point. The task of the investigator is to "reconstruct" the attractor from the limited data. Clearly that cannot always be done; if the attractor is too complicated, something will be lost. In some cases, however, it is possible to reconstruct the dynamics on the basic of limited data.

A technique introduced by us and put on a firm mathematical foundation by Takens made it possible to reconstruct a state space and look for chaotic attractors. The basic idea is that the evolution of any single component of a system is determined by the other components with which it interacts. Information about the relevant components is thus implicitly contained in the history of any single component. To reconstruct an "equivalent" state space, one simply looks at a single component and treats the measured values at fixed time delays (one second ago, two seconds ago and so on, for example) as though they were new dimensions.

The delayed values can be viewed as new coordinates, defining a single point in a multidimensional state space. Repeating the procedure and taking delays relative to different times generates many such points. One can then use other techniques to test whether or not these points lie on a chaotic attractor. Although this representation is in many respects arbitrary, it turns out that the important properties of an attractor are preserved by it and do not depend on the details of how the reconstruction is done.

The example we shall use to illustrate the technique has the advantage of being familiar and accessible to nearly everyone. Most people are aware of the periodic pattern of drops emerging from a dripping faucet. The time between successive drops can be quite regular, and more than one insomniac has been kept awake waiting for the next drop to fall. Less familiar is the behavior of a faucet at a somewhat higher flow rate. One can often find a regime where the drops, while still falling separately, fall in a never repeating patter, like an infinitely inventive drummer. (This is an experiment easily carried out personally; the faucets without the little screens work best.) The changes between periodic and random-seeming patterns are reminiscent of the transition between laminar and turbulent fluid flow. Could a simple chaotic attractor underlie this randomness?

The experimental study of a dripping faucet was done at the University of California at Santa Cruz by one of us (Shaw) in collaboration with Peter L. Scott, Stephen C. Pope and Philip J. Martein. The first form of the experiment consisted in allowing the drops from an ordinary faucet to fall on a microphone and measuring the time intervals between the resulting sound pulses. . . .

By plotting the time intervals between drops in pairs, one effectively takes a cross section of the underlying attractor. In the periodic regime, for example, the meniscus where the drops are detaching is moving in a smooth, repetitive manner, which could be represented by a limit cycle in the state space. But this smooth motion is inaccessible in the actual experiment; all that is recorded is the time intervals between the breaking off of the individual drops. This is like applying a stroboscopic light to regular motion around a loop. If the timing is right, one sees only a fixed point.

The exciting result of the experiment was that chaotic attractors were indeed found in the nonperiodic regime of the dripping faucet. It could have been the case that the randomness of the drops was due to unseen influences, such as small vibrations or air currents. If that was so, there would be no particular relation between one interval and the next, and the plot of the data taken in pairs would have shown only a featureless blob. The fact that any structure at all appears in the plots shows the randomness has a deterministic underpinning. In particular, many data sets show the horseshoelike shape that is the signature of the simple stretching and folding process discussed above. The characteristic shape can be thought of as a "snapshot" of a fold in progress, for example, a cross section partway around the Rössler attractor.

Other data sets seem more complicated; these may be cross sections of higher-dimensional attractors. The geometry of attractors above three dimensions is almost completely unknown at this time.

If a system is chaotic, how chaotic is it? A measure of chaos is the "entropy" of the motion, which roughly speaking is the average rate of stretching and folding, or the average rate at which information is produced. Another statistic is the "dimension" of the attractor. If a system is simple, its behavior should be described by a low-dimensional attractor in the state space, such as the examples given in this article. Several numbers may be required to specify the state of a more complicated system, and its corresponding attractor would therefore be higher-dimensional.

The technique of reconstruction, combined with measurements of entropy and dimension, makes it possible to reexamine the fluid flow originally studied by Gollub and Swinney. This was done by members of Swinney's group in collaboration with two of us (Crutchfield and Farmer). The reconstruction technique enabled us to make images of the underlying attractor. The images do not give the striking demonstration of a low-dimensional attractor that studies of other systems, such as the dripping faucet, do. Measurements of the entropy and dimension reveal, however, that irregular fluid motion near the transition in Couette flow can be described by chaotic attractors. As the rotation rate of the Couette cell increases so do the entropy and dimension of the underlying attractors.

In the past few years a growing number of systems have been shown to exhibit randomness due to a simple chaotic attractor. Among them are the convection pattern of fluid heated in a small box, oscillating concentration levels in a stirred-chemical reaction, the beating of chicken-heart cells and a large number of electrical and mechanical oscillators. In addition computer models of phenomena ranging from epidemics to the electrical activity of a nerve cell to stellar oscillations have been shown to possess this simple type of randomness. There are even experiments now under way that are searching for chaos in areas as disparate as brain waves and economics.

It should be emphasized, however, that chaos theory is far from a panacea. Many degrees of freedom can also make for complicated motions that are effectively random. Even though a given system may be known to be chaotic, the fact alone does not reveal very much. A good example is molecules bouncing off one another in a gas. Although such a system is known to be chaotic, that in itself does not make prediction of its behavior easier. So many particles are involved that all that can be hoped for is a statistical description, and the essential statistical properties can be derived without taking chaos into account.

There are other uncharted questions for which the role of chaos is unknown. What of constantly changing patterns that are spatially extended, such as the dunes of the Sahara and fully developed turbulence? It is not clear whether complex spatial patterns can be usefully described by a single attractor in a single state space. Perhaps, though, experience with the simplest attractors can serve as a guide to a more advanced picture, which may involve entire assemblages of spatially mobile deterministic forms akin to chaotic attractors.

The existence of chaos affects the scientific method itself. The classic approach to verifying a theory is to make predictions and test them against experimental data. If the phenomena are chaotic, however, long-term predictions are intrinsically impossible. This has to be taken into account in judging the merits of the theory. The process of verifying a theory thus becomes a much more delicate operation, relying on statistical and geometric properties rather than on detailed prediction.

Chaos brings a new challenge to the reductionist view that a system can be understood by breaking it down and studying each piece. This view has been prevalent in science in part because there are so many systems for which the behavior of the whole is indeed the sum of its parts. Chaos demonstrates, however, that a system can have complicated behavior that emerges as a consequence of simple, nonlinear interaction of only a few components.

The problem is becoming acute in a wide range of scientific disciplines, from describing microscopic physics to modeling macroscopic behavior of biological organisms. The ability to obtain detailed knowledge of a system's structure has undergone a tremendous advance in recent years, but the ability to integrate this knowledge has been stymied by the lack of a proper conceptual framework within which to describe qualitative behavior. For example, even with a complete map of the nervous system of a simple organism, such as the nematode studied by Sidney Brenner of the University of Cambridge, the organism's behavior cannot be deduced. Similarly, the hope that physics could be complete with an increasingly detailed understanding of fundamental physical forces and constituents is unfounded. The interaction of components on one scale can lead to complex global behavior on a larger scale that in general cannot be deduced from knowledge of the individual components.

Chaos is often seen in terms of the limitations it implies, such as lack of predictability. Nature may, however, employ chaos constructively. Through amplification of small fluctuations it can provide natural systems with access to novelty. A prey escaping a predator's attack could use chaotic flight control as an element of surprise to evade capture. Biological evolution demands genetic variability; chaos provides a means of structuring random changes, thereby providing the possibility of putting variability under evolutionary control.

Even the process of intellectual progress relies on the injection of new ideas and on new ways of connecting old ideas. Innate creativity may have an underlying chaotic process that selectively amplifies small fluctuations and molds them into macroscopic coherent mental states that are experienced as thoughts. In some cases the thoughts may be decisions, or what are perceived to be the exercise of will. In this light, chaos provides a mechanism that allows for free will within a world governed by deterministic laws.

## Note

This chapter originally appeared in *Scientific American* 255 (1986), 46–57. The figures in the original are omitted here.

# 21 Undecidability and Intractability in Theoretical Physics

**Stephen Wolfram**

There is a close correspondence between physical processes and computations. On one hand, theoretical models describe physical processes by computations that transform initial data according to algorithms representing physical laws. And on the other hand, computers themselves are physical systems, obeying physical laws. This chapter explores some fundamental consequences of this correspondence.[1]

The behavior of a physical system may always be calculated by simulating explicitly each step in its evolution. Much of theoretical physics has, however, been concerned with devising shorter methods of calculation that reproduce the outcome without tracing each step. Such shortcuts can be made if the computations used in the calculation are more sophisticated than those that the physical system can itself perform. Any computations must, however, be carried out on a computer. But the computer is itself an example of a physical system. And it can determine the outcome of its own evolution only by explicitly following it through: No shortcut is possible. Such computational irreducibility occurs whenever a physical system can act as a computer. The behavior of the system can be found only by direct simulation or observation: No general predictive procedure is possible. Computational irreducibility is common among the systems investigated in mathematics and computation theory. This paper suggests that it is also common in theoretical physics. Computational reducibility may well be the exception rather than the rule: Most physical questions may be answerable only through irreducible amounts of computation. Those that concern idealized limits of infinite time, volume, or numerical precision can require arbitrarily long computations, and so be formally undecidable.

A diverse set of systems are known to be equivalent in their computational capabilities, in that particular forms of one system can emulate any of the others. Standard digital computers are one example of such "universal computers": With fixed intrinsic instructions, different initial states or programs can be devised to simulate different systems. Some other examples are Turing machines, string transformation systems, recursively defined functions, and Diophantine equations.[2] One expects in fact that universal computers are as powerful in their computational capabilities as any physically realizable system can be, so that they can simulate any physical system.[3] This is the case if in all physical systems there is a finite density of information, which can be transmitted only at a finite rate in a finite-dimensional space.[4] No physically implementable procedure could then short cut a computationally irreducible process.

Different physically realizable universal computers appear to require the same order of magnitude times and information storage capacities to solve particular classes of finite problems.[5] One computer may be constructed so that in a single step it carries out the equivalent of two steps on another computer. However, when the amount of information $n$ specifying an instance of a problem becomes large, different computers use resources that differ only by polynomials in $n$. One may then distinguish several classes of problems.[6] The first, denoted $P$, are those such as arithmetical ones taking a time polynomial in $n$. The second, denoted *PSPACE*, are those that can be solved with polynomial storage capacity, but may require exponential time, and so are in practice effectively intractable. Certain problems are "complete" with respect to *PSPACE*, so that particular instances of them correspond to arbitrary *PSPACE* problems. Solutions to these problems mimic the operation of a universal computer with bounded storage capacity: A computer that solves *PSPACE*-complete problems for any $n$ must be universal. Many mathematical problems are *PSPACE*-complete.[6] (An example is whether one can always win from a given position in chess.) And since there is no evidence to the contrary, it is widely conjectured that $PSPACE \neq P$, so that *PSPACE*-complete problems cannot be solved in polynomial time. A final class of problems, denoted *NP*, consist in identifying, among an exponentially large collection of objects, those with some particular, easily testable property. An example would be to find an $n$-digit integer that divides a given $2n$-digit number exactly. A particular candidate divisor, guessed nondeterministically, can be tested in polynomial time, but a systematic solution may require almost all $O(2^n)$ possible candidates to be tested. A computer that could follow arbitrarily many computational paths in parallel could solve such problems in polynomial time. For actual computers that allow only boundedly many paths, it is suspected that no general polynomial time solution is possible.[5] Nevertheless, in the infinite time limit, parallel paths are irrelevant, and a computer that solves *NP*-complete problems is equivalent to other universal computers.[6]

The structure of a system need not be complicated for its behavior to be highly complex, corresponding to a complicated computation. Computational irreducibility may thus be widespread even among systems with simple construction. Cellular automata (CA)[7] provide an example. A CA consists of a lattice of sites, each with $k$ possible values, and each updated in time steps by a deterministic rule depending on a neighborhood of $R$ sites. CA serve as discrete approximations to partial differential equations, and provide models for a wide variety of natural systems. Figure 21.1 shows typical examples of their behavior. Some rules give periodic patterns, and the outcome after many steps can be predicted without following each intermediate step. Many rules,

**Figure 21.1**
Seven examples of patterns generated by repeated application of various simple cellular automaton rules. The last four probably are computationally irreducible and can be found only by direct simulation.

however, give complex patterns for which no predictive procedure is evident. Some CA are in fact known to be capable of universal computation, so that their evolution must be computationally irreducible. The simplest cases proved have $k = 18$ and $R = 3$ in one dimension,[8] or $k = 2$ and $R = 5$ in two dimensions.[9] It is strongly suspected that "class-4" CA are generically capable of universal computation: There are such CA with $k = 3$, $R = 3$ and $k = 2$, $R = 5$ in one dimension.[10]

Computationally, irreducibility may occur in systems that are not full universal computers. For inability to perform specific computations need not allow all computations to be short cut. Though class-3 CA and other chaotic systems may not be universal computers, most of them are expected to be computationally irreducible, so that the solution of problems concerning their behavior requires irreducible amounts of computation.

As a first example consider finding the value of a site in a CA after $t$ steps of evolution from a finite initial seed, as illustrated in figure 21.1. The problem is specified by giving the seed and the CA rule, together with the $\log t$ digits of $t$. In simple cases such as the first two shown in figure 21.1, it can be solved in the time $O(\log t)$ necessary to input this specification. However, the evolution of a universal computer CA for a polynomial in $t$ steps can implement any computation of length $t$. As a consequence, its evolution is computationally irreducible, and its outcome found only by an explicit simulation with length $O(t)$: exponentially longer than for the first two in figure 21.1.

One may ask whether the pattern generated by evolution with a CA rule from a particular seed will grow forever, or will eventually die out.[11] If the evolution is computationally irreducible, then an arbitrarily long computation may be needed to answer this question. One may determine by explicit simulation whether the pattern dies out after any specified number of steps, but there is no upper bound on the time needed to find out its ultimate fate.[12] Simple criteria may be given for particular cases, but computational irreducibility implies that no shortcut is possible in general. The infinite-time limiting behavior is formally undecidable: No finite mathematical or computational process can reproduce the infinite CA evolution.

The fate of a pattern in a CA with a finite total number of sites $N$ can always be determined in at most $k^N$ steps. However, if the CA is a universal computer, then the problem is *PSPACE*-complete, and so presumably cannot be solved in a time polynomial in $N$.[13]

One may consider CA evolution not only from finite seeds, but also from initial states with all infinitely many sites chosen arbitrarily. The value $a^{(t)}$ of a site after many time steps $t$ then in general depends on $2\lambda t \leq Rt$ initial site values, where $\lambda$ is the rate of information transmission (essentially Lyapunov exponent) in the CA.[9] In class-1 and -2 CA, information remains localized, so that $\lambda = 0$, and $a^{(t)}$ can be found by a length $O(\log t)$ computation. For class-3 and -4 CA, however, $\lambda > 0$, and $a^{(t)}$ requires an $O(t)$ computation.[14]

The global dynamics of CA are determined by the possible states reached in their evolution. To characterize such states one may ask whether a particular string of $n$ site values can be generated after evolution for $t$ steps from any (length $n + 2\lambda t$) initial string. Since candidate initial strings can be tested in $O(t)$ time, this problem is in the class *NP*. When the CA is a universal computer, the problem is in general *NP*-complete, and can presumably be answered essentially only by testing all $O(k^{n+2\lambda t})$ candidate initial strings.[15] In the limit $t \to \infty$, it is in general undecidable whether particular strings can appear.[16] As a consequence, the entropy or dimension of the limiting set of CA configurations is in general not finitely computable.

Formal languages describe sets of states generated by CA.[17] The set that appears after $t$ steps in the evolution of a one-dimensional CA forms a regular formal language: each possible state corresponds to a path through a graph with $\Xi^{(t)} < 2^{k^{Rt}}$ nodes. If, indeed, the length of computation to determine whether a string can occur increases exponentially with $t$ for computationally irreducible CA, then the "regular language complexity" $\Xi^{(t)}$ should also increase exponentially, in agreement with empirical data on certain class-3 CA,[17] and reflecting the "irreducible computational work" achieved by their evolution.

Irreducible computations may be required not only to determine the outcome of evolution through time, but also to find possible arrangements of a system in space. For example, whether an $x \times x$ patch of site values occurs after just one step in a two-dimensional CA is in general *NP*-complete.[18] To determine whether there is any complete infinite configuration that satisfies a particular predicate (such as being invariant under the CA rule) is in general undecidable[18]: It is equivalent to finding the infinite-time behavior of a universal computer that lays down each row on the lattice in turn.

There are many physical systems in which it is known to be possible to construct universal computers. Apart from those modeled by CA, some examples are electric circuits, hard-sphere gases with obstructions, and networks of chemical reactions.[19] The evolution of these systems is in general computationally irreducible, and so suffers from undecidable and intractable problems. Nevertheless, the constructions used to find universal computers in these systems are arcane, and if computationally complex problems occurred only there, they would be rare. It is the thesis of this chapter that such problems are in fact common.[20] Certainly there are many systems whose properties are in practice studied only by explicit simulation or exhaustive search: Few computational shortcuts (often stated in terms of invariant quantities) are known.

Many complex or chaotic dynamical systems are expected to be computationally irreducible, and their behavior effectively found only by explicit simulation. Just as it is undecidable whether a particular initial state in a CA leads to unbounded growth, to self-replication, or has some other outcome, so it may be undecidable whether a particular solution to a differential equation (studied say with symbolic dynamics) even enters a certain region of phase space, and whether, say, a certain $n$-body system is ultimately stable. Similarly, the existence of an attractor, say, with a dimension above some value, may be undecidable.

Computationally complex problems can arise in finding eigenvalues or extremal states in physical systems. The minimum energy conformation for a polymer is in general *NP*-complete with respect to its length.[21] Finding a configuration below a specified energy in a spin-glass with particular couplings is similarly *NP*-complete.[22] Whenever the stationary state of a physical system such as this can be found only by lengthy computation, the dynamic physical processes that lead to it must take a correspondingly long time.[5]

Global properties of some models for physical systems may be undecidable in the infinite-size limit (like those for two-dimensional CA). An example is whether a particular generalized Ising model (or stochastic multidimensional CA[23]) exhibits a phase transition.

Quantum and statistical mechanics involve sums over possibly infinite sets of configurations in systems. To derive finite formulas one must use finite specifications for these sets. But it may be undecidable whether two finite specifications yield equivalent configurations. So, for example, it is undecidable whether two finitely specified four-manifolds or solutions to the Einstein equations are equivalent (under coordinate reparametrization).[24] A theoretical model may be considered as a finite specification of the possible behavior of a system. One may ask for example whether the consequences of two models are identical in all circumstances, so that the models are equivalent. If the models involve computations more complicated than those that can be carried out by a computer with a fixed finite number of states (regular language), this question is in general undecidable. Similarly, it is undecidable what is the simplest such model that describes a given set of empirical data.[25]

This chapter has suggested that many physical systems are computationally irreducible, so that their own evolution is effectively the most efficient procedure for determining their future. As a consequence, many questions about these systems can be answered only by very lengthy or potentially infinite computations. But some questions answerable by simpler computations may still be formulated.

## Acknowledgments

This work was supported in part by the U. S. Office of Naval Research under Contract No. N00014-80-C-0657. I am grateful for discussions with many people, particularly C. Bennett, G. Chaitin, R. Feynman, E. Fredkin, D. Hillis, L. Hurd, J. Milnor, N. Packard, M. Perry, R. Shaw, K. Steiglitz, W. Thurston, and L. Yaffe.

### Notes

This chapter originally appeared in *Physical Review Letters* 54 (1985), 735–738.

1. For a more informal exposition see: S. Wolfram, Sci. Am. 251, 188 (1984). A fuller treatment will be given elsewhere.

2. E.g., *The Undecidable: Basic Papers on Undecidable Propositions, Unsolvable Problems, and Computable Functions*, edited by M. Davis (Raven, New York, 1965), or J. Hopcroft and J. Ullman, *Introduction to Automata Theory, Languages, and Computations* (Addison-Wesley, Reading, Mass., 1979).

3. This is a physical form of the Church-Turing hypothesis. Mathematically conceivable systems of greater power can be obtained by including tables of answers to questions insoluble for these universal computers.

4. Real-number parameters in classical physics allow infinite information density. Nevertheless, even in classical physics, the finiteness of experimental arrangements and measurements, implemented as coarse graining in statistical mechanics, implies finite information input and output. In relativistic quantum field theory, finite density of information (or quantum states) is evident for free fields bounded in phase space [e.g., J. Bekenstein, Phys. Rev. D 30, 1669 (1984)]. It is less clear for interacting fields, except if space-time is ultimately discrete [but cf. B. Simon, *Functional Integration and Quantum Physics* (Academic, New York, 1979), Sec. III.9]. A finite information transmission rate is implied by relativistic causality and the manifold structure of space-time.

5. It is just possible, however, that the parallelism of the path integral may allow quantum mechanical systems to solve any *NP* problem in polynomial time.

6. M. Garey and D. Johnson, *Computers and Intractability: A Guide to the Theory of NP-Completeness* (Freeman, San Francisco, 1979).

7. See S. Wolfram, Nature 311, 419 (1984); *Cellular Automata*, edited by D. Farmer, T. Toffoli, and S. Wolfram, Physica 10D, Nos. 1 and 2 (1984), and references therein.

8. A. R. Smith, J. Assoc. Comput. Mach. 18, 331 (1971).

9. E. R. Banks, Massachusetts Institute of Technology Report No. TR-81, 1971 (unpublished). The "Game of Life," discussed in E. R. Berlekamp, J. H. Conway, and R. K. Guy, *Winning Ways for Your Mathematical Plays* (Academic, New York, 1982), is an example with $k = 2$, $R = 9$. N. Margolus, Physica (Utrecht) 10D, 81 (1984), gives a reversible example.

10. S. Wolfram, Physica (Utrecht) 10D, 1 (1984), and to be published.

11. This is analogous to the problem of whether a computer run with particular input will ever reach a "halt" state.

12. The number of steps to check ("busy-beaver function") in general grows with the seed size faster than any finite formula can describe (Ref. 2).

13. Cf. C. Bennett, to be published.

14. Cf. B. Eckhardt, J. Ford, and F. Vivaldi, Physica (Utrecht) 13D, 339 (1984).

15. The question is a generalization of whether there exists an assignment of values to sites such that the logical expression corresponding the $t$-step CA mapping is true (cf. V. Sewelson, private communication).

16. L. Hurd, to be published.

17. S. Wolfram, Commun. Math. Phys. 96, 15 (1984).

18. N. Packard and S. Wolfram, to be published. The equivalent problem of covering a plane with a given set of tiles is considered in R. Robinson, Invent. Math. 12, 177 (1971).

19. E.g., C. Bennett, Int. J. Theor. Phys. 21, 905 (1982); E. Fredkin and T. Toffoli, Int. J. Theor. Phys. 21, 219 (1982); A. Vergis, K. Steiglitz, and B. Dickinson, "The Complexity of Analog Computation" (unpublished).

20. Conventional computation theory primarily concerns possibilities, not probabilities. There are nevertheless some problems for which almost all instances are known to be of equivalent difficulty. But other problems are known to be much easier on average then in the worst case. In addition, for some *NP*-complete problems the density of candidate solutions close to the actual one is very large, so approximate solutions can easily be found [S. Kirkpatrick, C. Gelatt, and M. Vecchi, Science 220, 671 (1983)].

21. Compare *Time Warps, String Edites, and Macromolecules*, edited by D. Sankoff and J. Kruskal (Addison-Wesley, Reading, Mass., 1983).

22. F. Barahona, J. Phys. A 13, 3241 (1982).

23. E. Domany and W. Kinzel, Phys. Rev. Lett. 53, 311 (1984).

24. See W. Haken, in *Word Problems*, edited by W. W. Boone, F. B. Cannonito, and R. C. Lyndon (North-Holland, Amsterdam, 1973).

25. G. Chaitin, Sci. Am. 232, 47 (1975), and IBM J. Res. Dev. 21, 350 (1977); R. Shaw, to be published.

# 22 Special Sciences (Or: The Disunity of Science as a Working Hypothesis)

Jerry Fodor

A typical thesis of positivistic philosophy of science is that all true theories in the special sciences should reduce to physical theories in the long run. This is intended to be an empirical thesis, and part of the evidence which supports it is provided by such scientific successes as the molecular theory of heat and the physical explanation of the chemical bond. But the philosophical popularity of the reductivist program cannot be explained by reference to these achievements alone. The development of science has witnessed the proliferation of specialized disciplines at least as often as it has witnessed their reduction to physics, so the widespread enthusiasm for reduction can hardly be a mere induction over its past successes.

I think that many philosophers who accept reductivism do so primarily because they wish to endorse the generality of physics *vis à vis* the special sciences: roughly, the view that all events which fall under the laws of any science are physical events and hence fall under the laws of physics.[1] For such philosophers, saying that physics is basic science and saying that theories in the special sciences must reduce to physical theories have seemed to be two ways of saying the same thing, so that the latter doctrine has come to be a standard construal of the former.

In what follows, I shall argue that this is a considerable confusion. What has traditionally been called "the unity of science" is a much stronger, and much less plausible, thesis than the generality of physics. If this is true it is important. Though reductionism is an empirical doctrine, it is intended to play a regulative role in scientific practice. Reducibility to physics is taken to be a *constraint* upon the acceptability of theories in the special sciences, with the curious consequence that the more the special sciences succeed, the more they ought to disappear. Methodological problems about psychology, in particular, arise in just this way: the assumption that the subject-matter of psychology is part of the subject-matter of physics is taken to imply that psychological theories must reduce to physical theories, and it is this latter principle that makes the trouble. I want to avoid the trouble by challenging the inference.

## 22.1

Reductivism is the view that all the special sciences reduce to physics. The sense of "reduce to" is, however, proprietary. It can be characterized as follows.[2]

Let

$$S_1x \rightarrow S_2x \tag{22.1}$$

be a law of the special science $S$. ((22.1) is intended to be read as something like "all $S_1$ situations bring about $S_2$ situations." I assume that a science is individuated largely by reference to its typical predicates, hence that if $S$ is a special science "$S_1$" and "$S_2$" are not predicates of basic physics. I also assume that the "all" which quantifies laws of the special sciences needs to be taken with a grain of salt; such laws are typically *not* exceptionless. This is a point to which I shall return at length.) A necessary and sufficient condition of the reduction of (22.1) to a law of physics is that the formulae (22.2) and (22.3) be laws, and a necessary and sufficient condition of the reduction of $S$ to physics is that all its laws be so reducible.[3]

$$S_1x \leftrightarrows P_1x \tag{22.2a}$$

$$S_2x \leftrightarrows P_2x \tag{22.2b}$$

$$P_1x \rightarrow P_2x. \tag{22.3}$$

"$P_1$" and "$P_2$" are supposed to be predicates of physics, and (22.3) is supposed to be a physical law. Formulae like (22.2) are often called "bridge" laws. Their characteristic feature is that they contain predicates of both the reduced and the reducing science. Bridge laws like (2) are thus contrasted with "proper" laws like (22.1) and (22.3). The upshot of the remarks so far is that the reduction of a science requires that any formula which appears as the antecedent or consequent of one of its proper laws must appear as the reduced formula in some bridge law or other.[4]

Several points about the connective "→" are in order. First, whatever other properties that connective may have, it is universally agreed that it must be transitive. This is important because it is usually assumed that the reduction of some of the special sciences proceeds via bridge laws which connect their predicates with those of intermediate reducing theories. Thus, psychology is presumed to reduce to physics via, say, neurology, biochemistry, and other local stops. The present point is that this makes no difference to the logic of the situation so long as the transitivity of "→" is assumed. Bridge laws which connect the predicates of $S$ to those of $S^*$ will satisfy the constraints upon the reduction of $S$ to physics so long as there are other bridge laws which, directly or indirectly, connect the predicates of $S^*$ to physical predicates.

There are, however, quite serious open questions about the interpretations of "→" in bridge laws. What turns on these questions is the respect in which reductivism is taken to be a physicalist thesis.

To begin with, if we read "→" as "brings about" or "causes" in proper laws, we will have to have some other connective for bridge laws, since bringing about and causing are presumably *a*symmetric, while bridge laws express symmetric relations. Moreover, if "→" in bridge laws is interpreted as any relation other than identity, the truth of reductivism will only guarantee the truth of a weak version of physicalism, and this would fail to express the underlying ontological bias of the reductivist program.

If bridge laws are not identity statements, then formulae like (22.2) claim at most that, by law, $x$'s satisfaction of a $P$ predicate and $x$'s satisfaction of an $S$ predicate are causally correlated. It follows from this that it is nomologically necessary that $S$ and $P$ predicates apply to the same things (i.e., that $S$ predicates apply to a subset of the things that $P$ predicates apply to). But, of course, this is compatible with a nonphysicalist ontology since it is compatible with the possibility that $x$'s satisfying $S$ should not itself *be* a physical event. On this interpretation, the truth of reductivism does *not* guarantee the generality of physics *vis à vis* the special sciences since there are some events (satisfactions of $S$ predicates) which fall in the domains of a special science ($S$) but not in the domain of physics. (One could imagine, for example, a doctrine according to which physical and psychological predicates are both held to apply to organisms, but where it is denied that the event which consists of an organism's satisfying a psychological predicate is, in any sense, a physical event. The up-shot would be a kind of psychophysical dualism of a non-Cartesian variety; a dualism of events and/or properties rather than substances.)

Given these sorts of considerations, many philosophers have held that bridge laws like (22.2) ought to be taken to express contingent event identities, so that one would read (22.2a) in some such fashion as "every event which consists of $x$'s satisfying $S_1$ is identical to some event which consists of $x$'s satisfying $P_1$ and vice versa." On this reading, the truth of reductivism would entail that every event that falls under any scientific law is a physical event, thereby simultaneously expressing the ontological bias of reductivism and guaranteeing the generality of physics *vis à vis* the special sciences.

If the bridge laws express event identities, and if every event that falls under the proper laws of a special science falls under a bridge law, we get the truth of a doctrine that I shall call "token physicalism." Token physicalism is simply the claim that all the events that the sciences talk about are physical events. There are three things to notice about token physicalism.

First, it is weaker than what is usually called "materialism." Materialism claims *both* that token physicalism is true *and* that every event falls under the laws of some science or other. One could therefore be a token physicalist without being a materialist, though I don't see why anyone would bother.

Second, token physicalism is weaker than what might be called "type physicalism," the doctrine, roughly, that every *property* mentioned in the laws of any science is a physical property. Token physicalism does not entail type physicalism because the contingent identity of a pair of events presumably does not guarantee the identity of the properties whose instantiation constitutes the events; not even where the event identity is nomologically necessary. On the other hand, if every event is the instantiation of a property, then type physicalism does ential token physicalism: two events will be identical when they consist of the instantiation of the same property by the same individual at the same time.

Third, token physicalism is weaker than reductivism. Since this point is, in a certain sense, the burden of the argument to follow, I shan't labour it here. But, as a first approximation, reductivism is the conjunction of token physicalism with the assumption that there are natural kind predicates in an ideally completed physics which correspond to each natural kind predicate in any ideally completed special science. It will be one of my morals that the truth of reductivism cannot be inferred from the assumption that token physicalism is true. Reductivism is a sufficient, but not a necessary, condition for token physicalism.

In what follows, I shall assume a reading of reductivism which entails token physicalism. Bridge laws thus state nomologically necessary contingent event identities, and a reduction of psychology to neurology would entail that any event which consists of the instantiation of a psychological property is identical with some event which consists of the instantiation of some neurological property.

Where we have got to is this: reductivism entails the generality of physics in at least the sense that any event which falls within the universe of discourse of a special science will also fall within the universe of discourse of physics. Moreover, any prediction which follows from the laws of a special science and a statement of initial conditions will also follow from a theory which consists of physics and the bridge laws, together with the statement of initial conditions. Finally, since "reduces to" is supposed to be an asymmetric relation, it will also turn out that physics is *the* basic science; that is, if reductivism is true, physics is the only science that is general in the sense just specified. I now want to argue that reductivism is too strong a constraint upon the unity of science, but that the relevantly weaker doctrine will preserve the desired consequences of reductivism: token physicalism, the generality of physics, and its basic position among the sciences.

## 22.2

Every science implies a taxonomy of the events in its universe of discourse. In particular, every science employs a descriptive vocabulary of theoretical and observation predicates such that events fall under the laws of the science by virtue of satisfying those predicates. Patently, not every true description of an event is a description in such a vocabulary. For example, there are a large number of events which consist of things having been transported to a distance of less than three miles from the Eiffel Tower. I take it, however, that there is no science which contains "is transported to a distance of less than three miles from the Eiffel Tower" as part of its descriptive vocabulary. Equivalently, I take it that there is no natural law which applies to events in virtue of their being instantiations of the property *is transported to a distance of less than three miles from the Eiffel Tower* (though I suppose it is conceivable that there is some law that applies to events in virtue of their being instantiations of some distinct but co-extensive property). By way of abbreviating these facts, I shall say that the property *is transported*... does not determine a *natural kind*, and that predicates which express that property are not natural kind predicates.

If I knew what a law is, and if I believed that scientific theories consist just of bodies of laws, then I could say that $P$ is a natural kind predicate relative to $S$ iff $S$ contains proper laws of the form $P_x \rightarrow \alpha_x$ or $\alpha_x \rightarrow P_x$; roughly, the natural kind predicates of a science are the ones whose terms are the bound variables in its proper laws. I am inclined to say this even in my present state of ignorance; accepting the consequence that it makes the murky notion of a natural kind viciously dependent on the equally murky notions *law* and *theory*. There is no firm footing here. If we disagree about what is a natural kind, we will probably also disagree about what is a law, and for the same reasons. I don't know how to break out of this circle, but I think that there are interesting things to say about which circle we are in.

For example, we can now characterize the respect in which reductivism is too strong a construal of the doctrine of the unity of science. If reductivism is true, then *every* natural kind is, or is co-extensive with, a physical natural kind. (Every natural kind *is* a physical natural kind if bridge laws express property identities, and every natural kind is co-extensive with a physical natural kind if bridge laws express event identities.) This follows immediately from the reductivist premise that every predicate which appears as the antecedent or consequent of a law of the special sciences must appear as one of the reduced predicates in some bridge, together with the assumption that the natural kind predicates are the ones whose terms are the bound variables in proper laws. If, in short, some physical law is related to each law of a special science in the way that (22.3) is related to (22.1), then every natural kind predicate of a special science is related to a natural kind predicate of physics in the way that (22.2) relates "$S_1$" and "$S_2$" to "$P_1$" and "$P_2$."

I now want to suggest some reasons for believing that this consequence of reductivism is intolerable. These are not supposed to be knock-down reasons; they couldn't be, given that the question whether reductivism is too strong is finally an *empirical* question. (The world could turn out to be such that every natural kind corresponds to a physical natural kind, just as it could turn out to be such that the property *is transported to a distance of less than three miles from the Eiffel Tower* determines a natural kind in, say, hydrodynamics. It's just that, as things stand, it seems very unlikely that the world *will* turn out to be either of these ways.)

The reason it is unlikely that every natural kind corresponds to a physical natural kind is just that (a) interesting generalizations (e.g., counter-factual supporting generalizations) can often be made about events whose physical descriptions have nothing in common, (b) it is often the case that *whether* the physical descriptions of the events subsumed by these generalizations have anything in common is, in an obvious sense, entirely irrelevant to the truth of the generalizations, or to their interestingness, or to their degree of confirmation or, indeed, to any of their epistemologically important properties, and (c) the special sciences are very much in the business of making generalizations of this kind.

I take it that these remarks are obvious to the point of self-certification; they leap to the eye as soon as one makes the (apparently radical) move of taking the special sciences at all seriously. Suppose, for example, that Gresham's "law" really is true. (If one doesn't like Gresham's law, then any true generalization of any conceivable future economics will probably do as well.) Gresham's law says something about what will happen in monetary exchanges under certain conditions. I am willing to believe that physics is general *in the sense that it implies that any event which consists of a monetary exchange* (hence any event which falls under Gresham's law) *has a true description in the vocabulary of physics and in virtue of which it falls under the laws of physics*. But banal considerations suggest that a description which covers all such events must be wildly disjunctive. Some monetary exchanges involve strings of wampum. Some involve dollar bills. And some involve signing one's name to a check. What are the chances that a disjunction of physical predicates which covers all these events (i.e., a disjunctive predicate which can form the right hand side of a bridge law of the form "$x$ is a monetary exchange $\leftrightarrows \ldots$") expresses a physical natural kind? In particular, what are the chances that such a predicate forms the antecedent or consequent of some proper law of physics? The point is that monetary exchanges have interesting things in common; Gresham's law, if true, says what one of these interesting things is. But what is interesting about monetary exchanges is surely not their commonalities under *physical* description. A natural kind like a monetary exchange *could* turn out to be co-extensive with a physical natural kind; but if it did, that would be an accident on a cosmic scale.

In fact, the situation for reductivism is still worse than the discussion thus far suggests. For, reductivism claims not only that all natural kinds are co-extensive with physical natural kinds, but that the co-extensions are nomologically necessary: bridge laws are *laws*. So, if Gresham's law is true, it follows that there is a (bridge) law of nature such that "$x$ is a monetary exchange $\rightleftarrows$ $x$ is $P$," where $P$ is a term for a physical natural kind. But, surely, there is no such law. If there were, then $P$ would have to cover not only all the systems of monetary exchange that there *are*, but also all the systems of monetary exchange that there *could be*; a law must succeed with the counterfactuals. What physical predicate is a candidate for "$P$" in "$x$ is a nomologically possible monetary exchange iff $P_x$"?

To summarize: an immortal econophysicist might, when the whole show is over, find a predicate in physics that was, in brute fact, co-extensive with "is a monetary exchange." If physics is general—if the ontological biases of reductivism are true—then there must *be* such a predicate. But (a) to paraphrase a remark Donald Davidson made in a slightly different context, nothing but brute enumeration could convince us of this brute co-extensivity, and (b) there would seem to be no chance at all that the physical predicate employed in stating the coextensivity is a natural kind term, and (c) there is still less chance that the co-extension would be lawful (i.e., that it would hold not only for the nomologically possible world that turned out to be real, but for any nomologically possible world at all).

I take it that the preceding discussion strongly suggests that economics is not reducible to physics in the proprietary sense of reduction involved in claims for the unity of science. There is, I suspect, nothing special about economics in this respect; the reasons why economics is unlikely to reduce to physics are paralleled by those which suggest that psychology is unlikely to reduce to neurology.

If psychology is reducible to neurology, then for every psychological natural kind predicate there is a co-extensive neurological natural kind predicate, and the generalization which states this co-extension is a law. Clearly, many psychologists believe something of the sort. There are departments of "psycho-biology" or "psychology and brain science" in universities throughout the world whose very existence is an institutionalized gamble that such lawful co-extensions can be found. Yet, as has been frequently remarked in recent discussions of materialism, there are good grounds for hedging these bets. There are no firm data for any but the grossest correspondence between types of psychological states and types of neurological states, and it is entirely possible that the nervous system of higher organisms characteristically achieves a given psychological end by a wide variety of neurological means. If so, then the attempt to pair neurological structures with psychological functions is foredoomed. Physiological psychologists of the stature of Karl Lashley have held precisely this view.

The present point is that the reductivist program in psychology is, in any event, *not* to be defended on ontological grounds. Even if (token) psychological events are (token) neurological events, it does not follow that the natural kind predicates of psychology are co-extensive with the natural kind predicates of any other discipline (including physics). That is, the assumption that every psychological event is a physical event does not guaranty that physics (or, *a fortiori*, any other discipline more general than psychology) can provide an appropriate vocabulary for psychological theories. I emphasize this point because I am convinced that the make-or-break commitment of many physiological psychologists to the reductivist program stems precisely from having confused that program with (token) physicalism.

What I have been doubting is that there are neurological natural kinds co-extensive with psychological natural kinds. What seems increasingly clear is that, even if there is such a co-extension, it cannot be lawlike. For, it seems increasingly likely that there are nomologically possible systems other than organisms (namely, automata) which satisfy natural kind predicates in psychology, and which satisfy no neurological predicates at all. Now, as Putnam has emphasized, if there are any such systems, then there are probably vast numbers, since equivalent automata can be made out of practically anything. If this observation is correct, then there can be no serious hope that the class of automata whose psychology is effectively identical to that of some organism can be described by *physical* natural kind predicates (though, of course, if token physicalims is true, that class can be picked out by some physical predicate or other). The upshot is that the classical formulation of the unity of science is at the mercy of progress in the field of computer simulation. This is, of course, simply to say that that formulation was too strong. The unity of science was intended to be an empirical hypothesis, defeasible by possible scientific findings. But no one had it in mind that it should be defeated by Newell, Shaw and Simon.

I have thus far argued that psychological reductivism (the doctrine that every psychological natural kind is, or is co-extensive with, a neurological natural kind) is not equivalent to, and cannot be inferred from, token physicalism (the doctrine that every psychological event is a neurological event). It may, however, be argued that one might as well take the doctrines to be equivalent since the only possible *evidence* one could have for token physicalism would also be evidence for reductivism: namely, the discovery of type-to-type psychophysical correlations.

A moment's consideration shows, however, that this argument is not well taken. If type-to-type psychophysical correlations would be evidence for token physicalism, so would correlations of other specifiable kinds.

We have type-to-type correlations where, for every *n*-tuple of events that are of the same psychological kind, there is a correlated *n*-tuple of events that are of the same neurological kind. Imagine a world in which such correlations are *not* forthcoming. What is found, instead, is that for every *n*-tuple of type identical psychological events, there is a spatio-temporally correlated *n*-tuple of type *distinct* neurological events. That is, every psychological event is paired with some neurological event or other, but psychological events of the same kind may be paired with neurological events of different kinds. My present point is that such pairings would provide as much support for token physicalism as type-to-type pairings do *so long as we are able to show that the type distinct neurological events paired with a given kind of psychological event are identical in respect of whatever properties are relevant to type-identification in psychology*. Suppose, for purposes of explication, that psychological events are type identified by reference to their behavioral consequences.[5] Then what is required of all the neurological events paired with a class of type homogeneous psychological events is only that they be identical in respect of their behavioral consequences. To put it briefly, type identical events do not, of course, have *all* their properties in common, and type distinct events must nevertheless be identical in *some* of their properties. The empirical confirmation of token physicalism does not depend on showing that the neurological counterparts of type identical psychological events are themselves type identical. What needs to be shown is only that they are identical in respect of those properties which determine which kind of *psychological* event a given event is.

Could we have evidence that an otherwise heterogeneous set of neurological events have these kinds of properties in common? Of course we could. The neurological theory might itself explain why an *n*-tuple of neurologically type distinct events are identical in their behavioral consequences, or, indeed, in respect of any of indefinitely many other such relational properties. And, if the neurological theory failed to do so, some science more basic than neurology might succeed.

My point in all this is, once again, not that correlations between type homogeneous psychological states and type heterogeneous neurological states would prove that token physicalism is true. It is only that such correlations might give us as much reason to be token physicalists as type-to-type correlations would. If this is correct, then the epistemological arguments from token physicalism to reductivism must be wrong.

It seems to me (to put the point quite generally) that the classical construal of the unity of science has really misconstrued the *goal* of scientific reduction. The point of reduction is *not* primarily to find some natural kind predicate of physics co-extensive with each natural kind predicate of a reduced science. It is, rather, to explicate the physical mechanisms whereby events conform to the laws of the special sciences. I have been arguing that there is no logical or epistemological reason why success in the second of these projects should require success in the first, and that the two are likely to come apart *in fact* wherever the physical mechanisms whereby events conform to a law of the special sciences are heterogeneous.

## 22.3

I take it that the discussion thus far shows that reductivism is probably too strong a construal of the unity of science; on the one hand, it is incompatible with probable results in the special sciences, and, on the other, it is more than we need to assume if what we primarily want is just to be good token physicalists. In what follows, I shall try to sketch a liberalization of reductivism which seems to me to be just strong enough in these respects. I shall then give a couple of independent reasons for supposing that the revised doctrine may be the right one.

The problem all along has been that there is an open empirical possibility that what corresponds to the natural kind predicates of a reduced science may be a heterogeneous and unsystematic disjunction of predicates in the reducing science, and we do not want the unity of science to be prejudiced by this possibility. Suppose, then, that we allow that bridge statements may be of the form

$$S_x \rightleftarrows P_1x \vee P_2x \vee \cdots \vee P_nx, \tag{22.4}$$

where "$P_1 \vee P_2 \vee \cdots \vee P_n$" is *not* a natural kind predicate in the reducing science. I take it that this is tantamount to allowing that at least some "bridge laws" may, in fact, not turn out to be laws, since I take it that a necessary condition on a universal generalization being lawlike is that the predicates which consitute its antecedent and consequent should pick out natural kinds. I am thus supposing that it is enough, for purposes of the unity of science, that every law of the special sciences should be reducible to physics by bridge statements which express true empirical generalizations. Bearing in mind that bridge statements are to be construed as a species of identity statements, (22.4) will be read as something like "every event which consists of $x$'s satisfying $S$ is identical with some event which consists of $x$'s satisfying some or other predicate belonging to the disjunction '$P_1 \vee P_2 \vee \cdots \vee P_n$.'"

Now, in cases of reduction where what corresponds to (22.2) is not a law, what corresponds to (22.3) will not be either, and for the same reason. Namely, the predicates appearing in the antecedent or consequent will, by hypothesis, not be natural kind predicates. Rather, what we will have is something that looks like (22.5).

Law of special science $X$: $S_1x \longrightarrow S_2x$ (22.5)

Bridge bi-conditionals

Disjunctive predicate of reducing science: $P_1x \vee P_2x \ldots P_nx$ $P_1^*x \vee P_2^*x \ldots P_m^*x$

That is, the antecedent and consequent of the reduced law will each be connected with a disjunction of predicates in the reducing science, and, if the reduced law is exceptionless, there will be laws of the reducing science which connect the satisfaction of each member of the disjunction associated with the antecedent to the satisfaction of some member of the disjunction associated with the consequent. That is, if $S_1x \to S_2x$ is exceptionless, then there must be some proper law of the reducing science which either states or entails that $P_1x \to P^*$ for some $P^*$, and similarly for $P_2x$ through $P_nx$. Since there must be such laws, it follows that each disjunct of "$P_1 \vee P_2 \vee \cdots \vee P_n$" is a natural kind predicate, as is each disjunct of "$P_1^* \vee P_2^* \vee \cdots \vee P_n^*$."

This, however, is where push comes to shove. For, it might be argued that if each disjunct of the $P$ disjunction is lawfully connected to some disjunct of the $P^*$ disjunction, it follows that (22.6) is itself a law.

$$P_1x \vee P_2x \vee \cdots \vee P_nx \rightarrow P_1^*x \vee P_2^*x \vee \cdots \vee P_n^*x. \tag{22.6}$$

The point would be that (22.5) gives us $P_1x \rightarrow P_2^*x$, $P_2x \rightarrow P_m^*x$, etc., and the argument from a premise of the form $(P \supset R)$ and $(Q \supset S)$ to a conclusion of the form $(P \vee Q) \supset (R \vee S)$ is valid.

What I am inclined to say about this is that it just shows that "it's a law that ———" defines a non-truth functional context (or, equivalently for these purposes, that not all truth functions of natural kind predicates are themselves natural kind predicates). In particular, that one may not argue from "it's a law that $P$ brings about $R$" and "it's a law that $Q$ brings about $S$" to "it's a law that $P$ or $Q$ brings about $R$ or $S$." (Though, of course, the argument from those premises to "$P$ or $Q$ brings about $R$ or $S$" *simpliciter* is fine.) I think, for example, that it is a law that the irradiation of green plants by sunlight causes carbohydrate synthesis, and I think that it is a law that friction causes heat, but I do not think that it is a law that (either the irradiation of green plants by sunlight or friction) causes (either carbohydrate synthesis or heat). Correspondingly, I doubt that "is either carbohydrate synthesis or heat" is plausibly taken to be a natural kind predicate.

It is not strictly mandatory that one should agree with all this, but one denies it at a price. In particular, if one allows the full range of truth functional arguments inside the context "it's a law that ———," then one gives up the possibility of identifying the natural kind predicates of a science with those predicates which appear as the antecedents or the consequents of its proper laws. (Thus (22.6) would be a proper law of physics which fails to satisfy that condition.) One thus inherits the need for an alternative construal of the notion of a natural kind, and I don't know what that alternative might be like.

The upshot seems to be this. If we do not require that bridge statements must be laws, then either some of the generalizations to which the laws of special sciences reduce are not themselves lawlike, or some laws are not formulable in terms of natural kinds. Whichever way one takes (22.5), the important point is that it is weaker than standard reductivism: it does not require correspondences between the natural kinds of the reduced and the reducing science. Yet it is physicalistic on the same assumption that makes standard reductivism physicalistic (namely, that the bridge statements express true token identities). But these are precisely the properties that we wanted a revised account of the unity of science to exhibit.

I now want to give two reasons for thinking that this construal of the unity of science is right. First, it allows us to see how the laws of the special sciences could reasonably have exceptions, and, second, it allows us to see why there are special sciences at all. These points in turn.

Consider, again, the model of reduction implicit in (22.2) and (22.3). I assume that the laws of basic science are strictly exceptionless, and I assume that it is common knowledge that the laws of the special sciences are not. But now we have a painful dilemma. Since "$\rightarrow$" expresses a relation (or relations) which must be transitive, (22.1) can have exceptions only if the bridge laws do. But if the bridge laws have exceptions, reductivism loses its ontological bite, since we can no longer say that every event which consists of the instantiation of an $S$ predicate is identical with some event which consists of the instantiation of a $P$ predicate. In short, given the reductionist model, we cannot consistently assume that the bridge laws and the basic laws are exceptionless while assuming that the special laws are not. But we cannot accept the violation of the bridge laws unless we are willing to vitiate the ontological claim that is the main point of the reductivist program.

We can get out of this (*salve* the model) in one of two ways. We can give up the claim that the special laws have exceptions or we can give up the claim that the basic laws are exceptionless. I suggest that both alternatives are undesirable. The first because it flies in the face of fact. There is just no chance at all that the true, counter-factual supporting generalizations of, say, psychology, will turn out to hold in strictly each and every condition where their antecedents are satisfied. Even where the spirit is willing, the flesh is often weak. There are always going to be behavioral lapses which are physiologically explicable but which are uninteresting from the point of view of psychological theory. The second alternative is only slightly better. It may, after all, turn out that the laws of basic science have exceptions. But the question arises whether one wants the unity of science to depend upon the assumption that they do.

On the account summarized in (22.5), however, everything works out satisfactorily. A nomologically sufficient condition for an exception to $S_1x \rightarrow S_2x$ is that the bridge statements should identify some occurrence of the satisfaction of $S_1$ with an occurrence of the satisfaction of a $P$ predicate which is not itself lawfully connected to the satisfaction of any $P^*$ predicate. (I.e., suppose $S_1$ is connected to a $P'$ such that there is no law which connects $P'$ to any predicate which bridge statements associate with $S_2$. Then any instantiation of $S_1$ which is contingently identical to an instantiation of $P'$ will be an event which constitutes an exception to $S_1x \rightarrow S_2x$.) Notice that, in this case, we need assume no exceptions to the laws of the *reducing* science since, by hypothesis, (22.6) *is not a law*.

In fact, strictly speaking, (22.6) has no status in the reduction at all. It is simply what one gets when one universally quantifies a formula whose antecedent is the physical disjunction corresponding to $S_1$ and whose consequent is the physical disjunction corresponding to $S_2$. As such, it will be true when $S_1 \rightarrow S_2$ is exceptionless and false otherwise. What does the work of expressing the physical mechanisms whereby $n$-tuples of events conform, or fail to conform, to $S_1 \to S_2$ is not (22.6) but the laws which severally relate elements of the disjunction $P_1 \vee P_2 \vee \cdots \vee P_n$ to elements of the disjunction $P_1^* \vee P_2^* \vee \cdots \vee P_n^*$. When there *is* a law which relates an event that satisfies one of the $P$ disjuncts to an event which satisfies one of the $P^*$ disjuncts, the pair of events so related conforms to $S_1 \to S_2$. When an event which satisfies a $P$ predicate is *not* related by law to an event which satisfies a $P^*$ predicate, that event will constitute an exception to $S_1 \to S_2$. The point is that none of the laws which effect these several connections need themselves have exceptions in order that $S_1 \to S_2$ should do so.

To put this discussion less technically: we could, if we liked, *require* the taxonomies of the special sciences to correspond to the taxonomy of physics by insisting upon distinctions between the natural kinds postulated by the former wherever they turn out to correspond to distinct natural kinds in the latter. This would *make* the laws of the special sciences exceptionless if the laws of basic science are. But it would also lose us precisely the generalizations which we want the special sciences to express. (If economics were to posit as many *kinds* of monetary systems as there are kinds of physical realizations of monetary systems, then the generalizations of economics *would* be exceptionless. But, presumably, only vacuously so, since there would be no generalizations left to state. Gresham's law, for example, would have to be formulated as a vast, open disjunction about what happens in monetary $\text{system}_1$ or monetary $\text{system}_n$ under conditions which would themselves defy uniform characterization. We would not be able to say what happens in monetary systems *tout court* since, by hypothesis, "is a monetary system" corresponds to no natural kind predicate of physics.)

In fact, what we do is precisely the reverse. We allow the generalizations of the special sciences to *have* exceptions, thus preserving the natural kinds to which the generalizations apply. But since we know that the *physical* descriptions of the natural kinds may be quite heterogeneous, and since we know that the physical mechanisms which connect the satisfaction of the antecedents of such generalizations to the satisfaction of their consequents may be equally diverse, we expect both that there will be exceptions to the generalizations and that these exceptions will be "explained away" at the level of the reducing science. This is one of the respects in which physics really is assumed to be bedrock science; exceptions to *its* generalizations (if there are any) had better be random, because there is nowhere "further down" to go in explaining the mechanism whereby the exceptions occur.

This brings us to why there are special sciences at all. Reductivism as we remarked at the outset, flies in the face of the facts about the scientific institution: the existence of a vast and interleaved conglomerate of special scientific disciplines which often appear to proceed with only the most token acknowledgment of the constraint that their theories must turn out to be physics "in the long run." I mean that the acceptance of this constraint, *in practice*, often plays little or no role in the validation of theories. Why is this so? Presumably, the reductivist answer must be *entirely* epistemological. If only physical particles weren't so small (if only brains were on the *out*side, where one can get a look at them), *then* we would do physics instead of paleontology (neurology instead of psychology; psychology instead of economics; and so on down). There is an epistemological reply; namely, that even if brains were out where they can be looked *at*, as things now stand, we wouldn't know what to look *for*: we lack the appropriate theoretical apparatus for the psychological taxonomy of neurological events.

*If* it turns out that the functional decomposition of the nervous system corresponds to its neurological (anatomical, biochemical, physical) decomposition, then there are only epistemological reasons for studying the former instead of the latter. But suppose there is no such correspondence? Suppose the functional organization of the nervous system crosscuts its neurological organization (so that quite different neurological structures can subserve identical psychological functions across times or across organisms). Then the existence of psychology depends not on the fact that neurons are so sadly small, but rather on the fact that neurology does not posit the natural kinds that psychology requires.

I am suggesting, roughly, that there are special sciences not because of the nature of our epistemic relation to the world, but because of the way the world is put together: not all natural kinds (not all the classes of things and events about which there are important, counterfactual supporting generalizations to make) are, or correspond to, physical natural kinds. A way of stating the classical reductionist view is that things which belong to different physical kinds *ipso facto* can have no projectible descriptions in common; that if *x* and *y* differ in those descriptions by virtue of which they fall under the proper laws of physics, they must differ in those descriptions by virtue of which they fall under any laws at all. But why should we believe that this is so? Any pair of entities, however different their physical structure, must nevertheless converge in indefinitely many of their properties. Why should there not be, among those convergent properties, some whose lawful inter-relations support the generalizations of the special sciences? Why, in short, should not the natural kind predicates of the special sciences *cross-classify* the physical natural kinds?[6]

Physics develops the taxonomy of its subject-matter which best suits its purposes: the formulation of exceptionless laws which are basic in the several senses discussed above. But this is not the only taxonomy which may be required if the purposes of science in general are to be served: e.g., if we are to state such true, counterfactual supporting generalizations as there are to state. So, there are special sciences, with their specialized taxonomies, in the business of stating some of these generalizations. If science is to be unified, then all such taxonomies must apply *to the same things*. If physics is to be basic science, then each of these things had better be a physical thing. But it is not further required that the taxonomies which the special sciences employ must themselves reduce to the taxonomy of physics. It is not required, and it is probably not true.

## Acknowledgment

I wish to express my gratitude to Ned Block for having read a version of this chapter and for the very useful comments he made.

### Notes

This chapter originally appeared in *Synthese* 28 (1974), 97–115.

1. I shall usually assume that sciences are about events, in at least the sense that it is the occurrence of events that makes the laws of a science true. But I shall be pretty free with the relation between events, states, things and properties. I shall even permit myself some latitude in construing the relation between properties and predicates. I realize that all these relations are problems, but they aren't my problem in this paper. Explanation has to *start* somewhere, too.

2. The version of reductionism I shall be concerned with is a stronger one than many philosophers of science hold; a point worth emphasizing since my argument will be precisely that it is too strong to get away with. Still, I think that what I shall be attacking is what many people have in mind when they refer to the unity of science, and I suspect (though I shan't try to prove it) that many of the liberalized versions suffer from the same basic defect as what I take to be the classical form of the doctrine.

3. There is an implicit assumption that a science simply *is* a formulation of a set of laws. I think this assumption is implausible, but it is usually made when the unity of science is discussed, and it is neutral so far as the main argument of this paper is concerned.

4. I shall sometimes refer to "the predicate which constitutes the antecedent or consequent of a law." This is shorthand for "the predicate such that the antecedent or consequent of a law consists of that predicate, together with its bound variables and the quantifiers which bind them." (Truth functions of elementary predicates are, of course, themselves predicates in this usage.)

5. I don't think there is any chance at all that this is true. What is more likely is that type-identification for psychological states can be carried out in terms of the "total states" of an abstract automaton which models the organism. For discussion, see Block and Fodor (1972).

6. As, by the way, the predicates of natural languages quite certainly do. For discussion, see Chomsky (1965).

### Bibliography

Block, N. and Fodor, J., "What Psychological States Are Not," *Philosophical Review* 81 (1972): 159–181.

Chomsky, N., *Aspects of the Theory of Syntax*, MIT Press, Cambridge, 1965.

# 23 Supervenience

**David Chalmers**

What is the place of consciousness in the natural order? Is consciousness physical? Can consciousness be explained in physical terms? To come to grips with these issues, we need to build a framework; in this chapter, I build one. The centerpiece of this framework is the concept of *supervenience*: I give an account of this concept and apply it to clarify the idea of reductive explanation. Using this account, I sketch a picture of the relationship between most high-level phenomena and physical facts, one that seems to cover everything except, perhaps, for conscious experience.

## 23.1 Supervenience

It is widely believed that the most fundamental facts about our universe are physical facts, and that all other facts are dependent on these. In a weak enough sense of "dependent," this may be almost trivially true; in a strong sense, it is controversial. There is a complex variety of dependence relations between high-level facts and low-level facts in general, and the kind of dependence relation that holds in one domain, such as biology, may not hold in another, such as that of conscious experience. The philosophical notion of supervenience provides a unifying framework within which these dependence relations can be discussed.

The notion of supervenience formalizes the intuitive idea that one set of facts can fully determine another set of facts.[1] The physical facts about the world seem to determine the biological facts, for instance, in that once all the physical facts about the world are fixed, there is no room for the biological facts to vary. (Fixing all the physical facts will simultaneously fix which objects are alive.) This provides a rough characterization of the sense in which biological properties supervene on physical properties. In general, supervenience is a relation between two sets of properties: *B*-properties—intuitively, the *high-level* properties—and *A*-properties, which are the more basic *low-level* properties.

For our purposes, the relevant A-properties are usually the physical properties: more precisely, the fundamental properties that are invoked by a completed theory of physics. Perhaps these will include mass, charge, spatio-temporal position; properties characterizing the distribution of various spatio-temporal fields, the exertion of various forces, and the form of various waves; and so on. The precise nature of these properties is not important. If physics changes radically, the relevant class of properties may be quite different from those I mention, but the arguments will go through all the same. Such high-level properties as juiciness, lumpiness, giraffehood, and the like are excluded, even though there is a sense in which these properties are physical. In what follows, talk of physical properties is implicitly restricted to the class of fundamental properties unless otherwise indicated. I will sometimes speak of "microphysical" or "low-level physical" properties to be explicit.

The *A-facts* and *B-facts* about the world are the facts concerning the instantiation and distribution of A-properties and B-properties.[2] So the physical facts about the world encompass all facts about the instantiation of physical properties within the spatio-temporal manifold. It is also useful to stipulate that the world's physical facts include its basic physical laws. On some accounts, these laws are already determined by the totality of particular physical facts, but we cannot take this for granted.

The template for the definition of supervenience is the following:

B-properties *supervene* on A-properties if no two possible situations are identical with respect to their A-properties while differing in their B-properties.

For instance, biological properties supervene on physical properties insofar as any two possible situations that are physically identical are biologically identical. (I use "identical" in the sense of indiscernibility rather than numerical identity here. In this sense, two separate tables might be physically identical.) More precise notions of supervenience can be obtained by filling in this template. Depending on whether we take the "situations" in question to be individuals or entire worlds, we arrive at notions of *local* and *global* supervenience, respectively. And depending on how we construe the notion of possibility, we obtain notions of *logical* supervenience, *natural* supervenience, and perhaps others. I will flesh out these distinctions in what follows.

### Local and Global Supervenience

B-properties supervene *locally* on A-properties if the A-properties of an *individual* determine the B-properties of that individual—if, that is, any two possible individuals that instantiate the same A-properties instantiate the same B-properties. For example, shape supervenes locally on physical properties: any two objects with the same physical properties will necessarily have the same shape. Value does not supervene locally on physical properties, however: an exact physical replica of the Mona Lisa is not worth as much as the Mona Lisa. In general, local supervenience of a property on the physical fails if that property is somehow context-dependent—that is, if an object's possession of that property depends not only on the object's physical constitution but also on its environment and its history. The Mona Lisa is more valuable than its replica because of a difference in their historical context: the Mona Lisa was painted by Leonardo, whereas the replica was not.[3]

B-properties supervene *globally* on A-properties, by contrast, if the A-facts about the entire *world* determine the B-facts: that is, if there are no two possible worlds identical with respect to their A-properties, but differing with respect to their B-properties.[4] A world here is to be thought of as an entire universe; different possible worlds correspond to different ways a universe might be.

Local supervenience implies global supervenience, but not vice versa. For example, it is plausible that biological properties supervene globally on physical properties, in that any world physically identical to ours would also be biologically identical. (There is a small caveat here, which I discuss shortly.) But they probably do not supervene locally. Two physically identical organisms can arguably differ in certain biological characteristics. One might be *fitter* than the other, for example, due to differences in their environmental contexts. It is even conceivable that physically identical organisms could be members of different species, if they had different evolutionary histories.

The distinction between global and local supervenience does not matter too much when it comes to conscious experience, because it is likely that insofar as consciousness supervenes on the physical at all, it supervenes locally. If two creatures are physically identical, then differences in environmental and historical contexts will not prevent them from having identical experiences. Of course, context can affect experience indirectly, but only by virtue of affecting internal structure, as in the case of perception. Phenomena such as hallucination and illusion illustrate the fact that it is internal structure rather than context that is *directly* responsible for experience.

### Logical and Natural Supervenience

A more important distinction for our purposes is between *logical* (or conceptual) supervenience, and mere *natural* (or nomic, or empirical) supervenience.

B-properties supervene *logically* on A-properties if no two *logically possible* situations are identical with respect to their A-properties but distinct with respect to their B-properties. I will say more about logical possibility later in this chapter. For now, one can think of it loosely as possibility in the broadest sense, corresponding roughly to conceivability, quite unconstrained by the laws of our world. It is useful to think of a logically possible world as a world that it would have been in God's power (hypothetically!) to create, had he so chosen.[5] God could not have created a world with male vixens, but he could have created a world with flying telephones. In determining whether it is logically possible that some statement is true, the constraints are largely *conceptual*. The notion of a male vixen is contradictory, so a male vixen is logically impossible; the notion of a flying telephone is conceptually coherent, if a little out of the ordinary, so a flying telephone is logically possible.

It should be stressed that the logical supervenience is not defined in terms of deducibility in any system of formal logic. Rather, logical supervenience is defined in terms of logically possible *worlds* (and individuals), where the notion of a logically possible world is independent of these formal considerations. This sort of possibility is often called "broadly logical" possibility in the philosophical literature, as opposed to the "strictly logical" possibility that depends on formal systems.[6]

At the global level, biological properties supervene logically on physical properties. Even God could not have created a world that was physically identical to ours but biologically distinct. There is simply no logical space for the biological facts to independently vary. When we fix all the physical facts about the world—including the facts about the distribution of every last particle across space and time—we will in effect also fix the macroscopic shape of all the objects in the world, the way they move and function, the way they physically interact. If there is a living kangaroo in this world, then *any* world that is physically identical to this world will contain a physically identical kangaroo, and that kangaroo will automatically be alive.

We can imagine that a hypothetical superbeing—Laplace's demon, say, who knows the location of every particle in the universe—would be able to straightforwardly "read off" all the biological facts, once given all the microphysical facts. The microphysical facts are enough for such a being to construct a model of the microscopic structure and dynamics of the world throughout space and time, from which it can straightforwardly deduce the macroscopic structure and dynamics. Given all that information, it has all the information it needs to determine which systems are alive, which systems belong to the same species, and so on. As long as it possesses the biological concepts and has a full specification of the microphysical facts, no other information is relevant.

In general, when B-properties supervene logically on A-properties, we can say that the A-facts *entail* the B-facts, where one fact entails another if it is logically impossible for the first to hold without the second. In such cases, Laplace's demon could read off the B-facts from a specification of the A-facts, as long as it possesses the B-concepts in question. (I will say much more about the connections between these different ways of understanding logical supervenience later in the chapter; the present discussion is largely for illustration.) In a sense, when logical supervenience holds, *all there is* to the B-facts being as they are is that the A-facts are as they are.

There can be supervenience without logical supervenience, however. The weaker variety of supervenience arises when two sets of properties are systematically and perfectly *correlated* in the natural world. For example, the pressure exerted by one mole of a gas systematically depends on its temperature and volume according to the law $pV = KT$, where $K$ is a constant (I pretend for the purposes of illustration that all gases are ideal gases). In the actual world, whenever there is a mole of gas at a given temperature and volume, its pressure will be determined: it is empirically impossible that two distinct moles of gas could have the same temperature and volume, but different pressure. It follows that the pressure of a mole of gas supervenes on its temperature and volume in a certain sense. (In this example, I am taking the class of A-properties to be much narrower than the class of physical properties, for reasons that will become clear.) But this supervenience is weaker than logical supervenience. It is *logically* possible that a mole of gas with a given temperature and volume might have a different pressure; imagine a world in which the gas constant *K* is larger or smaller, for example. Rather, it is just a fact about *nature* that there is this correlation.

This is an example of *natural* supervenience of one property on others: in this instance, pressure supervenes naturally on temperature, volume, and the property of being a mole of gas. In general, B-properties supervene naturally on A-properties if any two *naturally possible* situations with the same A-properties have the same B-properties.

A naturally possible situation is one that could actually occur in nature, without violating any natural laws. This is a much stronger constraint than mere logical possibility. The scenario with a different gas constant is logically possible, for example, but it could never occur in the real world, so it is not naturally possible. Among naturally possible situations, any two moles of gas with the same temperature and volume will have the same pressure.

Intuitively, natural possibility corresponds to what we think of as real *empirical* possibility—a naturally possible situation is one that could come up in the real world, if the conditions were right. These include not just actual situations but counterfactual situations that might have come up in the world's history, if boundary conditions had been different, or that might come up in the future, depending on how things go. A mile-high skyscraper is almost certainly naturally possible, for example, even though none has actually been constructed. It is even naturally possible (although wildly improbable) that a monkey could type *Hamlet*. We can also think of a naturally possible situation as one that conforms to the laws of nature of our world.[7] For this reason, natural possibility is sometimes called *nomic* or *nomological* possibility,[8] from the Greek term *nomos* for "law."

There are a vast number of logically possible situations that are not naturally possible. Any situation that violates the laws of nature of our world falls into this class: a universe without gravity, for example, or with different values of fundamental constants. Science fiction provides many situations of this sort, such as antigravity devices and perpetual-motion machines. These are easy to imagine, but almost certainly could never come to exist in our world.

In the reverse direction, any situation that is naturally possible will be logically possible. The class of natural possibilities is therefore a subset of the class of logical possibilities. To illustrate this distinction: both a cubic mile of gold and a cubic mile of uranium-235 seem to be logically possible, but as far as we know, only the first is naturally possible—a (stable) cubic mile of uranium-235 could not exist in our world.

Natural supervenience holds when, among all naturally possible situations, those with the same distribution of A-properties have the same distribution of B-properties: that is, when the A-facts about a situation *naturally necessitate* the B-facts. This happens when the same clusters of A-properties in our world are always accompanied by the same B-properties, and when this correlation is not just coincidental but *lawful*: that is, when instantiating the A-properties will always bring about the B-properties, wherever and whenever this happens. (In philosophical terms, the dependence must support counterfactuals.) This co-occurrence need not hold in every logically possible situation, but it must hold in every naturally possible situation.

It is clear that logical supervenience implies natural supervenience. If any two logically possible situations with the same A-properties have the same B-properties, then any two naturally possible situations will also. The reverse does not hold, however, as the gas law illustrates. The temperature and volume of a mole of gas determine pressure across naturally but not logically possible situations, so pressure depends naturally but not logically on temperature and volume. Where we have natural supervenience without logical supervenience, I will say that we have *mere* natural supervenience.

For reasons that will become clear, it is hard to find cases of natural supervenience on the set of *physical* properties without logical supervenience, but consciousness itself can provide a useful illustration. It seems very likely that consciousness is naturally supervenient on physical properties, locally or globally, insofar as in the natural world, any two physically identical creatures will have qualitatively identical experiences. It is not at all clear that consciousness is logically supervenient on physical properties, however. It seems *logically* possible, at least to many, that a creature physically identical to a conscious creature might have no conscious experiences at all, or that it might have conscious experiences of a different kind. (Some dispute this, but I use it for now only as an illustration.) If this is so, then conscious experience supervenes naturally but not logically on the physical. The necessary connection between physical structure and experience is ensured only by the laws of nature, and not by any logical or conceptual force.

The distinction between logical and natural supervenience is vital for our purposes.[9] We can intuitively understand the distinction as follows. If B-properties supervene logically on A-properties, then once God (hypothetically) creates a world with certain A-facts, the B-facts come along for free as an automatic consequence. If B-properties merely supervene naturally on A-properties, however, then after making sure of the A-facts, God has to do more work in order to make sure of the B-facts: he has to make sure there is a law relating the A-facts and the B-facts. (I borrow this image from Kripke 1972.) Once the law is in place, the relevant A-facts will automatically bring along the B-facts; but one could, in principle, have had a situation where they did not.

One also sometimes hears talk of *metaphysical* supervenience, which is based on neither logical nor natural necessity, but on "necessity *tout court*," or "metaphysical necessity" as it is sometimes known (drawing inspiration from Kripke's [1972] discussion of *a posteriori* necessity). I will argue later that the metaphysically possible worlds are just the logically possible worlds (and that metaphysical possibility of statements is logical possibility with an *a posteriori* semantic twist), but for now it is harmless to assume there is a notion of metaphysical supervenience, to be spelled out by analogy with the notions of logical and natural supervenience above. A notion of "weak" supervenience is also mentioned occasionally, but seems too weak to express an interesting dependence relation between properties.[10]

The logical–natural distinction and the global–local distinction cut across each other. It is reasonable to speak of both global logical supervenience and local logical supervenience, although I will more often be concerned with the former. When I speak of logical supervenience without a further modifier, global logical supervenience is intended. It is also coherent to speak of global and local natural supervenience, but the natural supervenience relations with which we are concerned are generally local or at least localizable, for the simple reason that evidence for a natural supervenience relation generally consists in local regularities between clusters of properties.[11]

### A Problem with Logical Supervenience

A technical problem with the notion of logical supervenience needs to be dealt with. This problem arises from the logical possibility of a world physically identical to ours, but with additional nonphysical stuff that is not present in our own world: angels, ectoplasm, and ghosts, for example. There is a *conceivable* world just like ours except that it has some extra angels hovering in a non-physical realm, made of ectoplasm. These angels might have biological properties of their own, if they reproduced and evolved. Presumably the angels could have all sorts of beliefs, and their communities might have complex social structure.

The problem these examples pose is clear. The angel world is physically identical to ours, but it is biologically distinct. If the angel world is logically possible, then according to our definition biological properties are not supervenient on physical properties. But we certainly *want* to say that biological properties are supervenient on physical properties, at least in *this* world if not in the angel world (assuming there are no angels in the actual world!). Intuitively, it seems undesirable for the mere logical possibility of the angel world to stand in the way of the determination of biological properties by physical properties in our own world.

This sort of problem has caused some (e.g., Haugeland 1982; Petrie 1987) to suggest that logical possibility and necessity are too strong to serve as the relevant sort of possibility and necessity in supervenience relations, and that a weaker variety such as natural possibility and necessity should be used instead. But this would render useless the very useful distinction between logical and natural supervenience outlined above, and would also ignore the fact that there is a very real sense in which the biological facts about our world are logically determined by the physical facts. Others (e.g., Teller 1989) have bitten the bullet by stipulating that worlds with extra nonphysical stuff are not logically or metaphysically possible, despite appearances, but this makes logical and metaphysical possibility seem quite arbitrary. Fortunately, such moves are not required. It turns out that it is possible to retain a useful notion of logical supervenience compatible with the possibility of these worlds, as long as we fix the definition appropriately.[12]

The key to the solution is to turn supervenience into a thesis about *our* world (or more generally, about particular worlds). This accords with the intuition that biological facts are logically determined by the physical facts in our world, despite the existence of bizarre worlds where they are not so determined. According to a revised definition, B-properties are logically supervenient on A-properties if the B-properties in our world are logically determined by the A-properties in the following sense: in any possible world with the same A-facts, the same B-facts will hold.[13] The existence of possible worlds with *extra* B-facts will thus not count against logical supervenience in our world, as long as *at least* the B-facts true in our world are true in all physically identical worlds. And this they generally will be (with an exception discussed below). If there is a koala eating in a gum tree in this world, there will be an identical koala eating in a gum tree in any physically identical world, whether or not that world has any angels hanging around.

There is a minor complication. There is a certain sort of biological fact about our world that does not hold in the angel world: the fact that our world has no living ectoplasm, for example, and the fact that all living things are based on DNA. Perhaps the angel world might even be set up with ectoplasm causally dependent on physical processes, so that wombat copulation on the physical plane sometimes gives rise to baby ectoplasmic wombats on the nonphysical plane. If so, then there might be a wombat that is childless (in a certain sense) in our world, with a counterpart that is not childless in the physically identical angel world. It follows that the property of being childless does not supervene according to our definition, and nor do the world-level properties such as that of having no living ectoplasm. Not all the facts about our world follow from the physical facts alone.

To analyze the problem, note that these facts all involve negative existence claims, and so depend not only on what is going on in our world but on what is not. We cannot expect these facts to be determined by any sort of localized facts, as they depend not just on local goings-on in the world but on the world's limits. Supervenience theses should apply only to *positive* facts and properties, those that cannot be negated simply by enlarging a world. We can define a positive fact in *W* as one that holds in every world that contains *W* as a proper part;[14] a positive property is one that if instantiated in a world *W*, is also instantiated by the corresponding individual in all worlds that contain *W* as a proper part.[15] Most everyday facts and properties are positive—think of the property of being a kangaroo, or of being six feet tall, or of having a child. Negative facts and properties will always involve negative existence claims in one form or another. These include explicitly negative existential facts such as the nonexistence of ectoplasm, universally quantified facts such as the fact that all living things are made of DNA, negative relational properties such as childlessness, and superlatives such as the property of being the most fecund organism in existence.

In future, the supervenience relations with which we are concerned should be understood to be restricted to positive facts and properties. When claiming that biological properties supervene on physical properties, it is only the positive biological properties that are at issue. All the properties with which we are concerned are positive—local physical and phenomenal properties, for instance—so this is not much of a restriction.

The definition of global logical supervenience of B-properties on A-properties therefore comes to this: for any logically possible world *W* that is A-indiscernible from our world, then the B-facts true of our world are true of *W*. We need not build in a clause about positiveness, but it will usually be understood that the only relevant B-facts and properties are positive facts and properties. Similarly, B-properties supervene locally and logically on A-properties when for every actual individual *x* and every logically possible individual *y*, if *y* is A-indiscernible from *x*, then the B-properties instantiated by *x* are instantiated by *y*. More briefly and more generally: B-properties supervene logically on A-properties if the B-facts about actual situations are entailed by the A-facts, where situations are understood as worlds and individuals in the global and local cases respectively. This definition captures the idea that supervenience claims are usually claims about our world, while retaining the key role of logical necessity.[16]

### Supervenience and Materialism

Logical and natural supervenience have quite different ramifications for ontology: that is, for the matter of what there is in the world. If B-properties are logically supervenient on A-properties, then there is a sense in which once the A-facts are given, the B-facts are a free lunch. Once God (hypothetically) made sure that all the physical facts in our world held, the biological facts came along for free. The B-facts merely redescribe what is described by the A-facts. They may be *different* facts (a fact about elephants is not a microphysical fact), but they are not *further* facts.

With mere natural supervenience, the ontology is not so straightforward. Contingent lawful connections connect distinct features of the world. In general, if B-properties are merely naturally supervenient on A-properties in our world, then there *could* have been a world in which our A-facts held without the B-facts. As we saw before, once God fixed all the A-facts, in order to fix the B-facts he had more work to do. The B-facts are something over and above the A-facts, and their satisfaction implies that there is something new in the world.

With this in mind we can formulate precisely the widely held doctrine of *materialism* (or *physicalism*), which is generally taken to hold that everything in the world is physical, or that there is nothing over and above the physical, or that the physical facts in a certain sense exhaust all the facts about the world. In our language, materialism is true if all the positive facts about the world are globally logically supervenient on the physical facts. This captures the intuitive notion that if materialism is true, then once God fixed the physical facts about the world, all the facts were fixed.

(Or at least, all the positive facts were fixed. The restriction to positive facts is needed to ensure that worlds with extra ectoplasmic facts do not count against materialism in our world. Negative existential facts such as "There are no angels" are not strictly logically supervenient on the physical, but their nonsupervenience is quite compatible with materialism. In a sense, to fix the negative facts, God had to do more than fix the physical facts; he also had to declare, "That's all." If we wanted, we could add a second-order "That's all" fact to the supervenience base in the definition of materialism, in which case the positive-fact constraint could be removed.)

According to this definition, materialism is true if all the positive facts about our world are entailed by the physical facts.[17] That is, materialism is true if for any logically possible world *W* that is physically indiscernible from our world, all the positive facts true of our world are true of *W*. This is equivalent in turn to the thesis that any world that is physically indiscernible from our world contains a copy of our world as a (proper or improper) part, which seems an intuitively correct definition.[18] (This matches the definition of physicalism given by Jackson [1994], whose criterion is that every minimal physical duplicate of our world is a duplicate *simpliciter* of our world.[19])

...Some may object to the use of logical possibility rather than possibility *tout court* or "metaphysical possibility." Those people may substitute metaphysical possibility for logical possibility in the definition above. Later, I will argue that it comes to the same thing.

### Notes

This chapter originally appeared in *The Conscious Mind*, pp. 32–42, New York, Oxford University Press, 1996.

1. The idea of supervenience was introduced by Moore (1922). The name was introduced in print by Hare (1952). Davidson (1970) was the first to apply to the notion to the mind–body problem. More recently, a sophisticated theory of supervenience has been developed by Kim (1978, 1984, 1993), Horgan (1982, 1984c, 1993), Hellman and Thompson (1975), and others.

2. I use "A-fact" as shorthand for "instantiation of an A-property." The appeal to facts makes the discussion less awkward, but all talk of facts and their relations can ultimately be cashed out in terms of patterns of co-instantiation of properties; I give the details in notes, where necessary. In

particular, it should be noted that the identity of the individual that instantiates an A-property is irrelevant to an A-fact as I am construing it; all that matters is the instantiation of the property. If the identity of an individual were partly constitutive of an A-fact, then any A-fact would entail facts about that individual's essential properties, in which case the definition of supervenience would lead to counterintuitive consequences.

3. I assume, perhaps artificially, that individuals have precise spatiotemporal boundaries, so that their physical properties consist in the properties instantiated in that region of space-time. If we are to count spatially distinct objects as physically identical for the purposes of local supervenience, any properties concerning absolute spatiotemporal position must be omitted from the supervenience base (although one could avoid the need to appeal to spatially distinct objects by considering only merely *possible* objects with the same position). Also, I always talk as if the same sort of individual instantiates low-level and high-level properties, so that a table, for example, instantiates microphysical properties by virtue of being characterized by a distribution of such properties. Perhaps it would be more strictly correct to talk of microphysical properties as being instantiated only by microphysical entities, but my way of speaking simplifies things. In any case, the truly central issues will all involve global rather than local supervenience.

4. There are various ways to specify precisely what it is for two worlds to be identical with respect to a set of properties; this will not matter much to the discussion. Perhaps the best is to say that two worlds are identical with respect to their A-properties if there is a one-to-one mapping between the classes of individuals instantiating A-properties in both worlds, such that any two corresponding individuals instantiate the same A-properties. For the purposes of global supervenience we then need to stipulated that the mappings by which two worlds are seen to be both A- and B-indiscernible are compatible with each other; that is, no individual is mapped to one counterpart under the A-mapping but to another under the B-mapping. The definition of global supervenience takes this form: Any two worlds that are A-identical (under a mapping) are B-identical (under an extension of that mapping).

A more common way to do this is to stipulate that A-identical worlds must contain exactly the same individuals, instantiating the same properties, but as McLaughlin (1995) points out, this is unreasonably strong: it ensures that such things as the cardinality of the world and essential properties of individuals supervene on any properties whatsoever. The definition I propose gets around this problem, by ensuring that *only* patterns of A-properties and nothing further enter into the determination relation.

5. With one exception: God could not have created a world that was not created by God, even though a world not created by God is presumably logically possible! I will ignore this sort of complication.

6. The relationship of this sort of possibility to deducibility in formal systems is a subtle one. It is arguable that the axioms and inference rules of specific formal systems are justified precisely in terms of a prior notion of logical possibility and necessity.

7. The intuitive notion of natural possibility is conceptually prior to the definition in terms of laws of nature: a regularity qualifies as a law just in case it holds in all situations that could come up in nature; that is, in all situations that are naturally possible in the intuitive sense. As it is

sometimes put, for something to count as a law it must hold not just in actual but in counterfactual situations, and the more basic notion of natural possibility is required to determine which counterfactual situations are relevant.

8. The terms "physical necessity" and "causal necessity" are also often used to pick out roughly this brand of necessity, but I do not wish to beg the question of whether all the laws of nature are physical or causal.

9. The important distinction between logical and natural supervenience is frequently glossed over or ignored in the literature, where the modality of supervenience relations is often left unspecified. Natural (or nomological) supervenience without logical supervenience is discussed by van Cleve (1990), who uses it to explicate a variety of emergence. Seager (1991) spells out a related distinction between what he calls *constitutive* and *correlative* supervenience. These correspond in a straightforward way to logical and natural supervenience, although Seager does not analyze the notions in quite the same way.

10. Weak supervenience requires only that "no B-difference without an A-difference" holds within a world, rather than across worlds (see Kim 1984 for details). The lack of modal strength in this relation makes it too weak for most purposes. At best, it may have a role in expressing conceptual constraints on nonfactual discourse (as in Hare 1984), although as Horgan (1993) points out, even these constraints seem to involve cross-world dependence. Seager (1988) appeals to weak supervenience to express a kind of systematic within-world correlation that is not strictly necessary, but natural supervenience serves this purpose much better.

11. Global natural supervenience without localized regularity is a coherent notion on a non-Humean account of laws, although perhaps not on a Humean (regularity-based) account. Even on a non-Humean account, though, it is hard to see what the evidence for such a relation could consist in.

12. Horgan (1982), Jackson (1994), and Lewis (1983b) address a related problem in the context of defining materialism.

13. The revised definition can be spelled out more precisely along the lines of note 4. Let $B(W)$ be the class of individuals with B-properties in a world $W$. We can say that $W'$ is *B-superior* to $W$ if there is an injection $f : B(W) \rightarrow B(W')$ (i.e., a one-to-one mapping from $B(W)$ onto a subset of $B(W')$) such that for all $a \in B(W)$, $f(a)$ instantiates every B-property that $a$ does. Then B-properties supervene logically on A-properties in $W$ if every world that is A-indiscernible from $W$ is B-superior to $W$, where the relevant B-mappings are again constrained to be extensions of the A-mappings.

To see that the constraint is necessary, imagine that our world has a countably infinite number of psychologically identical minds, of which one is realized in ectoplasm and the rest are physically realized. Intuitively, the psychological does not supervene on the physical in this world, but every physically indiscernible world is psychologically superior. Although we expect the ectoplasm-free world to count against supervenience, there is a one-to-one mapping between the psychologies in that world and in our world. The problem is that this mapping does not respect physical correspondence, as it maps a physical entity to an ectoplasmic entity; so we need the further constraint.

14. For the purposes of this definition, the containment relation between worlds can be taken as primitive. Lewis (1983a) and Jackson (1993) have noted that it is fruitless to analyze this sort of notion forever. Something needs to be taken as primitive, and the containment relation seems to be as clear as any. Some might prefer to speak, not of worlds that contain *W* as a proper part, but of worlds that contain a qualitative duplicate of *W* as a proper part; this works equally well.

15. Note that by this definition, there are positive facts that are not instantiations of positive properties. Think of instantiations of the property of being childless-or-a-kangaroo, for example. Perhaps positive facts should be defined more strictly as instantiations of positive properties, but as far as I can tell the weaker definition has no ill effects.

16. Arguably, the logical supervenience of properties in our world should be a *lawful* thesis. If it were the case that there *would* have been nonphysical living angels if things had gone a little differently in our world (perhaps a few different random fluctuations), even though the laws of nature were being obeyed, then it would be a mere accident of history that biological properties are logically supervenient on physical properties. One gets a stronger and more interesting metaphysical thesis by replacing the references to our world and actual individuals in the definitions of logical supervenience by a reference to naturally possible worlds and individuals. This rules out scenarios like the one above. As a bonus, it allows us to determine whether or not uninstantiated properties, such as that of being a mile-high skyscraper, are logically supervenient. On the previous definition, all such properties supervene vacuously.

This yields the following definition: B-properties are logically supervenient on A-properties iff for any naturally possible situation *X* and any logically possible situation *Y*, if *X* and *Y* are A-indiscernible then *Y* is B-superior to *X* (with the usual constraint). Or more briefly: for any naturally possible situation, the B-facts about that situation are entailed by the A-facts. This modification makes no significant difference to the discussion in the text, so I omit it for simplicity's sake. The discussion can easily be recast in terms of the stricter definition simply by replacing relevant references to "our world" by references to "all naturally possible worlds" throughout the text, usually without any loss in the plausibility of the associated claims.

The result resembles the standard definition of "strong" local supervenience (Kim 1984), in which there are two modal operators. According to that definition, B-properties supervene on A-properties if necessarily, for each *x* and each B-property *F*, if *x* has *F*, then there is an A-property *G* such that *x* has *G*, and necessarily if any *y* has *G*, it has *F*. (The A-properties *G* may be thought of as complexes of simpler A-properties, if necessary.) The angel issue makes it clear that the first modal operator should always be understood as natural necessity, even when the second is logical necessity. The standard definition of global supervenience (that A-indiscernible worlds are B-indiscernible) is less well off, and needs to be modified along the lines I have suggested. A parallel definition of "metaphysical" supervenience can be given if necessary. Of course, the angel problems do not arise for natural supervenience, as there is no reason to believe that ectoplasm is naturally possible, so the straightforward definition of natural supervenience is satisfactory.

17. Arguably, we should use the stronger definition of logical supervenience, so that materialism is true if all the positive facts about all *naturally possible* worlds are entailed by the physical facts about those worlds. Take an ectoplasm-free world in which nonphysical ectoplasm is nevertheless a natural possibility—perhaps it would have evolved if a few random fluctuations had gone

differently. It seems reasonable to say that materialism is false in such a world, or at least that it is true only in a weak sense.

18. To gain the equivalence, we need the plausible principle that if world *A* is a proper part of world *B*, then some positive fact holds in *B* that does not hold in *A*; that is, there is some fact that holds in *B* and in all larger worlds that does not hold in *A*.

19. It also comes to much the same thing as definitions by Horgan (1982) and Lewis (1983b), but unlike these it does not rely on the somewhat obscure notion of "alien property" to rule out ectoplasmic worlds from the range of relevant possible worlds.

## Bibliography

Davidson, D. (1970). "Mental Events." In L. Foster and J. Swanson (eds). *Experience and Theory.* London: Duckworth.

Hare, R. M. (1952). *The Language of Morals.* Oxford: Clarendon Press.

——— (1984). "Supervenience," *Proceedings of the Aristotelian Society*, suppl., 58, pp. 1–16.

Haugeland, J. (1982). "Weak Supervenience," *American Philosophical Quarterly*, 19, pp. 93–103.

Hellman, G., and Thompson, F. (1975). "Physicalism: Ontology, determination and reduction," *Journal of Philosophy* 72, pp. 551–564.

Horgan, T. (1982). "Supervenience and Microphysics," *Pacific Philosophical Quarterly*, 63, pp. 29–43.

——— (1984). "Supervenience and Cosmic Hermeneutics," *Southern Journal of Philosophy*, suppl., 22, 19–38.

——— (1993). "From Supervenience to Superdupervenience: Meeting the demands of a material world," *Mind* 102, pp. 555–586.

Jackson, F. (1993). "Armchair Metaphysics." In J. O'Leary-Hawthorne and M. Michael (eds), *Philosophy of Mind.* Dordrecht: Kluwer.

——— (1994). "Finding the Mind in the Natural World." In R. Casati, B. Smith, and G. White (eds), *Philosophy and the Cognitive Sciences*, Vienna: Holder-Pichler-Tempsky.

Kim, Jaegwon (1978). "Supervenience and Nomological Incommensurables," *American Philosophical Quarterly*, 15, pp. 149–156.

——— (1984). "Concepts of Supervenience," *Philosophy and Phenomological Resarch*, 45, pp. 153–176.

——— (1993). *Supervenience and Mind.* Cambridge: Cambridge University Press.

Kripke, Saul (1971). "Naming and Necessity," pp. 253–355 and 763–769 in *Semantics of Natural Language*, D. Davidson and G. Harman (eds). Dordrecht: D. Reidel Publishing Company.

——— (1980). *Naming and Necessity.* Cambridge, MA: Harvard University Press.

Lewis, David (1983a). "Extrinsic Properites," *Philosophical Studies*, 44, pp. 197–200.

——— (1983b). "New Work for a Theory of Universals," *Australian Journal of Philosophy*, 61, pp. 343–377.

McLaughlin, B. P. (1995). "Varieties of Supervenience," in E. E. Savellos and U. D. Yalein (eds), *Supervenience: New Essays*. Cambridge: Cambridge University Press.

Moore, G. E. (1922). *Philosophical Studies*. London: Routledge and Kegan Paul.

Petrie, B. (1987). "Global Supervenience and Reduction," *Philosophy and Phenomenological Research* 48, pp. 119–130.

Seager, W. E. (1988). "Weak Supervenience and Materialism," *Philosophy and Phenomological Research*, 48, pp. 697–709.

——— (1991). *Metaphysics and Consciousness*. London: Routledge.

van, Cleve, James (1990). "Mind-dust or magic? Panpsychism versus emergence," *Philosophical Perspectives* 4, pp. 215–226.

# 24 The Nonreductivist's Troubles with Mental Causation

Jaegwon Kim

## 24.1 A Bifurcated World or a Layered One?

Mind–body dualism in the classic Cartesian style envisages two non-overlapping domains of particulars ("substances") that are, by and large, equal in ontological standing. Mental items are thought to share a certain defining property ("thinking" or "consciousness," according to Descartes) that excludes the defining property shared by the items on the physical side ("extension," according to Descartes). And associated with each domain is a distinct family of properties, mental properties for one and physical properties for the other, in terms of which the particulars within that domain can be exhaustively characterized. We are thus presented with a bifurcated picture of reality: the world consists of two metaphysically independent spheres existing side by side.

But not everyone who accepts a picture like this thinks that the two domains are entirely unrelated; although there are notable exceptions, such as Leibniz and Malebranche, many substantival dualists, including of course Descartes, have held that, in spite of their separateness and independence, the domains are causally connected: mental events can be, and sometimes are, causes and effects of physical events, and changes in a mind can be causes or effects of changes in a body. This means that events of both kinds can occur as *links in the same causal chain*: if you pick a physical event and trace its causal ancestry or posterity, you may run into mental events, and similarly if you start off with a mental event. It follows then that under Cartesian causal dualism there can be *no complete physical theory of physical phenomena.* For it allows physical occurrences that cannot be causally explained by invoking physical antecedents and laws alone. Any comprehensive theory of the physical world must, on Cartesian interactionism, include references to non-physical causal agents and laws governing their behaviour. We can say then that *Cartesian interactionism violates the causal closure of the physical domain.* Of course, it violates the causal closure of the mental domain as well; Cartesianism implies that no scientific theory could hope to achieve complete coverage unless it encompassed both the physical and mental realms—unless, that is, we had a unified theory of both mental and physical phenomena.

The ontological picture that has dominated contemporary thinking on the mind problem is strikingly different from the Cartesian picture. The Cartesian model of a bifurcated world has been replaced by that of a layered world, a hierarchically stratified structure of "levels" or "orders" of entities and their characteristic properties. It is generally thought that there is a bottom level, one consisting of whatever micro-physics is going to tell us are the most basic physical particles out of which all matter is composed (electrons, neutrons, quarks, or whatever). And these objects, whatever they are, are characterized by certain fundamental physical properties and relations (mass, spin, charm, or whatever). As we ascend to higher levels, we find structures that are made up of entities belonging to the lower levels, and, moreover, the entities at any given level are thought to be characterized by a set of properties distinctive of that level. Thus, at a certain level, we will find lumps of $H_2O$ molecules, with such properties as transparency, power to dissolve sugar and salt, a characteristic density and viscosity, etc. At still higher levels we will find cells and organisms with their "vital" properties, and farther up organisms with consciousness and intentionality. Beyond them, there are social groups of organisms, and perhaps groups consisting of such groups.[1] Sometimes, one speaks in terms of "levels of description," "levels of analysis," or "levels of language"; the layered model is often implicit in such talk.

Thus, the world as portrayed in the new picture consists of an array of levels, each level consisting of two components: a set of *entities* constituting the domain of particulars for that level and a set of *properties* defined over this domain. What gives this array structure is the mereological relation of *being part of*: entities belonging to a given layer are mereologically composed of entities belonging to the lower levels, and this relation generates a hierarchical ordering of the levels. As earlier noted, this multi-tiered picture usually carries the assumption that there is a bottom tier, a layer of entities that have no physically significant parts.

The characterization thus far of the layered model leaves one important question unanswered: how are the properties characteristic of entities at a given level related to those that characterize entities of adjacent levels? Given that entities at distinct levels are ordered by the part–whole relation, is it the case that properties associated with different levels are also ordered by some distinctive and significant relationship?[2]

That is the crucial question answers to which have defined various currently contested positions on certain metaphysical and methodological issues including, most notably, the mind body problem. The classic positivist answer is that the distinctive properties of entities at a given level are *reducible to*, or *reductively explainable in terms of*, the properties and relations characterizing entities at lower levels. That is "reductionism." But reductionism has had a rough time of it for the past few decades, and has been eclipsed by its major rivals, "eliminativism" and "non-reductivism." These positions agree in their claim that higher-level properties are in general not reducible to lower-level ones, but differ on the status of irreducible higher properties. Non-reductivism maintains that they can be real and genuine properties of objects and events of this world, constituting an ineliminable part of its true ontology. Eliminativism, on the other hand, holds that they are useless danglers that must be expunged from the correct picture of reality. Thus, the split between non-reductivism and eliminativism hinges on the significance of reducibility: the former, unlike the latter, rejects reducibility as a test of legitimacy for higher-level properties, and holds that such properties can form an autonomous domain, a domain for an independent "special science" that is irreducible to the sciences about lower-level phenomena. "Emergentism," which was influential during the first half of this century, was the first systematic articulation of this non-reductivist approach.

At first blush, the layered model many appear to hold promise as an elegant way of averting violation of the causal closure of the physical: causal interactions could perhaps be confined within each level, in a way that respected the autonomy and closedness of the causal processes at the fundamental physical level. In particular, on the non-reductivist version of the layered model, it may be possible to view causal chains at a given level, like the properties distinctive of that level, as forming an autonomous and self-contained realm immune to causal intrusions from neighbouring levels. Moreover, this picture may not preclude the assignment of some special status to physical causation: in spite of the causal-nomological autonomy at each level, causal relations at higher levels may in some sense depend, or supervene, on the causal-nomological processes occurring at the lower levels (McLaughlin 1989, Fodor 1989)—just as, on the non-reductivist view, the irreducibility of higher-level properties is thought to be consistent with their supervenience on the lower-level properties and relations.

Let us narrow our focus on the mind–body problem. "Non-reductive physicalism," a position that can deservedly be called "the received view" of today, is non-reductivism applied to the mind–body case. It consists of the two characteristic theses of non-reductivism: its ontology is physical monism, the thesis that physical entities and their mereological aggregates are all that there is; but its "ideology" is anti-reductionist and dualist, consisting in the claim that psychological properties are irreducibly distinct from the underlying physical and biological properties. Its dualism is reflected in the belief that, though physically irreducible, psychological properties are genuine properties nonetheless, as real as underlying physical-biological properties. And there is a corollary about psychology: psychology is an autonomous special science independent of the physical and biological sciences in its methodology and in the concepts and explanations it generates.[3]

As we saw, Cartesian interactionism involves violation of the causal closure of the physical, and that was one cause of its downfall. I shall argue that non-reductive physicalism, and its more generalized companion, emergentism, are vulnerable to similar difficulties; in particular, it will be seen that the physical causal closure remains very much a problem within the stratified ontology of non-reductivism. Non-reductive physicalism, like Cartesianism, founders on the rocks of mental causation.

## 24.2 Non-Reductive Physicalism and "Physical Realization"

The basic ontological thesis of non-reductive physicalism confers on the physical a certain kind of primacy: all concrete existents are physical—there are no non-physical particulars, no souls, no Cartesian mental substances, and no "vital principles" or "entelechies." Stated as a thesis about properties, physical primacy in this sense comes to this: all mental properties are instantiated in physical particulars. Thus, although there can be, and presumably are, objects and events that have only physical properties, there can be none with mental properties alone; mentality must be instantiated in physical systems.

There appears to be no generally accepted account of exactly what it means to say that something is "physical." Minimally perhaps a physical entity must have a determinate location in space and time; but that may not be enough. Perhaps an entity is physical just in case it has some physical property or other. But what makes a property a physical property? Perhaps, the best answer we could muster is what Hellman and Thompson (1975) have offered: explain "physical" by reference to current theoretical physics. This strategy can be extended to higher-level sciences, chemistry and biology; when we reach psychological properties, however, the question as to what should be regarded as our reference scientific theories is itself an unsettled philosophical issue centrally involved in the debates about reductionism and mental causation. For mental properties, then, we must look to vernacular psychology and its characteristic intentional idioms of belief, desire, and the rest, and their intentional analogues in systematic psychology. Nothing in the discussion to follow will depend on precise general definitions of "physical" and "mental."

In any case, many physicalists, including most non-reductive physicalists, are not willing to rest with the ontological primacy of the physical in the sense explained, just as many substantival dualists are not content merely to posit two separate and independent domains. Most non-reductive physicalists want to go beyond the claim that mental properties are had by physical systems; they want to defend a thesis of *primacy, or basicness, for physical properties in relation to mental properties*. The main idea here is that, in spite of their irreducibility, mental properties are in some robust sense dependent on or determined by physical-biological properties. This means that the property dualism of non-reductive physicalism is an attenuated dualism: it is a dualism with dependency relations between the two domains, just as Cartesian dualism is a dualism with causal connections between its two domains. For many non-reductive physicalists, therefore, irreducibility is not the last word on the mind–body relationship. The irreducibility claim is a negative thesis; non-reductive physicalists want a positive account of the relationship between the two sets of properties. Here, the catch-words are "dependence" and "determination." Hellman and Thompson, who have proposed an elegant form of non-reductive physicalism, describe their project as follows:

> Of late there has been a growing awareness, however, that reductionism is an unreasonably strong claim. Along with this has come recognition that reductionism is to be distinguished from a purely ontological thesis concerning the sorts of entities of which the world is constituted.... Although a purely ontological thesis is a necessary component of physicalism it is insufficient in that it makes no appeal to the power of physical law.... We seek to develop principles of *physical determination* that spell out rather precisely the underlying physicalist intuition that the physical facts determine all the facts. (pp. 551–552)

Hellman and Thompson speak of physical facts as determining all the facts; presumably, that happens only because what objects have which physical properties (including relations) determines what mental properties these objects have.

But how do we capture this relation of determination, or dependence, in a way that escapes the threat of reductionism? Answering this question has been one of the principal projects of non-reductive physicalists in the past two decades. Two ideas have been prominent: "supervenience" and "physical realization." Hellman and Thompson themselves gave an account of the determination relation that is very close to what is now commonly known as "global supervenience": once the physical character of a world is fixed, its entire character is thereby fixed.[4] The idea that mental states are "physically realized" (or "instantiated" or "implemented") gained currency, in the late 1960s,[5] chiefly through an argument ("the multiple realization argument") that helped to defeat reductive physicalism and install non-reductivism in its current influential position. The reputed trouble for reductionism was that the realization is "multiple," namely that any given mental state is realizable in a variety of widely diverse physical structures, and that this makes its reductive identification with any single physical property hopeless. The two approaches, one based on "supervenience" and the other on "physical realization," are not mutually exclusive: as we shall see, on reasonable readings of the terms involved the claim that mental states are physically realized arguably entails the claim that they are physically supervenient. Whether the converse entailment holds is a question that depends, among other things, on the strength of the supervenience relation involved.

Many non-reductive physicalists avail themselves of both supervenience and realization, and some do so explicitly; for example, LePore and Loewer characterize non-reductive physicalism by three principles, one of which is "global supervenience," the thesis that "If two nomologically possible worlds are exactly alike with respect to fundamental physical facts.... then they are exactly alike with respect to all other facts" (1989, pp. 177–178). But they go on to say: "The relationship between psychological and neurophysiological properties is that the latter *realise* the former." What is it for a property to "realize" another? LePore and Loewer explain:

> Exactly what is it for one of an event's properties to *realize* another? The usual conception is that *e*'s being *P* realizes *e*'s being *F* iff *e* is *P* and *e* is *F* and there is a strong connection of some sort between *P* and *F*. We propose to understand this connection as a necessary connection which is *explanatory*. The existence of an explanatory connection between two properties is stronger than the claim that $P \rightarrow M$ is physically necessary since not every physically necessary connection is explanatory. (1989, p. 179)

It is clear that if all mental properties are realized in this sense by physical properties, the global supervenience of the mental on the physical is assured. In fact, their Physical Realization Thesis, as we might call it, entails a stronger form of supervenience, the "strong supervenience" of mental properties on physical ones (Kim 1984a).

For the purposes of my arguments in this paper I wish to focus on those versions of non-reductive physicalism that make use of "physical realization" to explain the psycho-physical property relationship. This for two reasons: first, there are a variety of non-equivalent supervenience relations, and this makes it difficult to formulate a reasonably uniform and perspicuous argument concerning supervenience-based versions; and, second, many philosophers have been converted to non-reductive physicalism by "the multiple realization argument" briefly alluded to earlier. In consequence, for many non-reductive physicalists, the Physical Realization Thesis is one of their early and basic commitments. It is often a sound expository strategy to formulate an argument in a stark and perspicuous fashion even if this involves making use of fairly strong premisses, and then worry about how the argument might be qualified and fine-tuned to accommodate weaker assumptions. It may well be that there are versions of non-reductive physicalism to which my considerations do not apply, at least not directly;[6] I believe, though, that they are relevant to many of the more popular and influential versions of it.

Let us return to the concept of realization. LePore and Loewer, as we saw above, explain "realization" for properties of events; however, there is no need to confine the relation to events, and we will assume that the realization relation, without substantive changes, applies to properties of objects ("substances") as well. In any case, according to LePore and Loewer, *P* realizes *M* just in case (1) $P \rightarrow M$ holds with nomological necessity (LePore and Loewer say "physical necessity") and (2) *P* "explains" *M*. I think LePore and Loewer are correct in suggesting these two kinds of conditions; however, their specific conditions need improvement. Consider first their explanatory requirement: I believe the idea behind this requirement is correct, but we should look for an objective metaphysical relation between *P* and *M*, not an essentially epistemic relation like explanation; that is, we should view the explanatory relation between the two properties as being supported by a metaphysical realization relation. Here I am taking a realist attitude about explanation: if $P$ explains $M$, that is so because some objective metaphysical relation holds between $P$ and $M$. That $P$ explains $M$ cannot be a brute, fundamental fact about $P$ and $M$. In causal explanations the required relation of course is the causal relation. In the case of realization, the key concepts, I suggest, are those of "causal mechanism" and "microstructure." When $P$ is said to "realize" $M$ in system $s$, $P$ must specify a micro-structural property of $s$ that provides a causal mechanism for the implementation of $M$ in $s$; moreover, in interesting cases—in fact, if we are speak meaningfully of "implementation" of $M$—$P$ will be a member of a family of physical properties forming a network of nomologically connected micro-structural states that provides a micro-causal mechanism, in systems appropriately like $s$, for the nomological connections among a broad system of mental properties of which $M$ is an element. These underlying micro-states will form an explanatory basis for the higher properties and the nomic relations among them; but the realization relation itself must be distinguished from the explanatory relation. Thus, my difference with LePore and Loewer in regard to their condition (2) is quite small: I agree that something like their explanatory condition should in general hold for the realization relation; however, it should not be regarded as constitutive of it.

What of the condition (1), to the effect that $P \rightarrow M$ be nomically necessary? I believe this condition is acceptable with the following proviso: in each system $s$ in which $M$ is physically realized, there must be a determinate set, finite or infinite, of physical properties, $P_1, P_2, \ldots$, each of which realizes $M$ in the sense explained, so that we may consider the disjunctive property, $P_1^{\vee}\, P_2^{\vee} \ldots$, as a nomic co-extension of $M$. Since I will not be making use of this requirement in this paper, I will not offer an argument for it here.[7]

## 24.3 Non-Reductive Physicalism as Emergentism

The non-reductive physicalism I have in mind, therefore, consists of the following theses:

1. Physical Monism All concrete particulars are physical.

2. Anti-Reductionism Mental properties are not reducible to physical properties.

3. The Physical Realization Thesis All mental properties are physically realized; that is, whenever an organism, or system, instantiates a mental property $M$, it has some physical property $P$ such that $P$ realizes $M$ in organisms of its kind.

And we must add a further thesis that is implicit in the above three and is usually taken for granted:

4. Mental Realism Mental properties are real properties of objects and events; they are not merely useful aids in making predictions or fictitious manners of speech.

I believe that these four basic tenets of non-reductive physicalism bring the position very close to "emergentism"—so close, in fact, that non-reductive physicalism of this variety is best viewed as a form of emergentism. I shall briefly explain why this is so (for more details, see Kim 1992).

Emergentists in general accepted a purely materialist ontology of concrete physical objects and events. For example, Samuel Alexander, one of the principal theoreticians of the emergence school, argues that there are no mental events over and above neural processes:

> We thus become aware, partly by experience, partly by reflection, that a process with the distinctive quality of mind or consciousness is in the same place and time with a neural process, that is, with a highly differentiated and complex process of our living body. We are forced, therefore, to go beyond the mere correlation of the mental with these neural processes and to identify them. There is but one process which, being of a specific complexity, has the quality of consciousness.... It has then to be accepted as an empirical fact that a neural process of a certain level of development possesses the quality of consciousness and is thereby a mental process; and, alternately, a mental process is *also* a vital one of a certain order. (1927, pp. 5–6)

This is, almost word for word, just what is claimed by "token physicalism" or "the token-identity thesis," a form of non-reductive physicalism. It is no surprise then that Alexander calls his position a version of "the identity doctrine of mind and body."

Both the "layered" structure of the emergentist ontology and its fundamentally physicalist character are evident in the following passage from C. Lloyd Morgan, another leader of the movement:

> In the foregoing lecture the notion of a pyramid with ascending levels was put forward. Near its base is a swarm of atoms with relational structure and the quality we may call atomicity. Above this level, atoms combine to form new units, the distinguishing quality of which is molecularity; higher up, on one line of advance, are, let us say, crystals wherein atoms and molecules are grouped in new relations of which the expression is crystalline form; on another line of advance are organisms with a different kind of natural relations which give the quality of vitality; yet higher, a new kind of natural relatedness supervenes and to its expression the word "mentality" may...be applied. Vital*ism* and anim*ism* are excluded if they imply the insertion of Entelechy. (1923, p. 35)

Atoms and their mereological aggregates exhaust all of concrete existence; on "entelechies," or any other physically alien entities, are to be "inserted" at any point in the hierarchy of levels of existence, although new properties emerge to characterize the more complex structures of basic entities. There is no room in this picture for any concrete existent not fully decomposable into atoms and other basic physical particulars.

The emergentist doctrine that "emergent" properties are irreducible to the "basal conditions" out of which they emerge is familiar; to most of us, this irreducibility claim is constitutive of the emergentist metaphysical world-view. Although the emergentists' idea of reduction or reductive explanation diverges from the model of reduction implicit in current anti-reductionist arguments (see Kim 1992), the philosophical significance of the denial of reducibility between two property levels is the same: the higher-level properties, being irreducible, are genuine new additions to the ontology of the world. Alexander, for example, says this:

> Out of certain physiological conditions nature has framed a new quality mind, which is therefore not itself physiological though it lives and moves and has its being in physiological conditions. Hence it is that there can be and is an independent science of psychology.... No physiological constellation explains for us why it should be mind. (1927, p. 8)

This idea of irreducible higher properties lies at the basis of some recent versions of emergentism, such as one promoted by the noted neurophysiologist Roger Sperry, who writes:

> First, conscious awareness... is interpreted to be a dynamic emergent property of cerebral excitation. As such conscious experience becomes inseparably tied to the material brain process with all its structural and physiological constraints. At the same time the conscious properties of brain excitation are conceived to be something distinct and special in their own right.... Among other implications of the current view for brain research is the conclusion that a full explanation of the brain process at the conscious level will not be possible solely in terms of the biochemical and physiological data. (1969, p. 533–535)

For both emergentism and non-reductive physicalism, then, the doctrine of irreducible higher-level properties is the centre-piece of their respective positions; and their proponents take it to be what makes their positions distinctive and important. As net additions to the world, the emergent higher-level properties cannot be reduced or explained away; and as irreducible new features of the world, they form an autonomous domain, and, as Alexander says, make "an independent science of psychology" possible. This is exactly what current non-reductive physicalists have been urging for over two decades: the "special sciences" are autonomous and independent from the underlying physical and biological sciences.[8]

Let us now turn to the third basic tenet of non-reductive physicalism, the Physical Realization Thesis. This involves the claim that for a mental property to be instantiated in a system, that system must instantiate an appropriate physical property, and further that whenever any system instantiates this physical property, the mental property must of necessity be instantiated by it as well. Mental events and states require physical bases, and when required physical bases are present, they must occur. A precisely parallel thesis was part of the emergentist doctrine: the emergence of higher-level properties require appropriate "basal conditions," and when these basal conditions are present, they must of necessity emerge. For both the non-reductive physicalist and the emergentist, physical bases are by themselves sufficient for the appearance of the higher-level properties; as Morgan says, "no insertion of Entelechy," or any other non-physical agent, is required for the emergence of higher properties:

> Since it is pretty sure to be said that to speak of an emergent quality of life savours of vitalism, one should here parenthetically say, with due emphasis, that if vitalism connotes anything of the nature of Entelechy or Elan—any insertion into physico-chemical evolution of an alien influence which must be invoked to explain the phenomenon of life—then, so far from this being implied, it is explicitly rejected under the concept of emergent evolution. (1923, p. 12)

And Morgan goes on to stress the necessity of physical bases for all higher-level phenomena: "Thus, for emergent evolution, conscious events at level *C* (mind) involves specific physiological events at level *B* (life), and these involve specific physico-chemical events at level *A* (matter). No *C* without *B*, and no *B* without *A*. No mind without life; and no life without 'a physical basis'" (1923, p. 15).

There is little doubt, I think, that on these three crucial tenets there is a broad agreement between emergentism and non-reductive physicalism, and that it is fair, and illuminating, to view non-reductive physicalism as a form of emergentism. It isn't for nothing that non-reductive physicalists sometimes speak of higher-level properties as "emergent."[9]

As for the fourth thesis of non-reductive physicalism, that is, Mental Realism, it is clear that the quotations from the emergentists that we have seen are shot through with realism about mentality. To the emergentist, emergent evolution is a historical fact of paramount importance; through the process of emergent evolution, the world has reached its present state—more complex, richer, and fuller. Most physicalists who reject reductionism also reject mental eliminativism; this realist attitude is implicit in what we have called the Physical Realization Thesis (assuming realism about the physical). And it is all but explicit in the claim, accepted by both emergentists and many non-reductivists, that psychology is a legitimate special science (unless one adopts a universal anti-realism about all science); as such it must investigate a domain of real phenomena and systematize them by discovering laws and causal connections governing them.

## 24.4 Mental Realism and Mental Causation

But just what does the commitment to the reality of mental properties amount to? What is the significance of saying of anything that it is "real"? Alexander supplies an apt answer, in a marvellous paragraph in which he curtly dismisses epiphenomenalism: "[Epiphenomenalism] supposes something to exist in nature which has nothing to do, no purpose to serve, a species of *noblesse* which depends on the work of its inferiors, but is kept for show and might as well, and undoubtedly would in time be abolished" (1927, p. 8).

This we may call "Alexander's dictum": *To be real is to have causal powers.* I believe this principle, as applied to concrete existents and their properties, will be accepted by most non-reductive physicalists.

Emphasis on the causal role of emergent properties is pervasive in the emergentist literature. Here is a pair of revealing quotations:

> Just as the holistic properties of the organism have causal effects that determine the course and fate of its constituent cells and molecules, so in the same way, the conscious properties of cerebral activity are conceived to have analogous causal effects in brain function that control subset events in the flow pattern of neural excitation. In this holistic sense the present proposal may be said to place mind over matter, but not as any disembodied or supernatural agent. (Sperry 1969, p. 533)

> But when some new kind of relatedness is supervenient (say at the level of life), the way in which the physical events which are involved run their course is different in virtue of its presence—different from what it would have been if life had been absent. (Morgan 1923, p. 16)

What is striking about these paragraphs is the reference to "downward causation": both Morgan and Sperry seem to be saying that mentality, having emerged from physical-biological processes, takes on a causal life of its own and begins to *exercise causal influence "downward" to affect what goes on in the underlying physical-biological processes*. Whether the idea of such causation makes sense is one of the main questions I want to discuss in the balance of this paper. But let us first note what the non-reductive physicalist has to say about mental causation.

There is no question that the typical non-reductive physicalist has a strong commitment to the reality of mental causation. As we saw, our non-reductivist is not an eliminativist: why bother with mental properties unless you think that they are good for some causal work and can play a role in causal explanations? Fodor puts the point this way:

> I'm not really convinced that it matters very much whether the mental is the physical; still less that it matters very much whether we can prove it. Whereas, if it isn't literally true that my wanting is causally responsible for my reaching, and my itching is causally responsible for my scratching, and my believing is causally responsible for my saying, . . . if none of that is literally true, then practically everything I believe about anything is false and it's the end of the world. (1989, p. 77)

One could hardly declare one's yearnings for mental causation with more feeling than this! Non-reductive physicalists in general regard mental causation seriously; they have expended much energy and ingenuity trying to show they are entitled to mental causation (see e.g. LePore and Loewer 1987, 1989; Fodor 1989; Davidson 1980, Chapter 1). I don't think that they can have what they want, and showing this will be my main burden in the remainder of this paper. My argument, if correct, will also show that Fodor is wrong in feeling that he could have his wishes fulfilled without having to worry about the mind–body problem—about "whether the mental is the physical."

Here is my plan for the remainder of this paper: I shall first show that both emergentism and non-reductive physicalism are committed to downward causation—that is, mental-to-physical causation for the mind–body case. This should be no news to the emergentist: for in a sense downward causation is much of the point of the emergentist programme. What I shall show would still be of interest, even for emergentism, since the argument will make clear that downward causation is entailed by the basic tenets of emergentism, and of non-physical reductionism. I shall then argue that the idea of downward causation is highly problematic, and perhaps incoherent, given the basic physicalist commitments.

## 24.5 Non-Reductive Physicalism Is Committed to Downward Causation

It is easy to see just how downward causation follows from the basic principles of emergentism and non-reductive physicalism. First, as we have observed, the emergentist and the non-reductive physicalist are mental realists, and Mental Realism, via Alexander's dictum, entails causal powers for mental properties. In fact, whether or not they accept Alexander's dictum, most of them will want causal powers for mental properties. Now, mental properties, on both positions, are irreducible net additions to the world. And this must mean, on Alexander's dictum, that mental properties bring with them *new causal powers, powers that no underlying physical–biological properties can deliver.* For unless mentality made causal contributions that are genuinely novel, the claim that it is a distinct and irreducible phenomenon over and beyond physical–biological phenomena would be hollow and empty. To be real, Alexander has said, is to have causal powers; *to be real, new, and irreducible, therefore, must be to have new, irreducible causal powers.*

This fits in well with the autonomy thesis, alluded to earlier, concerning the science of psychology: as an empirical science, psychology must generate causal explanations of phenomena in its domain; and as an irreducible, autonomous science, the causal explanations it delivers must themselves be irreducible, representing causal connections in the world not captured by the underlying sciences. The autonomy thesis, therefore, makes sense only if causal relations involving mental events are novel and irreducible—that is, mental properties are endowed with genuinely new causal powers irreducible to those of underlying physical–biological properties.

If $M$ is a mental property, therefore, $M$ must have some new causal powers. This must mean, let us suppose, that $M$ manifests its causal powers by being causally efficacious with respect to another property, $N$; that is, a given instance of $M$ can cause $N$ to be instantiated on that occasion. We shall assume here a broadly nomological conception of causality, roughly in the following sense: an instance of $M$ causes an instance of $N$ just in case there is an appropriate causal law that invokes the instantiation of $M$ as a sufficient condition for the instantiation of $N$. There are three cases to be distinguished: (i) the property $N$ for which $M$ is a cause is a mental property; (ii) $N$ is a physical property; (iii) $N$ is a higher-level property in relation to $M$. (i) is mental-to-mental causation ("same-level causation"); (ii) is mental-to-physical causation (that is, "downward causation"); and (iii) is a possibility if there are properties, perhaps social properties, that emerge from, or are realized by, mental properties ("upward causation").

My argument will show that case (i) is possible only if case (ii) is possible; namely, that mental-to-mental causation presupposes mental-to-physical causation. It will be clear that the same argument shows case (iii) presupposes case (ii), and therefore case (i). So suppose $M$ is causally efficacious with respect to some mental property $M^*$, and in particular that a given instance of $M$ causes an instance of $M^*$. But $M^*$, *qua* mental property, is physically realized; let $P^*$ be its physical realization base. Now we seem to have two distinct and independent answers to the question "Why is this instance of $M^*$ present?" *Ex hypothesi*, it is there because an instance of $M$ caused it; that's why it's there. But there is another answer: it's there because $P^*$ physically realizes $M^*$ and $P^*$ is instantiated on this occasion. I believe these two stories about the presence of $M^*$ on this occasion create a tension and must be reconciled.[10]

Is it plausible to suppose that the joint presence of $M$ and $P^*$ is responsible for the instantiation of $M^*$? No; because that contradicts the claim that $M^*$ is physically realized by $P^*$. As we saw, this claim implies that $P^*$ alone is sufficient to bring about $M^*$, whether or not any other condition, earlier or later or at the same time, obtained (unless it is somehow connected with the occurrence of $P^*$ itself—we shall recur to this possibility below). And the supposition is also inconsistent with our initial assumption that the given instance of $M$ was a sufficient condition for that instance of $M^*$. Nor is it plausible to suppose that the occurrence of $M^*$ on this occasion was somehow overdetermined in that it has two distinct and independent origins in $M$ and $P^*$. For this, too, conflicts with the assumption that $M^*$ is a property that requires a physical realization base in order to be instantiated, and that this instance of $M^*$ is there because it is realized by $P^*$. In the absence of $P^*$, we must suppose that $M^*$ could not have been there—unless an alternate realization base had been present. In either case, every instance of $M^*$ must have some physical base that is by itself sufficient for $M^*$; and this threatens to pre-empt $M$'s claim to be the cause of this instance of $M^*$.

I believe the only coherent story we can tell here is to suppose that the $M$-instance caused $M^*$ to be instantiated *by causing $P^*$, $M^*$'s physical realization base, to be instantiated*. This of course is downward causation, from $M$ to $P^*$, a case of mental-to-physical causation. I believe the argument goes through with "physical realization" replaced with "emergence" (see Kim 1992). The gist of my argument is encapsulated in the following principle, which I believe will be accepted by most non-reductive physicalists:

The Causal Realization Principle If a given instance of $S$ occurs by being realized by $Q$, then any cause of this instance of $S$ must be a cause of this instance of $Q$ (and of course any cause of this instance of $Q$ is a cause of this instance of $S$).

A parallel principle stated for emergence will be accepted by many, if not all, emergentists. In any case, I think we apply this principle constantly in daily life: for example, we treat pain by intervening with bodily processes, and we communicate by creating vibrations in the air or making marks on paper. (Direct mental-to-mental causation between different individuals is generally considered disreputable and unscientific: it goes by such names as "ESP," "telepathy," and "mind-reading.")

But couldn't we avoid this commitment to downward causation by exploiting the fact that *M*, as a mental property, has its own physical realization base, say *P*? Why not say then that *M*'s causation of $P^*$ comes to merely this: *M* is physically realized by *P*, and *P* causes $P^*$. The more basic causal relation obtains between the two physical properties, *P* and $P^*$, and *M*'s causation of $M^*$ is ultimately grounded in the causal relation between their respective physical realization bases. I think this is a highly appealing picture,[11] but it is not something that our non-reductivists can avail themselves of. For the picture reduces the causal powers of *M* to those of its realization base *P*: *P* is doing all the causal work, and *M*'s causation of $P^*$, or of $M^*$, turns out to be derivative from *P*'s causal powers. Thus, *M* has no causal powers over and beyond those of *P*, and this is contrary to Alexander's dictum and the assumption that *M* is an irreducible property. I shall take up this point again in the next section.

What these reflections show is that within the stratified world of non-reductive physicalism and emergentism, "same-level" causation can occur only if "cross-level" causation can occur. It will not be possible to isolate and confine causal chains within levels; there will be inevitable leakage of causal influence from level to level.

## 24.6 What's Wrong with Downward Causation?

So does downward causation makes sense—that is, within the scheme of non-reductive physicalism? I think there are some severe problems. As we shall see, the tension arises out of an attempt to combine "upward determination" with "downward causation." The non-reductive physicalist wants both: mentality is determined by, and dependent on, the physical, and yet minds are to have causal powers, novel causal powers that can be exercised, if my argument is correct, only by causally affecting physical-biological processes in novel ways.

Suppose then that mental property *M* is causally efficacious with respect to physical property $P^*$, and in particular that a given instance of *M* causes a given instance of $P^*$. Given the Physical Realization Thesis, this instance of *M* is there because it is realized by a physical property, say *P*. Since *P* is a realization base for *M*, it is sufficient for *M*, and it follows that *P* is sufficient, as a matter of law, for $P^*$. Now, the question that must be faced is this: What reason is there for not taking *P* as the cause of $P^*$, bypassing *M* and treating it as an epiphenomenon?[12]

I believe this epiphenomenalist solution with regard to *M* cannot easily be set aside: we are looking for a causal explanation of why $P^*$ is instantiated at this time. We see that *M* was instantiated and we can invoke a law connecting *M*-instances with $P^*$-instances. But we also see that *P* was instantiated at the same time, and there is an appropriate law connecting *P*-instances with $P^*$-instances. So the situation is this: *P* appears to have at least as strong a claim as *M* as a direct cause of $P^*$ (that is, without *M* as an intervening link). Is there any reason for invoking *M* as a cause of $P^*$ at all? The question is not whether or not *P* should be considered a cause of $P^*$; on anyone's account, it should be. Rather, the question is whether *M* should be given a distinct causal role in this situation? I believe there are some persuasive reasons for refusing to do so.

First, there is the good old principle of simplicity: we can make do with *P* as $P^*$'s cause, so why bother with *M*? Notice that given the simultaneity of the instances of *M* and *P* respectively, it is not possible to think of the *M*-instance as a temporally intermediate link in the causal chain from *P* to $P^*$. Moreover, if we insist on *M* as a cause of $P^*$, we fall foul of another serious difficulty, "the problem of causal-explanatory exclusion" (see Kim 1989a, 1990b). For we would be allowing two distinct sufficient causes, simultaneous with each other, of a single event. This makes the situation look like one of causal over-determination, which is absurd. And *ex hypothesi*, it is not possible to regard *M* and *P* as forming a single jointly sufficient cause, each being individually necessary but insufficient. And given the assumed irreducibility of *M*, we cannot regard *M* as identical with *P*, or as a part of it. The exclusion problem, then, is this: given that *P* is a sufficient physical cause of $P^*$, how could *M also* be a cause, a sufficient one at that, of $P^*$? What causal work is left over for *M*, or any other mental property, to do? *M*'s claim as a cause of $P^*$ will be weakened especially if, as we would expect in real-life neurobiological research, there is a continuous causal chain, a mechanism, connecting *P* with $P^*$. It is clear that the exclusion problem cannot be resolved within the framework of non-reductive physicalism.

All these considerations, I want to suggest, point to something like the following as the natural picture for the layered physicalist world: all causal relations are implemented at the physical level, and the causal relations we impute to higher-level processes are derivative from and grounded in the fundamental nomic processes at the physical level.[13] This goes perhaps a bit, but not much, beyond what is directly implied by the supervenience thesis most non-reductive physicalists accept: if, as the supervenience thesis claims, all the facts are determined by physical facts, then all causal relations involving mental events must be determined by physical facts (presumably including facts about physical causation).

Consider, then, a somewhat bald way of stating this idea:

The Principle of Causal Inheritance If *M* is instantiated on a given occasion by being realized by *P*, then the causal powers of *this instance of M* are identical with (perhaps, a subset of) the causal powers of *P*.

In other words, higher states are to inherit their causal powers from the underlying states that realize them. Non-reductivists must reject this principle; they will say that higher-level causal powers are "determined by," but not identical with (or reducible to) the lower-level causal powers. What our considerations have made clear is that if "determined but not identical" means that these higher-causal powers are genuinely novel powers, then non-reductivists are caught in a web of seemingly insurmountable difficulties. And I challenge those non-reductivists who would reject this principle to state an alternative principle on just how the causal powers of a realized property are connected with those of its realization base; or to explain, if no such connections are envisaged, the significance of talk of realization.

The implications of the Causal Inheritance Principle are devastating to non-reductive physicalism: if the causal powers of *M* are identical with those of its realization bases, then *M* in effect contributes nothing new causally, and *M*'s claim to be a new, irreducible property is put in jeopardy. And if, as suggested, *M* is treated as an epiphenomenal dangler from its physical realization base, with no causal work of its own to do, the next step of the argument, as mandated by Alexander's dictum, will be that *M* ought to be "abolished." All this seems like an inescapable lesson of Alexander's dictum.

The case for scepticism about downward causation is strengthened when we see that, as in the case of Cartesian interactionist dualism, it breaches the causal closure of the physical domain. What is worse, when we see that *P*, *M*'s realization base, is there to serve as a full cause of $P^*$—and this will always be the case whenever a mental cause is invoked—the violation isn't even as well motivated as it is with Cartesian interactionism.

Most emergentists will have no problem with the failure of the physical causal closure; although they may have to tinker with their doctrines somewhere to ensure the overall consistency of their position, they are not likely to shed any tears over the fate of the closure principle. For many emergentists that precisely was the intended consequence of their position. I doubt, however, that contemporary non-reductive physicalists can afford to be so cavalier about the problem of causal closure: to give up this principle is to acknowledge that there can in principle be no complete physical theory of physical phenomena, that theoretical physics, insofar as it aspires to be a complete theory, must cease to be pure physics and invoke irreducibly non-physical causal powers—vital principles, entelechies, psychic energies, elan vital, or whatnot. If that is what you are willing to embrace, why call yourself a "physicalist"? Your basic theory of the world will have to be a mixed one, a combined physical-mental theory, just as it would be under Cartesian interactionism. And all this may put the layered view of the world itself in jeopardy; it is likely to require some serious rethinking.

At this juncture it seems highly plausible that the only solution to the exclusion problem and the problem of the physical causal closure lies in some form of reductionism that will permit us to discard, or at least moderate, the claim that mental properties are distinct from their underlying physical properties. It is this claim that forced us to posit for mentality novel and distinct causal powers, thereby leading us to the present predicament. To identify the causal powers of mentality with those of its underlying physical base is, in effect, to deny it a distinct ontological status, and consider it reduced.

But a question must leap to your mind at this point, if you are at all familiar with current wisdom in philosophy of mind: Doesn't the Physical Realization Thesis itself, given the phenomenon of "multiple realizability," rule out any form of reductionism? Hasn't "the multiple realization argument" (see, e.g., Fodor 1974) refuted reductionism once and for all? The entrenched, almost automatic, response is "yes" to both questions. I believe the correct answer is a qualified but firm "no." But defending that answer is something I must leave for another time.[14]

## Notes

This chapter originally appeared in J. Heil and A. Mele (eds.), *Mental Causation*, 189–210, New York, Oxford University Press, 1993.

1. For a highly useful and informative presentation of this layered picture, see Oppenheim and Putnam 1958.

2. When the layered model is described in terms of "levels of description" or "levels of language," there is a corresponding question about how the descriptive apparatus (predicates, concepts, sentences, etc.) of one level is related to that of another.

3. There is also the position of Donald Davidson in his "Mental Events" (1970) which accepts both physical ontological monism (usually formulated as a thesis about individual events) and property dualism of non-reductive physicalism as characterized here, while rejecting the corollary about the scientific status of psychology. Davidson's views on psychology are hinted at by the title of one of his papers on this issue, "Psychology as Philosophy" (1974b).

4. For a survey of supervenience relations see Kim 1990a.

5. As far as I know, Hilary Putnam first introduced the idea of "physical realization," in the early 1960s, to describe the relationship between "logical" and "structural" states of computing machines, extending it by analogy to the mental–physical case (see Putnam 1960). I believe that the idea really began catching on, in discussions of the mind–body problem, when it was used to formulate the influential "multiple realization argument" in the seminal Putnam 1967 paper.

6. In particular, Davidson's "anomalous monism" *sans* a supervenience thesis will be largely immune to my argument. But there are other difficulties with such a position; in particular, an account of the causal powers of mental properties appears hopeless under anomalous monism. See Kim 1989b.

7. For reasons for requiring this see Kim (forthcoming).

8. In spite of all this, there is an apparent difference between emergentism and non-reductive physicalism concerning the relationship between properties belonging to adjacent levels. As may be recalled, LePore and Loewer require that the physical ("realization") base must *explain* the mental property it realizes. However, emergentists will deny that the "basal conditions" can ever constitute an explanatory basis for any property emergent from them. Much of the difference can be traced, I believe, to differing conceptions of explanation and reduction involved; for further details see Kim 1992. This difference, whether real or only apparent, will not affect the applicability of the main argument of this paper to both emergentism and non-reductive physicalism.

9. E.g. Hellman and Thompson (1975, p. 555) say this: "what may be called 'emergence' of higher-order phenomena is allowed for without departing from the physical ontology."

10. This is essentially identical to the situation we face when we are given two distinct independent causes for one and the same event, each claimed to be a sufficient cause. See Kim 1989a.

11. This is closely similar to the model of "supervenient causation" I have suggested in earlier papers of mine, e.g. Kim 1984b. It has been called "causal reductionism" by Menzies (1988). Also see LePore and Loewer 1989.

12. To be precise, we should put this in terms of instances of these properties rather than the properties themselves. In what follows, liberties of this form are sometimes taken to avoid verbosity.

13. There is a strong indication that Fodor e.g. accepts this sort of principle in Fodor 1989.

14. I undertake a defence of this reply in Kim (forthcoming). For a sketch of an argument see Kim 1989b.

## Bibliography

Alexander, S. (1927). *Space, Time and Diety*, ii, 2nd edn., London: Macmillan.

Beckermann, A., Flohr, H., Kim, J. (1992) (eds.) *Emergence or Reduction? Essays on the Prospect of Nonreductive Physicalism*, Berlin: Walter de Gruyter.

Capitan, W., and Merrill, D. (1967) (eds.), *Art, Mind, and Religion*, Pittsburgh: University of Pittsburgh Press.

Davidson, D. (1970). "Mental Events," in Foster and Swanson (1970), pp. 79–101; reprinted in Davidson (1980).

——— (1974). "Psychology as Philosophy," in S. Brown (ed.) *Philosophy of Psychology*, London: Macmillan, pp. 41–52; reprinted in Davidson (1980).

——— (1980). *Essays on Actions and Events*, Oxford: Clarendon Press.

Feigl, H., Scriven, M., and Maxwell, G. (1958) (eds.), *Concepts, Theories, and the Mind-Body Problem*. Minnesota Studies in the Philosophy of Science, 2. Minneapolis: University of Minnesota Press.

Fodor, J. (1974). "Special Sciences, or the Disunity of Science as a Working Hypothesis," *Synthese* 28, pp. 97–115.

——— (1989). "Making Mind Matter More," *Philosophical Topics* 17, pp. 59–80.

Foster, L., and Swanson, J. (1970) (eds.), *Experience and Theory*. Amherst, MA: University of Massachusetts Press.

Hellman, G., and Thompson, F. (1975). "Physicalism: Ontology, Determination, Reduction," *Journal of Philosophy* 72, pp. 551–564.

Kim, Jaegwon (1984a). "Concepts of Supervenience," *Philosophy and Phenomological Research* 65, pp. 153–176.

——— (1984b). "Epiphenomenal and Supervenient Causation," *Midwest Studies in Philosophy* 9, pp. 257–270.

——— (1989a). "Mechanism, Purpose and Explanatory Exclusion," *Philosophical Perspectives* 3, pp. 77–108.

——— (1989b). "The Myth of Nonreductive Materialism," *Proceedings of the American Philosophical Association* 63, pp. 31–47.

——— (1990a). "Supervenience as a Philosophical Concept," *Metaphilosophy* 21, pp. 1–27.

——— (1990b). "Explanatory Exclusion and the Problem of Mental Causation," in Villanueva (1990), pp. 36–56.

——— (1992). "Downward Causation," in Beckermann, Flohr, and Kim (1992), pp. 119–138.

——— (1992). "Multiple Realization and the Metaphysics of Reduction," *Philosophy and Phenomenological Research* 52, pp. 1–26.

LePore, E., McLaughlin, B. P., and Loewer, B. (1987). "Mind Matters," *Journal of Philosophy* 84, pp. 630–642.

LePore, E., McLaughlin, B. P., and Loewer, B. (1989). "More on Making Mind Matters," *Philosophical Topics* 17, pp. 175–191.

McLaughlin, B. (1989). "Type Epiphenomenalism, Type Dualism, and the Causal Priority of the Physical," *Philosophical Perspectives* 3, pp. 109–135.

Menzies, P. (1988). "Against Causal Reductionism," *Mind* 97, pp. 551–574.

Morgan, C. (1923). *Emergent Evolution*, London: Williams and Norgate.

Oppenheim, P., and Putnam, H. (1958). "Unity of Science as a Working Hypothesis," in Fiegl et al. (1958), pp. 3–36.

Putnam, H. (1960). "Minds and Machines" in Hook (1960), pp. 138–164; reprinted in Putnam 1975b.

——— (1967). "Psychological Predicates," in Capitan and Merrill (1967), pp. 37–48.

——— (1975b). *Philosophical Papers*, ii, Cambridge: Cambridge University Press.

Sperry, R. (1969). "A Modified Concept of Consciousness," *Psychological Review* 76, pp. 532–536.

Villanueva, E. (1990) (ed.). *Information, Semantics, and Epistemology*, Oxford: Basil Blackwell.

# Annotated Bibliography

The resources listed here provide broad but not exhaustive coverage of the discussion of emergence in contemporary philosophy and science. Most titles are self-explanatory. Brief comments place some sources in perspective.


Andersen, P. B., C. Emmeche, N. O. Finnemann, and P. Voetmann Christiansen. (2000). *Downward Causation: Minds, Bodies, and Matter*. Aarhus, Denmark: Aarhus University Press.

Anderson, P. W. (1981). Some general thoughts about broken symmetry. In N. Boccara (ed.), *Symmetries and Broken Symmetries in Condensed Matter Physics*, pp. 11–20. Paris: IDSET. (A survey of how broken symmetry underlies many cases of emergence in physics. The article also criticizes Hakens's account of self-organization and is relevant to the approach in Batterman 2002.)

Anderson, P. W. (1984). *Basic Notions of Condensed Matter Physics*. Menlo Park, Calif.: W. Benjamin.

Anderson, P. W. (1995). Historical overview of the twentieth century physics. In L. M. Brown, A. Pais, and B. Pippard (eds.), *Twentieth Century Physics*, pp. 2017–2032. New York: American Institute of Physics Press.

Anderson, P. W., and D. L. Stein. (1984). Broken symmetry, emergent properties, dissipative structures, life: Are they related. In Anderson (1984), pp. 262–284. (A useful overview of the concept of broken symmetry.)

Anthony, Louise M. (1999). Making room for the mental: Comments on Kim's "Making Sense of Emergence." *Philosophical Studies* 95: 37–44. (An early response to the Kim article in part I of this anthology.)

Auyang, Sunny. (1998). *Foundations of Complex-System Theories*. Cambridge: Cambridge University Press. (Chapter 6 contains a clear exposition of a number of concepts in physics relating to emergence.)

Ayala, F. J., and T. Dobzhansky (eds.). (1974). *Studies in the Philosophy of Biology: Reduction and Related Problems*. Berkeley: University of California Press.

Baas, N. A. (1994). Emergence, hierarchies, and hyperstructures. In Langton, C. G. (ed.), *Artificial Life III*, Santa Fe Studies in the Sciences of Complexity, Proc. Volume XVII, pp. 515–537. Redwood

City: Addison-Wesley. (This article presents a mathematical framework for emergence, and discusses some examples. This view is applied in chapter 17.)

Baas, N. A., and C. Emmeche. (1997). On emergence and explanation. *Intellectica* 25: 67–83.

Baas, N. A., Michael W. Olesen, and Steen Rasmussen. (1996). Generation of higher-order emergent structures. Working paper 96-08-057, Santa Fe Institute, Santa Fe, N.M.

Bak, Per, Chao Tang, and Kurt Weisenfeld. (1987). Self-organized criticality: An explanation of 1/f noise. *Physical Review Letters* 59: 381–384. (The original source of the claim that self-organized criticality is a universal phenomenon.)

Batterman, Robert. (2002). *The Devil in the Details.* New York: Oxford University Press. (Chapter 8 presents Batterman's account of emergence in the context of universality in physics.)

Beckerman, Ansgar, Hans Flohr, and Jaegwon Kim (eds.). (1992). *Emergence or Reduction? Essays on the Prospects of Nonreductive Physicalism.* Berlin: Walter de Gruyter. (A valuable early collection of articles in the title area.)

Beckner, Morton. (1974). Reduction, hierarchies, and organicism. In Ayala and Dobzhansky (1974), pp. 163–176.

Bedau, Mark A. (1997). Weak emergence. *Philosophical Perspectives* 11: 375–399. Malden: Blackwell. (The original article describing weak emergence.)

Bedau, Mark A. (1997). Emergent models of supple dynamics in life and mind. *Brain and Cognition* 34: 5–27. (Develops an emergent notion of functionalism, and applies it to life and mind.)

Bedau, Mark A. (2003). Artificial life: organization, adaptation, and complexity from the bottom up. *Trends in Cognitive Science* 7: 505–512. (An overview of the different kinds of emergent phenomena explored in contemporary artificial life.)

Bergman, Gustav. (1944). Holism, historicism, and emergence. *Philosophy of Science* 11: 209–221.

Blitz, David. (1992). *Emergent Evolution.* Dordrecht: Kluwer Academic Publishers. (An historical account of early to mid–twentieth century attempts to capture a particular type of diachronic emergence.)

Boden, Margeret (ed.). (1996). *The Philosophy of Artificial Life.* New York: Oxford University Press.

Boogerd, F. C., F. J. Bruggeman, R. C. Richardson, A. Stephan, and H. V. Westerhoff. (2005). Emergence and its place in nature: a case study of biochemical networks. *Synthese* 145: 131–164.

Bonabeau, Eric, Jean-Louis Dessalles, and Alain Grumbach. (1995). Characterizing emergent phenomena (1): A critical review. *Revue Internationale de Systémique* 9: 327–346.

Bonabeau, Eric, Jean-Louis Dessalles, and Alain Grumbach. (1995). Characterizing emergent phenomena (2): A conceptual framework. *Revue Internationale de Systémique* 9: 347–371.

Broad, C. D. (1925). *The Mind and Its Place in Nature.* London: Routledge and Kegan Paul. (Contains a classic account of the predictivist approach to emergence.)

Bunge, Mario. (1977). Levels and reduction. *American Journal of Physiology* 233: R75–R82.

Campbell, D. T. (1974). Downward causation in hierarchically organised biological systems. In Ayala and Dobzhansky (1974). (An early influential discussion.)

Castellani, Elena. (2002). Reductionism, emergence, and effective field theories. *Studies in History and Philosophy of Modern Physics* 33: 251–267. (A philosophically oriented description of how renormalization, an influential area of contemporary physics, approaches theory reduction.)

Clayton, Philip. (2004). *Mind and Emergence: From Quantum to Consciousness*. Oxford: Oxford University Press.

Clayton, Philip, and Paul Davies (eds.). (2006). *The Re-Emergence of Emergence: The Emergentist Hypothesis from Science to Religion*. Oxford: Oxford University Press.

Crutchfield, James. (2002). What lies between order and chaos? In J. Casti (ed.), *Art and Complexity*, pp. 31–45. Oxford: Oxford University Press.

Crutchfield, James. (1994). The calculi of emergence: computation, dynamics, and induction, *Physica D* 75: 11–54.

Cunningham, B. (2001). The re-emergence of emergence. *Philosophy of Science* 68: S62–S75.

Darley, Vince. (1994). Emergent phenomena and complexity. In R. Brooks and P. Meas (eds.), *Artificial Life VI, Proceedings of the Fourth International Workshop on the Synthesis and Simulation of Living Systems*, pp. 411–416. Cambridge, Mass.: MIT Press. (An early attempt to formalize underivability without simulation).

Emmeche, C., S. Køppe, and F. Stjernfelt. (1997). Explaining emergence: towards an ontology of levels. *Journal for General Philosophy of Science* 28: 83–119.

Feitz, B., M. Crommelinck, and P. Goujon (eds.). (2006). *Self-Organization and Emergence in Life Sciences*. Heidelberg: Springer. (A collection of previously unpublished essays by Gérard Weisbuch, Vincent Bauchau, Hugues Bersisi, René Thomas, Phillippe Lefévre, Cheng Tu, Marcus Missal, Marc Crommelinck, Francisco Varela, Henrí Atlan, Irun Cohen, Gertrudis Van de Vijver, Laurence Bouquiaux, François Duchesneau, Phillippe Goujon, Paul Mengel, Jean-Claude Heudin, Pierre Livet, Robert Brandon, Marc Maesschaick, Valérie Kokoszka, Paul Thompson, Robert Richardson, and Bernard Feitz.)

Flake, Garry William. (1998). *The Computational Beauty of Nature: Computer Explorations of Fractals, Chaos, complex systems, and Adaptation*. Cambridge, Mass.: MIT Press. (An accessible introduction to computational modeling).

Frigg, Roman. Self-organised criticality—what it is and what it isn't. (2002). *Studies in the History and Philosophy of Science* 34: 613–632. (A clear and straightforward set of criticisms of the claims that self-organized criticality is a universal theory.)

Gillett, Carl. (2006). Samuel Alexander's emergentism: Or, higher causation for physicalists. *Synthese* 153: 261–296.

Goldenfeld, Nigel, and Leo Kadanoff. (1999). Simple lessons from complexity. *Science* 284: 87–89. (An accessibly written short essay containing a number of interesting examples of how complexity requires us to think differently about physics.)

Grantham, Todd. (2006). Is macroevolution more than successive rounds of microevolution? *Paleontology* 50: 75–85. (Uses Bedau's notion of weak emergence to argue that a species' geographical range size is an emergent property, and thus that macroevolutionary phenomena cannot be fully explained by microevolutionary processes.)

Gross, Dominique, and Barry McMullin. (2001). Is it the right *ansatz*? *Artificial Life* 7: 355–365. (A critical discussion of chapter 17.)

Haldane, John. (1996). The mystery of emergence. *Proceedings of the Aristotelian Society* 96: 261–267.

Haken, Hermann. (2000). *Information and Self-Organization: A Macroscopic Approach to Complex Systems*. Second enlarged edition. Berlin: Springer. (An exposition of some areas of physics from the information-theoretic perspective. Difficult.)

Harré, R. (1972). *The Philosophies of Science*. London: Oxford University Press. (Chapter 5 is a clear early reference to nominal emergence.)

Healey, Richard. (1991). Holism and nonseparability. *Journal of Philosophy* 88: 393–421. (An exploration of the role of entangled states in quantum theory.)

Hofstadter, D. (1980). *Gödel, Escher, Bach: An Eternal Golden Braid*. Harmondsworth: Penguin Books. (An influential and inimitable book that develops a vivid picture of emergence in cognitive science.)

Holland, John. (1999). *Emergence: From Chaos to Order*. Perseus Group Books. (This book recounts the intellectual trajectory of a founding father in the science of complex systems and machine learning.)

Hooker, C. A. (2004). Asymptotics, reduction, and emergence. *British Journal for the Philosophy of Science* 55: 435–479.

Humphreys, Paul. (1996). Understanding in the not-so-special-sciences. *Southern Journal of Philosophy Supplement: Proceedings of the 1995 Spindel Conference* 34: 99–114.

Humphreys, Paul. (1997). Aspects of emergence. *Philosophical Topics* 24: 53–70.

Humphreys, Paul. (1997). Emergence, not supervenience. *Philosophy of Science* 64: S337–S345. (Argues that supervenience relations are an ineffective device for representing emergence.)

Humphreys, Paul. (2006). Emergence. In Donald Borchert (ed.), *The Encyclopedia of Philosophy. Second edition*. New York: MacMillan Reference Books.

Kauffman, S. A. (1996). *At Home in the Universe: The Search for the Laws of Self-Organization and Complexity*. New York: Oxford University Press. (A more accessible presentation of the central threads in Kauffman 1993.)

Kauffman, S. A. (1993). *The Origins of Order. Self-Organization and Selection in Evolution*. Oxford: Oxford University Press. (Presents a novel and widely known explanation of the emergence of order in biological systems as a result of intrinsic self-organization.)

Kauffman, S. A., and Philip Clayton. (2006). On emergence, agency, and organization. *Biology and Philosophy* 21: 501–521.

Kellert, Stephen H. (1994). *In the Wake of Chaos: Unpredictability in Dynamical Systems*. Chicago: University of Chicago Press. (An accessible introduction to chaos and dynamical systems theory for philosophers.)

Kim, Jaegwon. (1992). The layered model: metaphysical considerations. *Philosophical Explorations* 5: 2–20. (A discussion of the concept of levels in ontology.)

Kim, Jaegwon. (1993). *Supervenience and Mind*. Cambridge: Cambridge University Press. (Contains many influential articles on supervenience and related matters by the author.)

Kim, Jaegwon. (1997). Explanation, prediction, and reduction in emergentism. *Intellectica* 2: 45–57. (A very clear account, preceding the 1999 article reprinted in this collection, arguing for an alternative to Nagel reduction in terms of functionalizing properties. Emergent properties are those that cannot be reduced in this new sense.)

Kistler, Max (ed.). (2006). *New Perspectives on Reduction and Emergence in Physics, Biology, and Psychyology. Synthese* 151 (3, special issue). (Contains new articles by Max Kistler, C. Ulises Moulines, Stéphanie Ruphy, Alexander Rueger, Michel Morange, Ana Soto and Carlos Sonnenschein, Kenneth Schaffner, Luc Faucher, John Bickel, and Juib Looren de Jong.)

Klee, Robert. (1984). Microdeterminism and concepts of emergence. *Philosophy of Science* 51: 44–63. (An early taxonomy of emergence.)

Jackson, F., and P. Pettit. (1992). In defense of explanatory ecumenism. *Economics and Philosophy* 8: 1–21. (Argues for the explanatory autonomy of economics.)

Laughlin, Robert. (2006). *A Different Universe: Reinventing Physics from the Bottom Down*. New York: Basic Books. (An extended treatment of emergence in the physics of large agglomerations of matter, by one of the authors of chapter 14 in this anthology.)

Liu, Chuang. (1999). Explaining the emergence of cooperative phenomena. *Philosophy of Science* 66: S92–S106. (An account of emergence in the context of one branch of contemporary physics.)

Mainzer, Klaus. (2004). *Thinking in Complexity*. Fourth edition. Berlin: Springer. (A comprehensive survey of topics in complexity theory).

May, Robert M. (1986). When two and two do not make four: Non-linear processes in ecology. *Proceedings of the Royal Society of London* B228, 241–266. (The paper that made the logistic equation famous.)

Mayr, Ernst, and Stephen Weinberg. (1988). The limits of reductionism. *Nature* 331: 475–476. (An exchange of views on reduction resulting from the original publication of chapter 18 of this anthology.)

McLaughlin, Brian. (2003). Vitalism and emergence. In Thomas Baldwin (ed.), *The Cambridge History of Philosophy 1870–1945*, pp. 631–639. Cambridge: Cambridge University Press.

Morowitz, Harold J. (2004). *The Emergence of Everything*. New York: Oxford University Press. (A pioneer in the chemistry of the origin of life, Morowitz recounts a view of emergence that spans from the emergence of the solar system from the big bang to the emergence of human consciousness and technology.)

Nagel, Ernest. (1961). *The Structure of Science*. New York: Harcourt. (Chapter 11 is the classic source for Nagel-style reduction; chapter 19 of this anthology presents a later version of the position).

Newman, David V. (1996). Emergence and strange attractors. *Philosophy of Science* 63: 245–261. (An account of the relation between certain sorts of chaotic phenomena and emergence. Accurate and sophisticated, but does not deal with nonchaotic phenomena in complexity theory.)

Newman, David V. (2001). Chaos, emergence, and the mind–body problem. *Australasian Journal of Philosophy* 79: 180–196. (An elaboration and development of the view in Newman 1996.)

O'Connor, Timothy. (1994). Emergent properties. *American Philosophical Quarterly* 31: 91–104. (An account of emergence in terms of nonstructural properties).

O'Connor, Timothy. (2000). Agency, mind, and reductionism, chapter 6 of *Persons and Causes*. Oxford: Oxford University Press. (A later formulation of the view in O'Conner 1994.)

O'Connor, Timothy, and Wong, Hong Yu. (2005). The metaphysics of emergence. *Nous* 39: 658–678.

Polanyi, M. (1968). Life's irreducible structure. *Science* 160: 1308–1312.

Rasmussen, Steen, Chen, David Deamer, David C. Krakauer, Norman H. Packard, Peter F. Stadler, and Mark A. Bedau. (2004). Transitions from nonliving to living matter. *Science* 303: 963–965. (Describes recent attempts to understand how life can be made to emerge from nonliving materials.)

Rasmussen, Steen, Nils A. Baas, Bernd Mayer, and Martin Nilsson. (2001). Defense of the *ansatz* for dynamical hierarchies. *Artificial Life* 7: 367–373. (A reply to Gross and McMullin 2001.)

Redhead, Michael. (1995). *From Physics to Metaphysics*. Cambridge: Cambridge University Press. (Chapter 3 contains a reasonably simple account of how entangled states in quantum theory make reduction difficult.)

Rohrlich, Fritz. (1997). Cognitive emergence. *Philosophy of Science* 64: S346–S358. (A discussion of the need to introduce new technical vocabulary to capture novel phenomena.)

Rueger, Alexander. (2000a). Robust supervenience and emergence. *Philosophy of Science* 67: 466–489. (Uses dynamical systems theory to discuss a form of diachronic emergence that is compatible with supervenience in which invariance under small perturbations of control parameters is central).

Rueger, Alexander. (2000b). Physical emergence, diachronic and synchronic. *Synthese* 124: 297–322.

Rumelhard, David E., James L. Mclellend, et al. (1987). *Parallel Distributed Processing*, 2 volumes. Cambridge, Mass.: MIT Press. (Contains important discussions of emergence in the context of neural network models in cognitive science.)

Sawyer, R. Keith. (2001). Emergence in sociology. *American Journal of Sociology* 107: 551–585.

Sawyer, R. Keith. (2005). *Social Emergence: Societies as Complex Systemes*. Cambridge: Cambridge University Press.

Shimony, Abner. (1987). The methodology of synthesis: Parts and wholes in low-energy physics. In R. Kargon and P. Achinstein (eds.), *Kelvins Baltimore Lectures and Modern Theoretical Physics*. Cambridge, Mass.: MIT Press.

Shimony, Abner. (1993). Some proposals concerning parts and wholes. In A. Shimony, *Search for a Naturalistic World View*, Volume 2, pp. 218–227. Cambridge, Cambridge University Press.

Shoemaker, Sydney. (2002). Kim on emergence. *Philosophical Studies* 108: 53–63. (An important response to chapter 7 of this anthology.)

Silberstein, Michael, and John McGeever. (1999). The search for ontological emergence. *Philosophical Quarterly* 49: 182–200. (Describes the distinction between ontological and epistemological approaches to emergence.)

Silberstein, Michael. (2001). Converging on emergence: Consciousness, causation, and explanation. *Journal of Consciousness Studies* 8: 61–98.

Silberstein, Michael. (2002). Reduction, emergence, and explanation. In Peter Machamer and Michael Silberstein (eds.), *The Blackwell Guide to the Philosophy of Science*, pp. 80–107. Malden: Blackwell Publishers.

Sperry, R. W. (1970). An objective approach to subjective experience: further explanation of a hypothesis. *Psychological Review* 77: 585–590. (Classic expression of emergence in psychology that focuses on consciousness and raises key issues.)

Sperry, R. W. (1969). A modified concept of consciousness. *Psychological Review* 76: 532–536.

Sperry, R. W. (1986). Macro- versus micro-determination. *Philosophy of Science* 53: 265–275.

Stephan, Achim. (1999). Varieties of emergentism. *Evolution and Cognition* 5: 49–59.

Stephan, Achim. (2002). Emergentism, irreducibility, and downward causation. *Grazer Philosophische Studien* 65: 77–93.

Stoeckler, M. (1991). A short history of emergence and reduction. In E. Agazzi (ed.), *The Problem of Reductionism in Science*. Berlin: Springer.

Strogatz, Steven H. (1994). *Nonlinear Dynamics and Chaos, with Applications to Physics, Biology, Chemistry, and Engineering*. Reading, Mass.: Addison-Wesley Publishing Company. (A thorough but readable text on the title areas.)

Teller, Paul. (1986). Relational holism and quantum mechanics. *British Journal for the Philosophy of Science* 37: 71–81.

Teller, Paul. (1992). A contemporary look at emergence. In Beckerman et al. (1992), pp. 139–153. (Suggests that a property is emergent if it is not definable in terms of nonrelational lower level properties.)

van Cleve, James. (1990). Mind-dust or magic? Panpsychism versus emergentism. *Philosophical Perspectives* 4: 215–226. (An early suggestion that supervenience can represent emergence.)

Watkins, J. W. N. (1957). Historical explanation in the social sciences. *British Journal for the Philosophy of Science* 8: 89–10. (An assessment of methodological individualism.)

Webster, G., and B. Goodwin. (1996). *Form and Transformation. Generative and Relational Principles in Biology*. Cambridge: University of Cambridge Press. (Contains diverse concrete examples of interesting weak emergence in biological development.)

Wilson, Edward O. (1999). *Consilience: the unity of knowledge*. Vintage Publishers.

Wimsatt, William. (1976). Reductionism, levels of organization, and the mind–body problem. In G. G. Globus, G. Maxwell, and I. Savodnik (eds.), *Consciousness and the Brain: A Scientific and Philosophical Inquiry*, pp. 199–267. New York: Plenum.

Wimsatt, William. (1986). Forms of aggregativity. In A. Donagan, A. N. Perovich Jr., and M. V. Wedin (eds.), *Human Nature and Natural Knowledge*, pp. 259–291. Dordrecht: Reidel.

Wimsatt, William. (2000). Emergence as nonaggregativity and the biases of reductionisms. *Foundations of Science* 5: 269–297.

Wimsatt, William. (1994). The ontology of complex systems: levels of organization, perspectives, and causal thickets. *Canadian Journal of Philosophy* 40: 207–274.

Wolfram, Stephen. (1984). Universality and complexity in cellular automata. *Physica D* 10: 1–35. (The original classification of cellular automata).

Wolfram, Stephen. (2002). *A New Kind of Science*. Champaign, Ill.: Wolfram Media. (A thought-provoking but controversial book by the founder of the company that developed *Mathematica*.)

## About the Authors

**Philip W. Anderson** is Joseph Henry Professor of Physics at Princeton University. His principal work has been in condensed matter physics. In 1977 he was awarded the Nobel Prize in physics and in 1982 the National Medal of Science. A number of his papers are collected in *Basic Notions of Condensed Matter Physics*.

**Andrew Assad** is a research programmer in the Department of Electrical and Computer Engineering at the University of Illinois at Urbana–Champaign.

**Nils A. Baas** is a professor in the Department of Mathematical Sciences at the Norwegian University of Science and Technology.

**Mark A. Bedau** is a professor of philosophy and humanities at Reed College, editor-in-chief of the journal *Artificial Life*, a cofounder of ProtoLife Srl, and a cofounder of the European Center for Living Technology. His research interests include emergence, evolution and adaptation, the nature of life and intelligence, machine learning methods and their application to chemical screening, and the social and ethical implications of creating life from scratch.

**Mathieu S. Capcarrère** is a lecturer in computer science at the University of Kent.

**David Chalmers** is a professor of philosophy and director of the Centre for Consciousness at the Australian National University. His principal research interests are in the philosophy of mind and metaphysics. His article in this collection is drawn from his book *The Conscious Mind*.

**James P. Crutchfield** is a professor in the Center for Computational Science and Engineering at the University of California at Davis and an external faculty member at the Santa Fe Institute. His research interests include computational mechanics and evolutionary dynamics.

**Daniel C. Dennett** is Fletcher Professor of Philosophy and director of the Center for Cognitive Studies at Tufts University. Some of his contributions to the philosophy of mind and artificial intelligence can be found in his books *Brainstorms, Consciousness Explained,* and *Darwin's Dangerous Idea.*

**J. Doyne Farmer** is McKinsey Professor at the Santa Fe Institute. His research interests include dynamical systems and computational economics.

**Jerry Fodor** is State of New Jersey Professor of Philosophy at Rutgers University. His books include *The Language of Thought* and *The Mind Doesn't Work That Way.*

**Carl Hempel** was a key figure in the logical empiricist tradition, making seminal contributions to accounts of scientific explanation and confirmation theory. He taught at Yale, Princeton, and the University of Pittsburgh. His principal articles can be found in his *Aspects of Scientific Explanation and Other Essays.*

**John Holland** is a professor of psychology and a professor of electrical engineering and computer science at the University of Michigan at Ann Arbor. Inventor of novel adaptive machine learning methods such as genetic algorithms and classifier systems, he is among the founding fathers of the contemporary study of complex systems. His books include *Adaptation in Natural and Artificial Systems* and *Emergence: From Order to Chaos.*

**Paul Humphreys** is a professor of philosophy at the University of Virginia. His research interests include emergence and computational science. His most recent book is *Extending Ourselves: Computational Science, Empiricism, and Scientific Method.*

**Jaegwon Kim** is William Faunce Professor of Philosophy at Brown University. His research includes contributions to the philosophy of mind and metaphysics, especially in the exploration of supervenience. Among his books are *Supervenience and Mind, Philosophy of Mind,* and *Physicalism, Or Something Near Enough.*

**Robert B. Laughlin** is Bass Professor of Physics at Stanford University. His current research is in theoretical condensed matter physics. In 1998 he was awarded the Nobel Prize in physics. His recent nontechnical book is *A Different Universe: Remaking Physics from the Bottom Down.*

**Bernd Mayer** teaches in the Department of Theoretical Chemistry at the University of Vienna.

**Brian P. McLaughlin** is a professor of philosophy at Rutgers University. His principal areas of research are the philosophy of mind and philosophical psychology, and he is a coeditor of the *Oxford Handbook for the Philosophy of Mind.*

**Bernd Mayer** is a lecturer in the Department of Theoretical Chemistry at the University of Vienna.

**Ernest Nagel** was University Professor of Philosophy at Columbia University and an influential contributor to modern philosophy of science. His best-known work is *The Structure of Science.*

**Martin Nillson** (who now goes by the name Martin Nillson Jacobi) is an assistant professor in the Complex Systems Group at Chalmers University of Technology.

**Paul Oppenheim** was a contemporary of Carl Hempel, Kurt Grelling, Hilary Putnam, Albert Einstein, and Kurt Gödel.

**Norman H. Packard** is a cofounder and present CEO of ProtoLife Srl, and a cofounder and codirector of the European Center for Living Technology. Originally in the physics faculty at the University of Illinois at Urbana–Champaign, he became a cofounder and CEO of the Prediction Company in Santa Fe, New Mexico.

**David Pines** is a research professor of physics at the University of Illinois at Urbana-Champaign, a founding codirector of the Institute for Complex Adaptive Matter, and a member of the National Academy of Sciences. His research interests include emergent behavior in superconductivity and superfluidity. Among his publications is *The Many-Body Problem.*

**Steen Rasmussen** is a team leader for self-organizing systems at Los Alamos National Laboratories, an external research professor at the Santa Fe Institute, and a cofounder of the European Center for Living Technology.

**Edmund M. A. Ronald** is an affiliate researcher in the Center for Applied Mathematics at the École Polytechnique, Paris.

**Robert S. Shaw** is a senior research scientist at ProtoLife Srl. He Wrote a classic book in dynamical systems theory entitled *The Dripping Faucet as a Model Chaotic System.*

**Thomas Schelling** is Distinguished University Professor, Emeritus in the Department of Economics and School of Public Affairs at the University of Maryland. In 2005 he was awarded the Nobel Prize in economics. Some representative work can be found in his *Micromotives and Macrobehavior.*

**John Searle** is Slusser Professor of Philosophy at the University of California at Berkeley. Some of his contributions to the philosophy of language and the philosophy of mind can be found in his books *Speech Acts, The Rediscovery of the Mind,* and *Consciousness and Language.*

**Herbert Simon** was Mellon University Professor of Computer Science and Psychology at Carnegie Mellon University. His research interests included artificial intelligence, economics, computer science, and philosophy. In 1978 he was awarded the Nobel Prize in economics and in 1986 the National Medal of Science.

**Moshe Sipper** is an associate professor in the Department of Computer Science, Ben-Gurion University, Israel.

**Daniel L. Stein** is a professor of physics at the University of Arizona at Tucson.

**Stephen Weinberg** is Josey Welch Foundation Chair in Science and Regental Professor in the Department of Physics at the University of Texas at Austin. In 1979 he was awarded the Nobel Prize in physics and in 1991 the National Medal of Science. His books include *The Quantum Theory of Fields* and *Dreams of a Final Theory*.

**William C. Wimsatt** is a professor of philosophy and a member of the Committee on Evolutionary Biology at the University of Chicago.

**Stephen Wolfram** has held appointments at the California Institute of Technology, the Institute of Advanced Studies, and the University of Illinois at Urbana–Champaign. He is the developer of *Mathematica* and is currently president and CEO of Wolfram Research. His most recent book is *A New Kind of Science*.

# Index

Printed in the United States
by Baker & Taylor Publisher Services